Saline Soil-based Agriculture by Halotolerant Microorganisms

Manoj Kumar · Hassan Etesami
Vivek Kumar *Editors*

耐盐微生物与盐土农业

[印]马努基·库马尔　[伊朗]哈桑·埃泰萨米　[印]维韦克·库马尔　主编

徐宗昌　李义强　张成省　译著

中国农业科学技术出版社

版权合同登记号 01-2021-1303

图书在版编目（CIP）数据

耐盐微生物与盐土农业：Saline Soil-based Agriculture by Halotolerant Microorganisms /（印）马努基·库马尔（Manoj Kumar），（伊朗）哈桑·埃泰萨米（Hassan Etesami），（印）维韦克·库马尔（Vivek Kumar）主编；徐宗昌，李义强，张成省译著. —北京：中国农业科学技术出版社，2021.4

ISBN 978-7-5116-5249-2

Ⅰ.①耐… Ⅱ.①马…②哈…③维…④徐… Ⅲ.①盐土植物—研究 Ⅳ.①Q949.4

中国版本图书馆 CIP 数据核字（2021）第 049756 号

版权声明

First published in English under the title
Saline Soil-based Agriculture by Halotolerant Microorganisms
edited by Manoj Kumar, Hassan Etesami and Vivek Kumar
Copyright© Springer Nature Singapore Pte Ltd., 2019
This edition has been translated and published under licence from
Springer Nature Singapore Pte Ltd.

责任编辑	姚　欢
责任校对	贾海霞
责任印制	姜义伟　王思文

出 版 者	中国农业科学技术出版社 北京市中关村南大街 12 号　邮编：100081
电　　话	（010）82106631（编辑室）　　（010）82109702（发行部） （010）82109709（读者服务部）
传　　真	（010）82106631
网　　址	http：//www.castp.cn
经 销 者	全国各地新华书店
印 刷 者	北京科信印刷有限公司
开　　本	787mm×1 092mm　1/16
印　　张	18.5
字　　数	400 千字
版　　次	2021 年 4 月第 1 版　2021 年 4 月第 1 次印刷
定　　价	98.00 元

◄━━ 版权所有·翻印必究 ━━►

译者简介

徐宗昌，男，博士，助理研究员。主要以罗布麻为研究对象进行植物-微生物互作研究，以及开展耐盐基因挖掘与功能鉴定、主要活性成分逆境积累调控研究以及产业开发等工作。现主持国家自然科学基金青年基金1项，山东省博士基金1项，横向课题3项，参编著作1部，发表SCI论文10余篇。

李义强，男，博士，研究员，博士生导师。长期致力于农作物病虫害防治和农药残留领域的创新研究与技术应用，近几年对滩涂生物资源农业应用有较深入的研究。先后主持参加省部级以上项目10余项，7项成果获省部级奖励；第一（通信）作者发表论文50余篇；编写著作7部；制定标准10余项；获授权专利9件；获青岛市劳动模范、中国农业科学院十佳青年等荣誉称号。

张成省，男，博士，研究员，硕士生导师。主要研究方向是植物抗病、抗逆领域的创新研究与技术应用。基于微生物组学技术，揭示植物抗病、抗盐碱的微生态机制，进而研发新型生物农药和生物刺激素产品。主持国家和省部级科研项目10余项，获省部级科技成果奖励4项，第一作者或通信作者发表论文40篇，获授权发明专利6项，主编学术著作2部，起草国家和行业标准10项。

著者简介

Manoj Kumar，副教授，目前就职于印度贾肯德邦中央大学生命科学系，致力于高质量的科学研究，具有丰富的专业经验，主持多个行业级的研究项目，国际合作单位包括贾瓦哈拉尔·尼赫鲁大学、墨尔本大学、比勒陀利亚大学和顿杜尚大学等。重点研究方向：树木分子生物学、植物-微生物相互作用和土壤污染物的生物修复；在主要国际期刊上发表研究论文45篇和评论文章5篇，出版图书13部。

Hassan Etesami，博士，毕业于伊朗德黑兰大学农业与自然资源学院土壤科学系。研究领域为土壤生物学和生物技术，研究方向为生物肥料和生物防治剂，包括微生物生态学、生物肥料、土壤污染、生物（盐分、干旱、重金属和营养平衡）和生物（真菌病原体）胁迫的综合管理、植物-微生物相互作用、环境微生物学和生物修复等，在包括生物肥料和生物控制在内的各个领域发表了50余篇研究论文、评论论文，是27种国际期刊的审稿人。

Vivek Kumar，副教授，目前就职于印度德拉敦的斯瓦米喜马拉雅大学喜马拉雅生物科学学院，从事教学、研究和指导工作，研究领域包括微生物相互作用、可持续农业、环境微生物学和生物修复等。目前，发表了100多篇研究论文、评论文章，并在斯普林格出版社出版了多部图书。Vivek教授是几家著名国际期刊的评论员，也是荷兰科学研究组织（NWO）项目建议书的评估员。Vivek教授长达8年在科威特农业事务和渔业资源公共管理局土壤和水研究部担任友好专家，他首次报道并鉴定了由黏质沙雷氏菌（*Serratia marcescens*）引起的科威特枣树烂穗病。2002年他被印度微生物学家协会授予"农业微生物学年度青年科学家奖"。

译者前言

粮食安全是人类社会不可忽视的基本需求之一。根据联合国数据，全球人口规模在 2025 年底将超过 80 亿。一方面，为了满足不断增长的人口对口粮的需求，到 2025 年全球粮食的产量需要较目前增加 38% 以上；另一方面，目前全球 33% 的灌溉农业用地和 20% 的正常耕地正在遭受土壤盐渍化的威胁，随着气候的变化这一数据还在不断地扩大当中。当前，大多数农作物对盐分十分敏感，土壤中可溶性盐过度积累造成土壤盐渍化会严重制约农作物的生产能力。农作物遭受盐渍化环境胁迫后减产能够达到 20%~50%，严重时甚至绝产，土壤盐渍化对农作物生产的制约严重威胁着世界人口安全。因此土壤盐渍化的问题引起了各国政府和各界社会团体的重视，并积极倡导寻找减缓这一趋势或者通过技术手段提高农作物耐盐性的方案。

耐盐植物和耐盐微生物能够在盐胁迫条件下生存，发展出了各自的应对策略来适应恶劣的生存环境。因此，研究耐盐植物和耐盐微生物的耐盐机制并加以利用，促进提高农作物的耐盐能力是提高盐渍化土壤利用率并增加粮食产量的一条有效路径。

译者所在的滩涂生物资源保护利用创新团队是中国农业科学院烟草研究所 2015 年新成立的研究团队，针对滩涂农业领域，立足盐土农业特殊生态环境，在耐盐碱基因改良与植物种质创新、海洋微生物及活性物质农业应用、滨海滩涂改良与综合利用等方面开展共性、关键性技术的科技攻关及示范应用研究。正是在这样一个背景之下，我们团队倾注精力，完成了该书的译著工作。

本书概述了耐盐微生物促进植物生长的各种机制，以此提高植物的耐盐性；阐述了盐生植物根际具有丰富的耐盐微生物，是挖掘有益耐盐微生物的良好场所；并强调了盐生植物–微生物相互作用在盐渍土农业应用方面的一些最新进展，对其在以盐渍土为基础的农业发展中的应用进行了友好的展望。

希望书中介绍的内容能够给关心粮食安全、土壤安全、环境保护和关注耐盐植物微生物生物技术的企事业单位、机关团体和个人带来帮助。

译者

2021 年 1 月

著者前言

粮食安全是社会正常运转的基本要求之一，任何一个正常的社会都需要保证粮食安全。目前，自然环境的持续恶化和人口数量的不断增加对全球粮食生产产生了非常不利的影响，在全球人口呈爆发增长之前，世界的粮食生产很有可能就会发生短缺。在未来50年内，世界人口将达到100亿人，为了满足不断增长的人口对粮食的需求，主要粮食作物的产量需要大幅增加50%左右。世界人口不断增加，而土壤盐渍化导致农业可耕种土地不断减少，进一步加剧了全球的粮食安全问题。治理土壤盐渍化的成本是巨大的，全球每年大约需要投入120亿美元的治理成本，并且这种投入还在持续上升中。近年来，由于可耕作的肥沃土地数量的减少，盐土农业得到了迅速发展。但是限制盐土农业生产方式发展的一个重要原因就是相关农作物和树木的耐盐性低。在盐土环境中，耐盐性好的植物（作物）能够产生更多的产量。为了盐土农业的可持续发展，我们必须筛选培育具有耐盐能力的粮食作物和纤维植物，使这些植物能够在盐土地区成功生长。利用传统育种和基因工程技术培育耐盐作物一直以来都备受期待，但是收效甚微。引入能够促进作物生长的耐盐微生物可能是提高作物耐盐性的另外一种行之有效的方法。具有耐土壤盐渍化特性的微生物能够促进许多不同作物在盐渍土壤中的生长，这种方法可能在较难培育耐盐种质的地方获得成功。耐盐微生物的鉴定和利用，不仅能够满足我们对耐盐作物的需求，而且也减少了粮食生产对正常耕地的压力。不仅如此，耐盐微生物还提供了一种用于了解作物的耐盐性和适应机制的良好模型，进而可以对作物进行改良，以适应、缓解因气候变化而引起的各种逆境胁迫对农作物的伤害。最近的植物-细菌相互作用研究表明，植物能够获得根际和根内微生物群活动带来的益处。在各种逆境胁迫条件下，植物在不同生态系统中的生长和形态建成需要具有相关耐性的微生物。共生细菌存在于所有不同自然生态系统的植物中，这种共生关系可能是影响植物耐逆能力的关键因素。事实上，植物对环境的局部适应能力是由与之密切相关的细菌的遗传分化和进化所驱动。

盐生植物是极端耐盐的植物，通常在不适宜耕种的野生环境中可以通过保持负水势在高达5 g/L的盐浓度下存活和生长。而生活在盐生植物根际的耐盐微生物可能对盐生植物的耐盐性有协同作用。盐生植物的根际和根内是分离不同类型耐盐微生物的理想资源，这些耐盐微生物能够促进不同作物在盐胁迫条件下的生长。耐盐微生物作为生物肥料在农业中的应用已经有很多报道，因此耐盐微生物已具备替代农用化学品的可能性。目前，耐盐微生物对干旱半干旱地区盐碱地退耕还林的林木生长具有重要的作用。从盐生植物中分离的微生物在盐胁迫下也能够促进甜土植物（盐敏感作物）的生长。

  本书提出了耐盐微生物能够提高植物耐盐能力并促进植物生长的观点。盐生植物根际作为发掘耐盐微生物的有效场所，未来能够开发以耐盐微生物为主体的生物接种剂，并且在农业上进行应用，也为今后提高作物耐盐性、促进盐土农业的发展提供有价值的资源。总的来说，这一系列综述将着重介绍植物（盐生植物）-微生物相互作用在农业应用方面的一些最新进展，及其对农业生态系统（盐土农业）在可持续发展方面的贡献。

Manoj Kumar
Hassan Etesami
Vivek Kumar
2019 年

目 录

1 土壤盐分对农业可持续发展的挑战及细菌介导的作物盐害缓解 …………… 1
   1.1 前言 ……………………………………………………………………… 1
   1.2 盐分对植物生长的影响及植物的耐盐机理 …………………………… 4
   1.3 细菌介导的作物盐胁迫缓解 …………………………………………… 6
   1.4 结论 ……………………………………………………………………… 13
   参考文献 ……………………………………………………………………… 14

2 盐生植物生长促生菌在缓解植物受到的盐胁迫危害中的作用 …………… 27
   2.1 前言 ……………………………………………………………………… 27
   2.2 盐渍化 …………………………………………………………………… 29
   2.3 盐渍化对土壤、植物和微生物的影响 ………………………………… 31
   2.4 耐盐植物生长促生菌是缓解植物盐胁迫的新手段 …………………… 36
   2.5 耐盐植物生长促生菌——藻类：一个很有前途的缓解植物盐胁迫
       危害的联合体 …………………………………………………………… 45
   2.6 结论 ……………………………………………………………………… 46
   参考文献 ……………………………………………………………………… 47

3 耐盐根际菌：一种很有前途的盐渍土农业益生菌 ………………………… 61
   3.1 前言 ……………………………………………………………………… 62
   3.2 盐胁迫对植物生长的影响 ……………………………………………… 63
   3.3 对植物益生菌的需求 …………………………………………………… 64
   3.4 根际细菌可作为植物生长益生菌 ……………………………………… 66
   3.5 耐盐根际细菌 …………………………………………………………… 69
   3.6 耐盐细菌是盐渍土农业的植物益生菌 ………………………………… 69
   3.7 结论 ……………………………………………………………………… 75
   参考文献 ……………………………………………………………………… 76

4 盐生植物根际土壤中的生长促生菌对提高作物耐盐性研究进展 ………… 84
   4.1 前言 ……………………………………………………………………… 85
   4.2 盐胁迫对植物生理的影响 ……………………………………………… 87

4.3　盐分对根际微生物多样性的影响 ⋯⋯⋯⋯⋯⋯⋯⋯⋯⋯⋯⋯⋯⋯⋯ 87
　　4.4　根际植物生长促生菌的耐盐机制 ⋯⋯⋯⋯⋯⋯⋯⋯⋯⋯⋯⋯⋯⋯⋯ 88
　　4.5　从不同盐生植物根际分离植物生长促生菌 ⋯⋯⋯⋯⋯⋯⋯⋯⋯⋯⋯ 89
　　4.6　应激条件下植物生长促生菌的促生长机制 ⋯⋯⋯⋯⋯⋯⋯⋯⋯⋯⋯ 92
　　4.7　植物生长促生菌在耐盐作物品种开发中的应用 ⋯⋯⋯⋯⋯⋯⋯⋯⋯ 98
　　4.8　结论和展望 ⋯⋯⋯⋯⋯⋯⋯⋯⋯⋯⋯⋯⋯⋯⋯⋯⋯⋯⋯⋯⋯⋯⋯ 101
　　参考文献 ⋯⋯⋯⋯⋯⋯⋯⋯⋯⋯⋯⋯⋯⋯⋯⋯⋯⋯⋯⋯⋯⋯⋯⋯⋯⋯ 102

5　耐盐植物生长促生真菌和细菌提高盐胁迫下作物养分利用效率的替代
　　策略 ⋯⋯⋯⋯⋯⋯⋯⋯⋯⋯⋯⋯⋯⋯⋯⋯⋯⋯⋯⋯⋯⋯⋯⋯⋯⋯⋯⋯ 113
　　5.1　前言 ⋯⋯⋯⋯⋯⋯⋯⋯⋯⋯⋯⋯⋯⋯⋯⋯⋯⋯⋯⋯⋯⋯⋯⋯⋯⋯ 114
　　5.2　盐胁迫 ⋯⋯⋯⋯⋯⋯⋯⋯⋯⋯⋯⋯⋯⋯⋯⋯⋯⋯⋯⋯⋯⋯⋯⋯⋯ 118
　　5.3　盐胁迫下植物生长促生菌和菌根真菌介导的氮素利用有效性的
　　　　提高 ⋯⋯⋯⋯⋯⋯⋯⋯⋯⋯⋯⋯⋯⋯⋯⋯⋯⋯⋯⋯⋯⋯⋯⋯⋯⋯ 125
　　5.4　植物生长促生菌和丛枝菌根真菌介导的盐胁迫下磷利用有效性的
　　　　提高 ⋯⋯⋯⋯⋯⋯⋯⋯⋯⋯⋯⋯⋯⋯⋯⋯⋯⋯⋯⋯⋯⋯⋯⋯⋯⋯ 132
　　5.5　植物生长促生菌和丛枝根菌真菌介导的盐胁迫下钾利用有效性的
　　　　提高 ⋯⋯⋯⋯⋯⋯⋯⋯⋯⋯⋯⋯⋯⋯⋯⋯⋯⋯⋯⋯⋯⋯⋯⋯⋯⋯ 140
　　5.6　植物生长促生菌和丛枝菌根真菌介导的盐胁迫下微量元素利用
　　　　有效性的提高 ⋯⋯⋯⋯⋯⋯⋯⋯⋯⋯⋯⋯⋯⋯⋯⋯⋯⋯⋯⋯⋯⋯ 144
　　5.7　结论 ⋯⋯⋯⋯⋯⋯⋯⋯⋯⋯⋯⋯⋯⋯⋯⋯⋯⋯⋯⋯⋯⋯⋯⋯⋯⋯ 146
　　参考文献 ⋯⋯⋯⋯⋯⋯⋯⋯⋯⋯⋯⋯⋯⋯⋯⋯⋯⋯⋯⋯⋯⋯⋯⋯⋯⋯ 147

6　盐生植物内生细菌：如何帮助植物缓解盐胁迫危害 ⋯⋯⋯⋯⋯⋯⋯⋯⋯ 166
　　6.1　前言 ⋯⋯⋯⋯⋯⋯⋯⋯⋯⋯⋯⋯⋯⋯⋯⋯⋯⋯⋯⋯⋯⋯⋯⋯⋯⋯ 167
　　6.2　盐生植物及其相关微生物 ⋯⋯⋯⋯⋯⋯⋯⋯⋯⋯⋯⋯⋯⋯⋯⋯⋯ 168
　　6.3　盐生植物-微生物互作与非生物胁迫 ⋯⋯⋯⋯⋯⋯⋯⋯⋯⋯⋯⋯⋯ 169
　　6.4　从盐生植物中分离的内生微生物如何帮助植物缓解盐胁迫？ ⋯⋯⋯ 170
　　6.5　结论和展望 ⋯⋯⋯⋯⋯⋯⋯⋯⋯⋯⋯⋯⋯⋯⋯⋯⋯⋯⋯⋯⋯⋯⋯ 175
　　参考文献 ⋯⋯⋯⋯⋯⋯⋯⋯⋯⋯⋯⋯⋯⋯⋯⋯⋯⋯⋯⋯⋯⋯⋯⋯⋯⋯ 175

7　盐胁迫条件下嗜盐细菌对水稻品种生化特性的影响 ⋯⋯⋯⋯⋯⋯⋯⋯⋯ 182
　　7.1　前言 ⋯⋯⋯⋯⋯⋯⋯⋯⋯⋯⋯⋯⋯⋯⋯⋯⋯⋯⋯⋯⋯⋯⋯⋯⋯⋯ 183
　　7.2　植物材料 ⋯⋯⋯⋯⋯⋯⋯⋯⋯⋯⋯⋯⋯⋯⋯⋯⋯⋯⋯⋯⋯⋯⋯⋯ 184
　　7.3　细菌菌株 ⋯⋯⋯⋯⋯⋯⋯⋯⋯⋯⋯⋯⋯⋯⋯⋯⋯⋯⋯⋯⋯⋯⋯⋯ 184
　　7.4　讨论 ⋯⋯⋯⋯⋯⋯⋯⋯⋯⋯⋯⋯⋯⋯⋯⋯⋯⋯⋯⋯⋯⋯⋯⋯⋯⋯ 191
　　7.5　结论 ⋯⋯⋯⋯⋯⋯⋯⋯⋯⋯⋯⋯⋯⋯⋯⋯⋯⋯⋯⋯⋯⋯⋯⋯⋯⋯ 193

  参考文献 ··· 193
8 耐盐细菌海水发酵生产鼠李糖脂及其在草莓白粉病防治中的应用 ····· 197
  8.1 研究背景 ··· 197
  8.2 生物表面活性剂的低成本生产 ······································ 198
  8.3 生物表面活性剂菌株的筛选 ·· 199
  8.4 发酵工艺 ··· 200
  8.5 在盐渍土草莓上的应用 ··· 201
  8.6 成本评估 ··· 203
  8.7 结论 ··· 204
  参考文献 ··· 204

9 盐渍土有益耐盐微生物商品化的瓶颈及发展前景 ······················· 207
  9.1 前言 ··· 207
  9.2 盐分干扰作物的营养和生殖发育 ··································· 208
  9.3 农田中的盐分管理 ·· 209
  9.4 耐盐植物生长促生菌在农业中的应用，提高植物在盐胁迫环境
    中的存活率 ·· 211
  9.5 商业化阶段 ·· 215
  9.6 商业化的瓶颈 ··· 218
  参考文献 ··· 221

10 耐盐微生物在盐胁迫条件下促进植物生长的作用 ······················· 228
  10.1 前言 ··· 229
  10.2 耐盐微生物耐盐机理 ·· 231
  10.3 逆境农业促进植物生长和产量有效性的机制 ··················· 235
  10.4 耐盐微生物的协同应用 ··· 250
  10.5 耐盐微生物在环境科学中的作用 ································ 256
  10.6 展望 ··· 259
  参考文献 ··· 260

附表 主要名词中英文对照 ··· 280

1 土壤盐分对农业可持续发展的挑战及细菌介导的作物盐害缓解

Hassan Etesami, Fatemeh Noori

[摘要]

目前，由于气候变化、全球气温上升和环境胁迫因子增加等负面影响，导致农业生产力正在不断下降。因此，为了实现农业的可持续发展和保证世界人口的粮食安全，就必须使用适当的生态和环境友好的技术和方法来减少这些胁迫因子对农作物生长的不利影响。土壤盐渍化是影响全球农业生产力的一个重要问题。根据目前的研究结果，生长在盐胁迫条件下的农作物由于受到渗透胁迫、土壤理化性质差、作物营养失调和离子毒性胁迫等影响，导致农作物产量降低。在世界粮食需求不断增加的大背景下，减少由于盐渍胁迫造成的农作物产量损失逐渐成为一个研究热点。解决这个问题需要新的农业技术来改善盐渍土条件下农作物的生产力。有报道指出，一些耐盐植物的根际细菌能够通过影响植物根系发育、改善土壤团粒结构、增加农作物水分与养分吸收、减少Na^+吸收、减少应激乙烯的产生与负面影响，以及增加能够提高农作物耐盐性的相关基因的表达等机制来提高农作物对盐胁迫的耐受性。通过接种微生物来缓解植物盐胁迫是一种成本相对较低，并且能够在较短时间内达到目标、比较有效且环境友好的方法。这种方法有助于在盐胁迫条件下促进农业的可持续发展。

[关键词]

耐盐植物生长促生菌；盐胁迫；农作物；植物；微生物互作；盐渍土农业

1.1 前言

粮食安全是人类社会不可忽视的基本需求之一。根据联合国粮农组织

H. Etesami (✉), 伊朗德黑兰大学土壤科学系农业与自然资源学院，E-mail: hassanetesami@ut.ac.ir

F. Noori, 伊朗萨里农业科学与自然资源大学生物技术与植物育种系

(Food and Agriculture Organization，FAO) 的调查数据，到2030年，全世界对农产品的总需求将比现在高出60%。近半个多世纪以来，世界各国一直依靠增加农作物的产量来满足不断增长的粮食需求（Ladha 等，1998）。根据联合国的估计，到2050年，世界人口增长将达到89亿（Wood，2001），而用于农业生产的土壤肥力水平却在不断下降，可用来进行农作物生产的正常农业土地的比例正在迅速下降。因此，在盐渍土条件下提高作物的单位面积产量是提高粮食产量的途径之一。盐渍土是指土壤饱和浸出液的电导率在4 dS/m（大约相当于在25℃条件下40 mmol/L 的 NaCl 溶液）和可交换性 Na^+ 在15%以上的土壤。尽管有一些农作物在盐渍土程度较低的土壤中产量损失较小，但不可否认的是，在这种盐渍化土壤中大多数农作物的产量确实都降低了（Munns，2005）。当然，对于盐渍土的定义也有不同的看法。Shannon 和 Grieve（1998）认为，盐渍危害是指水和土壤溶液中可溶性盐和矿物质元素的浓度过高，导致植物根区盐分积累，植物难以从土壤溶液中吸收足够的水分。而在另一个定义中，盐渍危害是指可溶性阴离子和阳离子在植物体内过度积累并影响了植物的正常生长发育（Çavusoglu 和 Kabar，2010）。

土壤盐渍化是指土壤表层可溶性盐积累增加，表层土壤失去作为植物生长介质能力的过程。一般来说，土壤盐渍化分为初生盐渍化和次生盐渍化两个主要的过程。初生盐渍化过程是指盐分在土壤或地表水中长期累积的自然过程。而次生盐渍化的形成主要是人类活动的结果，比如灌溉。灌溉是导致土壤剖面中可溶性盐浓度增加的主要原因，可溶性盐浓度的增加会抑制植物生长，从而导致正常农业用地的盐渍化（Egamberdieva 等，2007；Manchanda 和 Garg，2008；Munns，2005）。微咸水灌溉农田、灌溉经验不足、长期干旱和降水量少、地表蒸发量大以及栽培管理不善是导致盐渍化土壤以每年10%的速度增加的主要原因（Jamil 等，2011）。据估计，世界上约有3亿 hm^2 的灌溉土地，其中约有90%的灌溉用水生产了全球36%的粮食（Rengasamy，2006）。在灌溉水短缺（或没有灌溉水）和降水量较少的地区，甚至在大多数温度较高的农业生产区，土壤盐渍化的程度都会增加（Othman 等，2006）。有预测研究表明，全球33%的灌溉农业用地和20%的正常耕地会受到土壤盐渍化的威胁。而气候变化可能会导致许多非灌溉地区出现更多的盐渍化景观（Othman 等，2006）。据估计，到2050年，世界上50%以上的可耕地都将受到盐渍化的威胁（Jamil 等，2011）。

在影响农作物生产和栽培的环境胁迫因子（如极端温度、大风、干旱、土壤盐渍化和涝害）中，盐渍化胁迫被认为是最具破坏性的环境胁迫因子之一。盐渍化胁迫能够降低农作物的生产能力和经济效益，并且产生严重的土壤侵蚀（Hu 和 Schmidhalter，2004），进而导致耕地面积的大幅度减少（Shahbaz 和

Ashraf，2013；Yamaguchi 和 Blumwald，2005）。由于大多数农作物对盐分敏感，因此土壤中可溶性盐过度积累造成的土壤盐渍化严重制约了农作物的生产能力和质量，农作物减产甚至能够达到20%~50%（Shrivastava 和 Kumar，2015）。除了能够大幅度地减少大多数农作物的产量和质量（特别是对盐分敏感的农作物），盐渍化还能够对耕作区内的土壤理化性质和生态平衡（如对生活在土壤中或土壤上的微生物、原生动物和线虫等生物的生长和多样性的影响）产生诸多不利影响（Hu 和 Schmidhalter，2004；Parida 和 Das，2005）。到2050年，全世界的人口将在现有人口的基础上增加23亿，而增加的这一部分人口需要的粮食消耗需要在目前粮食产能的基础上提高70%才能满足。在粮食需求不断增加和可耕作土地面积不断减少形势下，我们应该充分利用盐渍土的生产潜力来进行水稻、小麦和玉米等农作物的生产。在这种情况下，这些农作物的粮食产量到2050年估计会增加50%左右，才有可能满足不断增长的人口对粮食消耗的需求（Godfray 等，2010）。如前所述，生长在盐渍土环境下的农作物粮食产量显著降低，在这些受到土壤盐渍化威胁的耕作区，农民正试图通过使用更多的化肥等投入来弥补盐渍化造成的农作物产量损失。然而，大量的化肥投入使用不仅提高了农作物的生产成本，还存在着污染地表水、地下水以及土壤板结等环境污染的风险。另外，大量营养元素的过度投入反而会增加土壤中盐分离子的含量，进一步导致更多的土壤出现盐渍化现象。因此，为使盐渍化土壤在农业生产中更好地发挥作用，加强盐渍化土壤管理非常必要。目前，已经有很多研究提出了许多方法来改善盐渍土（盐渍土修复），从而增强这些土壤在粮食作物生产方面的利用潜力（Bai 等，2017；Bauder 等，2004；Etesami 和 Maheshwari，2018；Qadir 等，2000；Tejada 等，2006；Wang 等，2014）。尽管目前已经采用了很多方法诸如根区盐分淋溶、农用耕地管理方法优化、使用有机改良剂、耕作土壤管理经验交流、滴灌（或喷灌）、作物轮作、种植耐盐农作物、通过育种和基因工程培育耐盐品种等策略来改善盐渍化条件下的作物减产现象，但往往由于当地水资源的限制，这些手段的应用都或多或少地受到了制约。换言之，这些方法要起到良好的效果都需要漫长的时间或者消耗大量的成本（Araus 等，2008；Dwivedi 等，2010；Flowers，2004；Manchanda 和 Garg，2008；Shrivastava 和 Kumar，2015；Venkateswarlu 和 Shanker，2009）。因此，有必要开发成本较低并且操作相对简单的生物方法来治理盐渍化土壤，并且这些方法还需要能够在短期内见到成效。众所周知，与农作物共生的耐盐微生物在提高宿主作物耐盐性方面发挥着重要的作用（Etesami，2018；Etesami 和 Beattie，2017；Etesami 和 Maheshwari，2018；Shrivastava 和 Kumar，2015）。本章主要讨论盐渍化胁迫对农作物生长的影响，以及与植物共生的耐盐植物生长促生菌（Plant growth promoting bacteria，

PGPB）在提高植物耐盐能力中发挥的作用。

1.2 盐分对植物生长的影响及植物的耐盐机理

盐分作为一种主要的非生物胁迫因子在世界范围内限制了大多数农作物的生产能力，同时也影响了世界人口增长对粮食作物产量增长的需求。目前，全球有超过20%的耕地受到盐渍化胁迫的影响，并且这种影响造成的危害还在不断加剧。植物在漫长的进化过程中发生了分化，有些植物有较强的耐受盐分的能力，这类植物可以统称为盐生植物；另一些植物则无法耐受盐分并在长时间的盐分胁迫下导致死亡，这类植物称为甜土植物。

类属于甜土植物的大多数农作物种类对盐分胁迫都非常敏感（Flowers，2004；Munns和Tester，2008；Horneck等，2007；Ondrasek等，2009），由于汲取不到足够的营养，这些甜土农作物在面临盐渍化胁迫时的生产能力就会大大降低（Chinnusamy等，2005；Mantri等，2012）。

盐分胁迫几乎影响植物的种子萌发、营养生长和生殖生长的各个方面（Bano和Fatima，2009）。盐分胁迫能够导致农作物各种代谢和生理过程以及农作物的形态特征发生变化，根据农作物受到盐分胁迫的严重程度和持续时间，造成的生理过程变化包括气孔缩小（导致气孔导度降低）、生物膜系统合成中断、叶片生长发育减缓、叶绿素含量低、光系统Ⅱ光合作用效率降低；抑制硝酸还原酶（nitrate reductase，NR）活性而破坏光合功能、叶片早衰（影响光合作用等过程并导致生长迟缓）、引起细胞膨大并导致能量产生减少、对各种酶活性产生不利影响（如核酸代谢相关酶的活性受到抑制）；对种子萌发、植株活力、作物产量和根系生长产生不利影响；损害活性氧（reactive oxygen species，ROS）解毒过程，并促进衰老和落叶的发生；对光合作用、蛋白质合成、能量和脂质代谢、营养平衡、植物激素合成和细胞壁成熟等所有主要的生理过程都产生不利影响；导致植物体内激素失衡，如乙烯水平升高（乙烯在较高浓度下对植物正常生长产生不利影响，包括促进叶片衰老和落叶发生以及其他生命过程）；抗氧化酶活性变异、植株生理过程失调（如离层脱落、衰老和易感病）以及光合活性的降低（Arbona等，2005；Barnawal等，2014；Cramer和Nowak，1992；Dantas等，2005；Dolatabadian等，2011；Flowers，2004；Glick，2014；Hashem等，2015；Kang等，2014a；Munns，2002；Munns和Tester，2008；Nadeem等，2014；Netondo等，2004；Nunkaew，2015；Parida和Das，2005；Paul和Lade，2014；Prakash和Prathapasenan，1990；Rahnama等，2010；Shirokova等，2000；Tavakkoli等，2011），并最终导致作物产量的严重减产（James等，2011；

Rozema 和 Flowers，2008）。

总的来说，盐分胁迫对农作物的影响是包括生物化学、植物生理学和形态学过程（如植物生长、种子萌发、水和养分吸收）之间复杂相互作用的结果（Akbarimoghaddam 等，2011）。其中，作物种子萌发（根系发育阶段）和幼苗生长早期是植物对盐分胁迫最敏感的生长阶段，因为幼苗根系与土壤直接接触，因此对受到的盐分胁迫十分敏感（Bae 等，2006；Rahman 等，2000）。另外，由于盐渍化土壤的土壤溶液中充斥着大量的离子，导致渗透压的增加（Atak 等，2006；Neamatollahi 等，2009），较高的土壤溶液渗透压导致包括大豆（Essa，2002）、小麦（Egamberdieva 和 Kucharova，2009）、水稻（Xu 等，2011）、玉米（Khodarahmpour 等，2012）和蚕豆（Rabie 和 Almadini，2005）等在内的农作物种子的萌发率显著降低。

盐胁迫在早期也被称为高渗胁迫，通过渗透胁迫的形式抑制植物生长。因此在盐胁迫的早期，植物根系的吸水能力减弱。渗透胁迫能够使植物叶片失水，这导致受到盐分胁迫的植物组织中盐分浓度升高（Munns，2005）。盐胁迫在植物生长早期能够造成植物的生理缺水，并且由于组织中积累了较高的盐离子浓度，还能够造成植物体内的离子毒性，并引起叶片衰老。盐分胁迫还能够改变土壤的基本质地，导致土壤孔隙度降低和土壤的理化性质改变，造成植物水分吸收困难从而引起生理干旱（Munns 和 Tester，2008）。

盐分胁迫通过渗透作用造成土壤溶液渗透势降低、植物营养失衡（营养元素 N、Ca、K、P、Fe、Zn 等的缺乏）、离子毒害效应、氧化胁迫，或这些胁迫因素的组合，在生理生化水平和分子水平（Tester 和 Davenport，2003）对植物的生长和发育产生许多不利影响（Ashraf，2004；Bano 和 Fatima，2009；Grover 等，2011），最终导致作物生产能力的大幅下降。

土壤盐分胁迫显著降低了植物对养分的吸收，比如 P（P 与 Ca 容易形成沉淀物）（Bano 和 Fatima，2009）、N、K 和 Mg（Heidari 和 Jamshid，2010；James 等，2011）。在农业生产中，只有少数豆科植物能够在盐渍化土壤中相对正常地生长（Ashraf 和 McNeilly，2004）。但是，在盐分胁迫条件下，对盐分敏感的豆科植物的生长也会受到影响，固氮作用下降，导致通过固氮作用进入农业生产系统的 N 元素的减少，因此又导致化肥使用量的增加。盐胁迫尤其会对豆科植物与根瘤菌（*Rhizobial*）良好的共生作用造成危害。根瘤菌能够通过根瘤固氮酶复合物固定大气中的 N，使之能够被与之共生的豆科植物吸收利用（Quispel，1988）。比如，在盐胁迫条件下，根瘤菌侵染豆科植物和在植物根部结瘤的能力减弱，导致这些生理过程的失败，植物体内固氮酶的活性降低，从而导致豆科作物（如菜豆、大豆和蚕豆等）的固氮能力降低（Rabie 等，2005；Singleton 和

Bohlool，1984）。另外，盐胁迫还能够抑制与根瘤菌共生的豆科植物的根、根尖和根毛等组织的生长，从而减少根瘤菌进一步侵染根瘤发育的部位。有研究指出，盐胁迫引起的植物根系发育受阻是由于在盐胁迫条件下植物对 Ca 元素的吸收减少所导致的（Bouhmouch 等，2005）。

为了能够在盐渍土环境中生存，植物就必须适应周围的胁迫环境（Paul 和 Lade，2014）。在盐渍胁迫环境下，植物也进化出了多种机制用来提高耐盐性抵御盐分危害。在这些耐盐机制中，比较重要的有以下 7 种：①激素调节；②渗透调节物质的生物合成；③离子稳态和分区储存；④离子运输和吸收；⑤抗氧化酶的激活和抗氧化化合物的合成；⑥多胺的合成；⑦一氧化氮（NO）的产生（Gupta 和 Huang，2014）。除了这些耐盐机制外，目前报道的多种研究表明，与植物共生的微生物（细菌）群落也可以在盐胁迫条件下促进植物的生长和发育，在提高植物对多种环境胁迫（包括盐胁迫）的耐受性方面发挥重要的作用（Etesami 和 Beattie，2018；Etesami 和 Maheshwari，2018）。

1.3 细菌介导的作物盐胁迫缓解

与植物共生的有益微生物（包括根际微生物和内生菌）都能够在植物抵御盐胁迫方面发挥重要作用（Dodd 和 Pérez–Alfocea，2012；Etesami 和 Beattie，2017，2018；Etesami 和 Maheshwari，2018；Gill 等，2016；Hamilton 等，2016；Singh 等，2011；Vimal 等，2016）。植物生长促生菌一般定殖在许多植物的根际中并对植物的生长发育产生正向促进效果（Glick 等，2007；Van Loon 等，1998）。在植物抵抗盐胁迫的反应中，给目标植物接种植物生长促生菌可以有效诱导植物生物化学和形态学发生改变，进而增强植物对盐胁迫的耐受性，这种方式被定义为系统性诱导抗性（induced systemic tolerance，IST）（Glick，2014；Kaushal 和 Wani，2016；Yang 等，2009）。

以往对微生物的研究主要集中在其改善土壤质量和肥力等方面，而目前对微生物的研究则主要聚焦在其缓解植物遭受到的包括盐胁迫在内的各种非生物胁迫领域（Mapelliet 等，2013）。众所周知，高渗透强度和离子毒性作用对土壤微生物的活性有不利影响，然而，对于频繁暴露于盐胁迫条件下或与盐生植物共生的细菌在提高植物耐盐性等方面会有更加优良的效果。耐盐微生物能够在条件恶劣的土壤生境或者根际环境中生存和繁殖（Etesami 和 Beattie，2018；Garcia 和 Hernandez，1996；Paul 等，2005）。目前已经证明，耐盐细菌在盐胁迫条件下能够作为植物生长促进剂促进植物更好地生长（Etesami 和 Beattie，2018；Shrivastava 和 Kumar，2015）。在以土壤为生境的细菌中，植物根际细菌对盐胁迫的耐受性更强，这是由于植物根系的吸水作用导致根际周围的渗透压和离子强度都有所增强，因此在这种生境下的根际细菌就对盐胁迫具有更高的耐受性（Tripathi 等，1998）。这些植物生长促生细菌至少能够在3% NaCl 浓度下生存（Egamberdieva 等，2011）。由于在盐渍土生境下生存持久并具有很强的生存竞争力，因此这些植物生长促生细菌能够在植物根际生存（Mayak 等，2004b；Yasmin 等，2007）。从耐盐植物根际分离出来的耐盐细菌能够通过很多生理生化过程促进植物的生长，比如产生吲哚-3-乙酸（IAA）、增强磷酸盐的溶解程度、提高1-氨基环丙烷-1-羧酸脱氨酶的活性、产生氨以及氮的固定等（Etesami 和 Beattie，2018；Mapelli 等，2013）。在盐胁迫条件下，耐盐和嗜盐微生物可以增加细胞内渗透调节物质的含量（Zhou 等，2015）。大量的研究表明，耐盐植物生长促生细菌对植物的生长具有有利作用，具体内容见表1.1（Etesami 和 Beattie，2017，2018；Etesami 和 Maheshwari，2018）。在盐渍土生长环境中，耐盐植物生长促生细菌显著提高了各种农作物的生长和产量，这些农作物包括小麦（Barra 等，2016；Egamberdieva 和 Kucharova，2009；Nabti 等，2010；Nia 等，2012；Orhan，2016；Rajput 等，2013；Ramadoss 等，2013；Upadhyay 和 Singh，2015；Upadhyay 等，2012）、黄瓜（Egamberdieva 等，2011；Kang 等，2014a；Nadeem 等，2016）、甘薯（Yasmin 等，2007）、罗勒（*Ocimum basilicum*）（Heidari 等，2011）、番茄（Albacete 等，2008；Essghaier 等，2014；Mayak 等，2004a；Tank 和 Saraf，2010）、胡椒、油菜、豆类（Egamberdieva，2011）、莴苣（Barassi 等，2006；Yildirim 和 Taylor，2005；Yildirim 等，2011）、棉花（Yao 等，2010）、马铃薯（Shaterian 等，2005b）、大豆（*Glycine max* L.）（Kang 等，2014b；Naz 等，2009）、水稻（*Oryza sativa* L.）（Jha 等，2011）、白三叶草（Han 等，2014）、落花生（*Arachis hypogaea*）（Shukla 等，2012；Saravanakumar 和 Samiyappan，2007）、草莓（Esitken 等，2010）、蒺藜状苜蓿（*Medicago truncatula*）（Bianco 和 Defez，2009）和山羊豆（*Galega officinalis*）（Saravanakumar 和 Samiyappan，2007）等。

表1.1　盐胁迫下引发植物系统性耐受性的植物生长促生菌及其作用机制

植物生长促生菌	作用机制	作物种类	参考文献
假单胞菌属（*Pseudomonas* sp.）	增加吲哚-3-乙酸和蛋白质含量	罗勒（*Ocimum basilicum*）	Heidari 等（2011）
皮氏无色杆菌（*Achromobacter piechaudii*）	1-氨基环丙烷-1-羧酸脱氨酶	番茄	Zhang 等（2008）
生脂固氮螺菌（*Azospirillum lipoferum*）	增加脱落酸含量	玉米	Cohen 等（2009）

（续表）

植物生长促生菌	作用机制	作物种类	参考文献
皮氏无色杆菌（Achromobacter piechaudii）	1-氨基环丙烷-1-羧酸脱氨酶	辣椒和番茄	Mayak 等（2004b）
枯草芽孢杆菌（Bacillus subtilis）和节杆菌（Arthrobacter sp.）	增加可溶性总糖和脯氨酸含量	小麦	Upadhyay 等（2012）
类产碱假单胞菌（Pseudomonas pseudoalcaligenes）和短小芽孢杆菌（Bacillus pumilus）	增加渗透调节物质和提高抗氧化酶类物质活性	水稻	Jha 等（2011）
固氮螺菌属（Azospirillum）	增加氮元素含量	小麦	Nia 等（2012）
莱比托游动球菌（Planococcus rifietoensis）	产生吲哚-3-乙酸和1-氨基环丙烷-1-羧酸脱氨酶，提高吸收可溶性磷酸盐的能力	小麦	Rajput 等（2013）
溶血性链球菌（Streptococci haemolytic）和枯草芽孢杆菌（Bacillus subtilis）	积累渗透调节物质	鹰嘴豆	Essghaier 等（2014）
枯草芽孢杆菌（Bacillus subtilis）	增加叶绿素Ⅱ含量，并减少丙二醛含量	白三叶草	Han 等（2014）
荧光假单胞菌（Pseudomonas fluorescens）	增强电子传递速率	叙利亚松（Pinus halepensis）	Rincón 等（2008）
巴西固氮螺菌（Azospirillum brasilense）	增加叶子水分含量，并降低丙二醛含量	拟南芥	Cohen 等（2015）
橘黄假单胞菌（Pseudomonas aurantiaca）	产生吲哚-3-乙酸	小麦	Egamberdieva 和 Kucharova（2009）
黄芪叶杆菌（Phyllobacterium brassicacearum）	改善植物氮元素状况	拟南芥	Kechid 等（2013）
洋葱伯克霍尔德菌（Burkholderia cepacia）和乙酸钙不动杆菌（Acinetobacter calcoaceticus）	产生赤霉素	黄瓜	Kang 等（2014a）
恶臭假单胞菌（Pseudomonas putida）	产生赤霉素	黄豆	Kang 等（2014b）
枯草芽孢杆菌（Bacillus subtilis）	产生细胞分裂素	莴苣	Arkhipova 等（2007）
平凡假单胞菌（Pseudomonas trivialis）	1-氨基环丙烷-1-羧酸脱氨酶	山羊豆（Galega officinalis）	Egamberdieva 和 Jabborova（2013）

（续表）

植物生长促生菌	作用机制	作物种类	参考文献
荧光假单胞菌（*Pseudomonas fluorescens*）	1-氨基环丙烷-1-羧酸脱氨酶	落花生（*Arachis hypogaea*）	Saravanakumar 和 Samiyappan（2007）
生脂固氮螺菌（*Azospirillum lipoferum*）	1-氨基环丙烷-1-羧酸脱氨酶	小麦	Zaki 等（2004）
巴西固氮螺菌（*Azospirillum brasilense*）	1-氨基环丙烷-1-羧酸脱氨酶	玉米	Hamdia 等（2004）
肠杆菌（*Azospirillum*）	提高抗氧化酶类活性	番茄	Kim 等（2014）
偶氮螺菌（*Enterobacter* spp.）	增加脯氨酸含量	玉米	Kandowangko 等（2009）
巴西固氮螺菌（*Azospirillum brasilense*）	合成多胺	水稻	Cassan 等（2009）
芽孢杆菌属（*Bacillus*）	提高 K^+/Na^+ 比例	唐菖蒲	Damodaran 等（2014）
争论贪噬菌属（*Variovorax paradoxus*）	增强植物结瘤能力和增加氮元素含量	鹰嘴豆	Belimov 等（2009）

耐盐植物生长促生菌包括固氮螺菌（*Azospirillum*）、根瘤菌（*Rhizobium*）、芽孢杆菌（*Bacillus*）、假单胞菌（*Pseudomonas*）、泛菌属（*Pantoea*）、类芽孢杆菌（*Paenibacillus*）、肠杆菌属（*Enterobacter*）、伯克霍尔德菌（*Burkholderia*）、无色杆菌属（*Achromobacte*）、微杆菌属（*Microbacterium* sp.）、甲基杆菌属（*Methylobacterium*）、黄杆菌属（*Flavobacterium*）、节杆菌属（*Arthrobacter*）、沙雷氏菌属（*Serratia*）、分枝杆菌（*Mycobacterium*）、大洋芽孢杆菌属（*Oceanobacillus* sp.）、盐单胞菌属（*Halomonas* sp.）、微杆菌（*Exiguobacterium* sp.）、刘志恒氏菌（*Zhihengliuella*）、贪噬菌属（*Variovorax*）、短杆菌属（*Brachybacterium*）、葡萄球菌（*Staphylococcus*）、考克氏菌（*Kocuria*）等（Adesemoye 等，2008；Egamberdieva 和 Kucharova，2009；Egamberdieva 等，2011；Egamberdiyeva，2007；Egamberdiyeva 和 Islam，2008；Etesami 和 Beattie，2018；Rajput 等，2013；Shrivastava 和 Kumar，2013；Shukla 等，2012；Siddikee 等，2010）。它们能够通过各种机制来提高植物对盐胁迫的耐受性，并刺激植物生长和发育（Dimkpa 等，2009；Etesami 和 Maheshwari，2018；Glick，2014；Glick 等，2007；Grover 等，2011；Lugtenberg 和 Kamilova，2009；Lugtenberg 等，2013；Mayak 等，2004b；Shrivastava 和 Kumar，2015；Yang 等，2009）。植物生

长促生菌提高植物耐盐性的机制如下。

（1）植物激素的产生，如生长素、细胞分裂素和赤霉素等。众所周知，植物激素参与植物生长发育的各个阶段，并参与包括盐胁迫在内的各种逆境胁迫响应（Shaterian 等，2005a），强化各种植物细胞防御系统以保护植物免受胁迫危害（Shaterian 等，2005b）。在高浓度盐胁迫的条件下，受到危害的植物根和叶中的植物激素如生长素、赤霉素和玉米素等的产生会减少（Pérez-Alfocea 等，2010；Sakhabutdinova 等，2003），因此植物的发芽率会降低，生长和发育过程也会受到抑制（Sakhabutdinova 等，2003；Werner 和 Finkelstein，1995）。除了各种植物激素的合成受到抑制外，盐胁迫等逆境胁迫因子还导致植物激素从根到茎的极性运输困难（比如细胞分裂素）（Naqvi 和 Ansari，1974）。目前，有研究表明，盐胁迫对耐盐植物生长促生菌的激素合成没有明显影响（Albacete 等，2008；Egamberdieva 和 Kucharova，2009）。有研究发现，耐盐植物生长促生菌巴西固氮螺菌（*Azospirillum brasilense*）在 200 mmol/L 浓度 NaCl 胁迫下也能够产生 IAA（Nabti 等，2010）。另外，其他一些研究人员研究发现，耐盐植物生长促生菌如普城沙雷菌 RR2-5-10（*Serratia plymuthica* RR2-5-10）、窄食单胞菌 e-p10（*Stenotrophomonas rhizophila* e-p10）、假单胞菌 TSAU13（*Pseudomonas chlororaphis* TSAU13）和荧光假单胞菌 SPB2145（*Pseudomonas fluorescens* SPB2145）在浓度为 1.5% 的 NaCl 胁迫下同样也都能够合成 IAA（Gamberdieva，2011，2012）。共生细菌产生的激素能够被植物直接利用，参与植物根形态建成，包括植物根系根尖数量以及根系表面积的增加，从而增强植物对养分的吸收，进而改善植物在盐胁迫下的生长能力（Dodd 等，2010；Egamberdieva 和 Kucharova，2009；Etesami 和 Alikhani，2016；Etesami 等，2015a，2015b；Etesami 和 Beattie，2017；Kurepin 等，2015；Postma 和 Lynch，2011）。即通过增加植物根系根毛数量使植物获得更加庞大的根系系统（根表面积增大），能够增强植物从土壤中获得更多养分的能力，从而增强对盐胁迫的耐受能力（Boiero 等，2007；Egamberdieva 和 Kucharova，2009；Mantelin 和 Touraine，2004）。这样的研究事例有很多，例如在 100 mmol/L 的 NaCl 胁迫条件下在小麦上接种能够产生 IAA 的植物生长促生菌，比如橘黄假单胞菌 TSAU22（*Pseudomonas aurantiaca* TSAU22）、极端东方化假单胞菌 TSAU6（*Pseudomonas extremorientalis* TSAU6）和极端东方化假单胞菌 TSAU20（*Pseudomonas extremorientalis* TSAU20），与未接种植物生长促生菌的小麦植株相比，试验组小麦的根系和地上部分生物量分别增长了 40% 和 52%（Egamberdieva 和 Kucharova，2009）。植物在响应盐胁迫的过程中会产生脱落酸引起气孔关闭，减少水分蒸腾损失，并且调控根系分枝生长增加，以提高对水分的吸收能力，从而促进叶片生长（Tardieu 等，2010）。有研究指出，一些植物生长促生菌，比如生脂固氮螺菌（*Azospirillum lipoferum*），能够促进受到盐胁迫危害的植物体内脱落酸含量的增加（Cohen 等，2009）。一般来说，植物生长促生菌菌株能够通过产生激素类物质调节植物体内的激素平衡，从而缓解遭受盐胁迫危害的植物受到的盐害危害，并促进植物的生长（Postma 和 Lynch，2011）。

（2）通过产生富含铁载体的细胞（增加铁营养元素）、促进氮元素的获取和转化、促进无机磷酸盐和含钾矿物的溶解以实现植物吸收养分的增加，促进植株体内离子稳态平衡的建立（Bell 等，2015；Etesami 等，2017；Etesami 和 Maheshwari，2018）。众所周知，N、P 和 K 元素是植物生长发育所必需的大量营养元素，而生长环境中的高 Na^+ 含量则会抑制植物对这些营养元素的吸收（Etesami 和 Maheshwari，2018）。换言之，盐胁迫破坏了植物细胞内原有的离子稳态平衡环境，造成离子稳态失衡，从而导致生理活动紊乱（Nadeem 等，2009）。受到盐胁迫危害的植物可以通过限制 Na^+ 流入细胞、从木质部中回流 Na^+ 到达根部并在根部排出过多的 Na^+ 等方式来保护自身免受高盐胁迫的危害（Chinnusamy 等，2006）。根据以往报道的研究，许多植物生长促生菌可以通对 N、P 和 K 等元素的吸收来提高细胞内的 K^+/Na^+ 比，从而减少 Na^+ 在盐胁迫条件下的过度积累，以维持细胞内离子稳态平衡的状态（Belimov 等，2009；Bharti 等，2014；Damodaran 等，2014；Etesami 和 Maheshwari，2018；Kang 等，2014b；Mayak 等，2004b；Nadeem 等，2009；Tewari 和 Arora，2014）。

（3）减少因应激反应引起的乙烯大量产生。乙烯是一种植物内源激素，参与调节植物的生长和发育。根据植物细胞内乙烯的浓度，它可以发挥抑制性的作用（乙烯是一种植物根系伸长的抑制剂），也能够发挥刺激性的作用（比如促进种子萌发、根毛发育、茎伸长和果实成熟等）（Glick，2005；Penrose 等，2001；Siddikee 等，2011）。在盐胁迫条件下，植物因应激反应产生过量的乙烯会导致植物根和芽的生长受到抑制（Glick 等，2007）。1-氨基环丙烷-1-羧酸（1-aminocyclopropane-1-carboxylic acid，ACC）是乙烯生物合成的前体物质，ACC 脱氨酶产生菌能够将 ACC 催化降解为 α-酮丁酸盐和氨（氮营养元素的供体）。因此，植物生长促生菌通过降解 ACC 不仅能够降低乙烯的生物合成，缓解因乙烯过量合成带来的毒害作用，还能够增加植物对氮营养元素的吸收，从而促进盐胁迫下植物的生长。一般来说，含 ACC 脱氨酶的植物生长促生菌能够减少植物应激性乙烯的产生，对那些遭受盐胁迫危害的植物而言，能够促进这些植物侧根的生长，在侧根数量、长度和根重等方面有所增加（Shahzad 等，2010）。有研究指出，细菌 ACC 脱氨酶的活性与植物根系的生长发育具有直接关联性（Shaharoona 等，2006）。在盐胁迫条件下，细菌 ACC 脱氨酶能够介导植物形成较长的根系，这样植物就能够从深层土壤中吸收相对较多的水分和营养元素，从而提高自身营养，在遭受盐胁迫时就能够利用更多的水分，从而降低盐害对自身生长和发育的不利影响（Zahir 等，2008）。许多植物生长促生菌都能够合成 ACC 脱氨酶（Glick，2010），这其中包括耐盐植物生长促生菌（Etesami 和 Beattie，2018）。目前，有许多研究表明，植物生长促生菌可以通过产生 ACC 脱氨酶来促进植物的生长和发育，并提高其对盐胁迫的耐受能力（Ahmed 等，2004；Etesami 和 Beattie，2018；Glick，2014；Hamdia 等，2004；Mayak 等，2004a，2004b；Nadeem 等，2007，2009，2013；Paul 和 Sarma，2006；Penrose 和 Glick，2003；Saravanakumar 和 Samiyappan，2007；Zaki 等，2004）。

（4）代谢相容性物质也称为渗透压调节物质，比如脯氨酸、小分子糖、甘氨酸、甜菜碱、多元醇和胆碱等（Essghaier 等，2014）。这些调节物质是一些化学性质多样的有机化合物，物质分子具有极性但不带电荷，并且有良好的可溶性，这些调节物质在细胞中即使达到很高的浓度也不会干扰细胞的正常代谢（Ford，1984；Saxena 等，2013）。植物通过在细胞中积累这些渗透调节物质，以保持细胞在正常的生理条件下维持最大程度的膨胀性。遭受盐胁迫危害的植物通过细胞自身的渗透调节，可以保护自身减轻或者免受盐胁迫带来的危害（Evelin 等，2009；Gill 和 Tuteja，2010；Serraj 和 Sinclair，2002）。以往的研究也表明，植物生长促生菌，比如固氮螺菌（*Azospirillum*）、类产碱假单胞菌（*Pseudomonas pseudoalcaligenes*）、伯克霍尔德氏菌（*Burkholderia*）、节杆菌（*Arthrobacter*）和芽孢杆菌（*Bacillus*）等，都可以促进盐胁迫条件下植物细胞中渗透调节物质（如脯氨酸、甘氨酸、类甜菜碱类化合物、游离氨基酸和可溶性糖等）的增加，通过渗透调节来减缓植物遭受盐渍胁迫的危害（Bano 等，2013；Barka 等，2006；Bharti 等，2014；Cassan 等，2009；Damodaran 等，2014；Jha 等，2011；Kandowangko 等，2009；Paul 和 Nair，2008；Sarma 和 Saikia，2014；Shintu 和 Jayaram，2015；Sziderics 等，2007；Vardharajula 等，2011）。

（5）抗氧化酶的合成。活性氧（ROS）是细胞代谢的副产物，在正常的生长条件下，植物的各种组织和细胞中 ROS 的含量较低（Gill 和 Tuteja，2010）。众所周知，低水平的 ROS 可以作为信号分子激活植物的应激反应和防御系统（Pitzschke 等，2006）。然而，在盐胁迫的条件下，植物体内的 ROS 物质（比如羟基自由基、超氧自由基、单线态氧、过氧化氢和超氧化物等）会大量产生，甚至过量增加（Ahmad 和 Umar，2011；Andersend 等，2004；Apel 和 Hirt，2004；Chaves 等，2002；Groß 等，2013；Zhang 等，2016）。过量的 ROS 物质会导致细胞不同组分产生氧化损伤，比如 DNA 氧化损伤、脂质过氧化、膜结合蛋白受到破坏、蛋白质的生物合成过程受到抑制、生物膜流动性降低以及各种酶活性的丧失等，这些生理过程的异常会导致盐胁迫下植物细胞功能的中断与细胞死亡（Dietz 等，2016；Gill 和 Tuteja，2010；Gupta 和 Huang，2014）。换言之，盐胁迫会导致植物体内一种称为"氧化应激胁迫"的次级胁迫反应，这是由于产生了高于阈值水平的各种 ROS 物质所导致的细胞内稳态失衡所引起的（Gill 和 Tuteja，2010）。但是，植物体内含有抗氧化酶系统，比如过氧化氢酶（catalase，CAT）、谷胱甘肽过氧化物酶（glutathione peroxidase，GPX）、超氧化物歧化酶（superoxide dismutase，SOD）、抗坏血酸过氧化物酶（ascorbate peroxidase，APX）、单脱氢抗坏血酸还原酶（monodehydroascorbate reductase，MHDAR）、谷胱甘肽还原酶（glutathione reductase，GR）以及非酶类组分包括谷胱甘肽、半胱氨酸、生育酚和抗坏血酸等。植物可以利用体内的这些物质通过抗氧化防御系统消除或者中和逆境条件下植物细胞产生的 ROS，消除因氧化应激反应产生的过多 ROS 带来的不利影响，保护自身免受 ROS 的危害（Miller 等，2010）。目前，各种研究表明，在逆境胁迫条件下植物的抗氧化能力和体内抗氧化酶的浓度有强烈的相关性（Asada，1999；Gupta 等，2005；Sairam 和 Srivastava，2002）。通过产生抗氧化酶，一些植物生长促生菌也可以消除 ROS。据报道，植物对逆境胁迫的耐受性也与抗氧化酶的活性高低密切相关（Štajner 等，1997）。有大量的证据表明，植物生长促生菌可以通过调控植物抗氧化酶的活性来减轻多种植物由于盐胁迫而引起的氧化胁迫（比如 ROS 的产生）（Damodaran 等，2014；Han 和 Lee，2005；Kim 等，2014；Sandhya 等，2010；Wang 等，2012）。值得注意的是，目前与植物共生的植物生长促生菌所引起的植物体内抗氧化酶活性水平变化的机制尚不明确（Kaushal 和 Wani，2016），还需要进一步研究。

（6）挥发性化合物的产生（Kaushal 和 Wani，2016）。产生的这些挥发性化合物能够影响植物响应外界胁迫的信号传导途径（Bhattacharyya 等，2015）。

（7）改变植物-水分利用关系和增强植物细胞渗透调节（Creus 等，2004）。

（8）通过抗菌、空间和养分的争夺以及诱导植物的系统性抗性来控制各种植物病害（根、叶等广谱病原体）的发生（Compant 等，2005；Singh 等，2011）。

1.4 结论

盐生植物根际的微生物群落习惯称作根际微生物群落，与盐渍土微生物群落协同作用，以可持续以及环境友好的方式提高植物抗盐胁迫能力，从而保证作物在盐渍土环境中能够正常生长，并获得产量的提高。利用植物生长促生菌提高作物的耐盐能力开辟了农业生产应对盐胁迫的新技术。植物生长促生菌还可以促进农作物从盐渍土中吸收养分的增加，从而减少盐渍土农业生产对化肥的需求。这种方式有助于减少农用化学品的投入,也有利于盐渍化土壤的修复,因此可以维持盐胁迫条件下农业的可持续发展。但是我们也应该清醒地认识到,尽管目前在这方面开展了许多研究,但对植物生长促生菌和植物之间的协同作用提高作物耐盐能力的互作机制的了解仍然不够透彻。此外,耐盐植物生长促生菌作为生物肥料的应用还需要在大田条件下进行更多深入和细致的研究。

[致谢]

感谢德黑兰大学为这项研究提供了必要的设施和资金。

参考文献

Adesemoye AO, Obini M, Ugoji EO, 2008. Comparison of plant growth-promotion with *Pseudomonas aeruginosa* and *Bacillus subtilis* in three vegetables. Braz J Microbiol 39: 423-426.

Ahmad P, Umar S, 2011. Oxidative stress: role of antioxidants in plants. Studium Press, New Delhi.

Ahmed W, Shahroona B, Zahir ZA, Arshad M, 2004. Inoculation with ACC-deaminase containing rhizobacteria for improving growth and yield of wheat. Pak J Agric Sci 41: 119.

Akbarimoghaddam H, Galavi M, Ghanbari A, Panjehkeh N, 2011. Salinity effects on seed germination and seedling growth of bread wheat cultivars. Trakia J Sci 9: 43-50.

Albacete A, Ghanem ME, Martínez-Andújar C, Acosta M, Sánchez-Bravo J, Martínez V, Lutts S, Dodd IC, Pérez-Alfocea F, 2008. Hormonal changes in relation to biomass partitioning and shoot growth impairment in salinized tomato (*Solanum lycopersicum* L.) plants. J Exp Bot 59: 4119-4131.

Andersen L, Williams M, Serek M, 2004. Reduced water availability improves drought tolerance of potted miniature roses: is the ethylene pathway involved? J Hortic Sci Biotechnol 79: 1-13.

Apel K, Hirt H, 2004. Reactive oxygen species: metabolism, oxidative stress, and signal transduction. Annu Rev Plant Biol 55: 373-399.

Araus JL, Slafer GA, Royo C, Serret MD, 2008. Breeding for yield potential and stress adaptation in cereals. Crit Rev Plant Sci 27: 377-412.

Arbona V, Marco AJ, Iglesias DJ, López-Climent MF, Talon M, Gomez-Cadenas A, 2005. Carbohydrate depletion in roots and leaves of salt-stressed potted *Citrus clementina* L. Plant Growth Regul 46: 153-160.

Arkhipova TN, Prinsen E, Veselov SU, Martinenko EV, Melentiev AI, Kudoyarova GR, 2007. Cytokinin producing bacteria enhance plant growth in drying soil. Plant Soil 292: 305-315.

Asada K, 1999. The water-water cycle in chloroplasts: scavenging of active oxygens and dissipation of excess photons. Annu Rev Plant Biol 50: 601-639.

Ashraf M, 2004. Some important physiological selection criteria for salt tolerance in plants. Flora

Morphol Distrib Funct Ecol Plants 199: 361-376.

Ashraf M, McNeilly T, 2004. Salinity tolerance in Brassica oilseeds. Crit Rev Plant Sci 23: 157-174.

Atak M, Kaya MD, Kaya G, Çikili Y, Çiftçi CY, 2006. Effects of NaCl on the germination, seedling growth and water uptake of triticale. Turk J Agric For 30: 39-47.

Bae D, Yong K, Chun S, 2006. Effect of salt (NaCl) stress on germination and early seedling growth of four vegetables species. J Cent Eur Agric 7: 273-282.

Bai YC, Yan YY, Zuo WG, Gu CH, Xue WJ, Mei LJ, Shan YH, Feng K, 2017. Coastal mudflat saline soil amendment by dairy manure and green manuring. Int J 478 Agron 2017: 1-9.

Bano A, Fatima M, 2009. Salt tolerance in *Zea mays* (L). following inoculation with *Rhizobium* and *Pseudomonas*. Biol Fertil Soils 45: 405-413.

Bano Q, Ilyas N, Bano A, Zafar N, Akram A, Hassan F, 2013. Effect of *Azospirillum inoculationon* maize (*Zea mays* L.) under drought stress. Pak J Bot 45: 13-20.

Barassi CA, Ayrault G, Creus CM, Sueldo RJ, Sobrero MT, 2006. Seed inoculation with *Azospirillum mitigates* NaCl effects on lettuce. Sci Horticult 109: 8-14.

Barka EA, Nowak J, Clément C, 2006. Enhancement of chilling resistance of inoculated grapevine plantlets with a plant growth-promoting rhizobacterium, Burkholderia phytofirmans strain PsJN. Appl Environ Microbiol 72: 7246-7252.

Barnawal D, Bharti N, Maji D, Chanotiya CS, Kalra A, 2014. ACC deaminase-containing Arthrobacter protophormiae induces NaCl stress tolerance through reduced ACC oxidase activity and ethylene production resulting in improved nodulation and mycorrhization in *Pisum sativum*. J Plant Physiol 171: 884-894.

Barra PJ, Inostroza NG, Acuña JJ, Mora ML, Crowley DE, Jorquera MA, 2016. Formulation of bacterial consortia from avocado (*Persea americana* Mill.) and their effect on growth, biomass and superoxide dismutase activity of wheat seedlings under salt stress. Appl Soil Ecol 102: 80-91.

Bauder TA, Davis JG, Waskom RM, Cardon GE, Follett RH, Franklin WT, Heil RD, 2004. Managing saline soils. Service in action; no 0503.

Belimov AA, Dodd IC, Hontzeas N, Theobald JC, Safronova VI, Davies WJ, 2009. Rhizosphere bacteria containing 1-aminocyclopropane-1-carboxylate deaminase increase yield of plants grown in drying soil via both local and systemic hormone signalling. New Phytol 181: 413-423.

Bell CW, Asao S, Calderon F, Wolk B, Wallenstein MD, 2015. Plant nitrogen uptake drives rhizosphere bacterial community assembly during plant growth. Soil Biol Biochem 85: 170-182.

Bharti N, Barnawal D, Awasthi A, Yadav A, Kalra A, 2014. Plant growth promoting rhizobacteria alleviate salinity induced negative effects on growth, oil content and physiological status in *Mentha arvensis*. Acta Physiol Plant 36: 45-60.

Bhattacharyya D, Yu SM, Lee YH, 2015. Volatile compounds from *Alcaligenes faecalis* JBCS1294 confer salt tolerance in *Arabidopsis thaliana* through the auxin and gibberellin pathways and differential modulation of gene expression in root and shoot tissues. Plant Growth Regul 75: 297-306.

Bianco C, Defez R, 2009. *Medicago truncatula* improves salt tolerance when nodulated by an indole-3-acetic acid-overproducing Sinorhizobium meliloti strain. J Exp Bot 60: 3097-3107.

Boiero L, Perrig D, Masciarelli O, Penna C, Cassán F, Luna V, 2007. Phytohormone production by three strains of *Bradyrhizobium japonicum* and possible physiological and technological implications. Appl Microbiol Biotechnol 74: 874-880.

Bouhmouch I, Souad-Mouhsine B, Brhada F, Aurag J, 2005. Influence of host cultivars and Rhizobium species on the growth and symbiotic performance of *Phaseolus vulgaris* under salt stress. J Plant Physiol 162: 1103-1113.

Cassan F, Maiale S, Masciarelli O, Vidal A, Luna V, Ruiz O, 2009. Cadaverine production by *Azospirillum brasilense* and its possible role in plant growth promotion and osmotic stress mitigation. Eur J Soil Biol 45: 12-19.

Çavusoglu K, Kabar K, 2010. Effects of hydrogen peroxide on the germination and early seedling growth of barley under NaCl and high temperature stresses. EurAsian J Biosci 4: 70-79.

Chaves MM, 2002. How plants cope with water stress in the field? Photosynthesis and growth. Ann Bot 89: 907-916.

Chinnusamy V, Jagendorf A, Zhu J K, 2005. Understanding and improving salt tolerance in plants. Crop Sci 45: 437-448.

Chinnusamy V, Zhu J, Zhu J K, 2006. Salt stress signaling and mechanisms of plant salt tolerance. In: Genetic engineering. Springer, p. 141-177.

Cohen AC, Travaglia CN, Bottini R, Piccoli PN, 2009. Participation of abscisic acid and gibberellins produced by endophytic *Azospirillum* in the alleviation of drought effects in maize. Botany 87: 455-462.

Cohen AC, 2015. *Azospirillum brasilense* ameliorates the response of *Arabidopsis thaliana* to drought mainly via enhancement of ABA levels. Physiol Plant 153: 79-90.

Compant S, Duffy B, Nowak J, Clément C, Barka EA, 2005. Use of plant growth-promoting bacteria for biocontrol of plant diseases: principles, mechanisms of action, and future prospects. Appl Environ Microbiol 71: 4951-4959.

Cramer GR, Nowak RS, 1992. Supplemental manganese improves the relative growth, net assimilation and photosynthetic rates of salt-stressed barley. Physiol Plant 84: 600-605.

Creus CM, Sueldo RJ, Barassi CA, 2004. Water relations and yield in *Azospirillum*-inoculated wheat exposed to drought in the field. Can J Bot 82: 273-281.

Damodaran T, 2014. Rhizosphere and endophytic bacteria for induction of salt tolerance in gladiolus grown in sodic soils. J Plant Interact 9: 577-584.

Dantas BF, Ribeiro LS, Aragão CA, 2005. Physiological response of cowpea seeds to salinity stress. Rev Bras Sementes 27: 144-148.

Dietz KJ, Mittler R, Noctor G, 2016. Recent progress in understanding the role of reactive oxygen species in plant cell signaling. Plant Physiol 171: 1535-1539.

Dimkpa C, Weinand T, Asch F, 2009. Plant-rhizobacteria interactions alleviate abiotic stress conditions. Plant Cell Environ 32: 1682-1694.

Dodd IC, Pérez-Alfocea F, 2012. Microbial amelioration of crop salinity stress. J Exp Bot 63: 3415-3428.

Dodd IC, Zinovkina NY, Safronova VI, Belimov AA, 2010. Rhizobacterial mediation of plant hormone status. Ann Appl Biol 157: 361-379.

Dolatabadian A, Sanavy SAMM, Ghanati F, 2011. Effect of salinity on growth, xylem structure and anatomical characteristics of soybean. Notulae Sci Biol 3: 41.

Dwivedi S, Upadhyaya H, Subudhi P, Gehring C, Bajic V, Ortiz R, 2010. Enhancing abiotic stress tolerance in cereals through breeding and transgenic interventions. Plant Breed Rev 33: 31-114.

Egamberdieva D, 2011. Survival of *Pseudomonas extremorientalis* TSAU20 and *P. chlororaphis* TSAU13 in the rhizosphere of common bean (*Phaseolus vulgaris*) under saline conditions. Plant Soil Environ 57: 122-127.

Egamberdieva D, 2012. *Pseudomonas chlororaphis*: a salt-tolerant bacterial inoculant for plant growth stimulation under saline soil conditions. Acta Physiol Plant 34: 751-756.

Egamberdieva D, Jabborova D, 2013. Biocontrol of cotton damping-off caused by *Rhizoctonia solani* in salinated soil with rhizosphere bacteria. Asian Australas J Plant Sci Biotechnol 7: 31-38.

Egamberdieva D, Kucharova Z, 2009. Selection for root colonising bacteria stimulating wheat growth in saline soils. Biol Fertil Soils 45: 563-571.

Egamberdieva D, Gafurova L, Islam R, 2007. Salinity effects on irrigated soil chemical and biological properties in the Syr Darya Basin of Uzbekistan.

Egamberdieva D, 2011. Bacteria able to control foot and root rot and to promote growth of cucumber in salinated soils. Biol Fertil Soils 47: 197-205.

Egamberdieva D, Berg G, Lindström K, Räsänen LA, 2013. Alleviation of salt stress of symbiotic *Galega officinalis* L. (goat's rue) by co-inoculation of Rhizobium with root-colonizing *Pseudomonas*. Plant Soil 369: 453-465.

Egamberdiyeva D, 2007. The effect of plant growth promoting bacteria on growth and nutrient uptake of maize in two different soils. Appl Soil Ecol 36: 184-189.

Egamberdiyeva D, Islam KR, 2008. Salt-tolerant rhizobacteria: plant growth promoting traits and physiological characterization within ecologically stressed environments. In: Plant-bacteria interactions: strategies and techniques to promote plant growth. p. 257-281.

Esitken A, Yildiz HE, Ercisli S, Donmez MF, Turan M, Gunes A, 2010. Effects of plant growth promoting bacteria (PGPB) on yield, growth and nutrient contents of organically grown strawberry. Sci Horticult 124: 62-66.

Essa TA, 2002. Effect of salinity stress on growth and nutrient composition of three soybean (*Glycine max* L. Merrill) cultivars. J Agron Crop Sci 188: 86-93.

Essghaier B, Dhieb C, Rebib H, Ayari S, Boudabous ARA, Sadfi-Zouaoui N, 2014. Antimicrobial behavior of intracellular proteins from two moderately halophilic bacteria: strain J31 of *Terribacillus halophilus* and strain M3-23 of *Virgibacillus marismortui*. J Plant Pathol Microbiol 5: 1.

Etesami H, 2018. Can interaction between silicon and plant growth promoting rhizobacteria benefit in alleviating abiotic and biotic stresses in crop plants? Agric Ecosyst Environ 253: 98-112. https://doi.org/10.1016/j.agee.2017.11.007.

Etesami H, Alikhani HA, 2016. Co-inoculation with endophytic and rhizosphere bacteria allows reduced application rates of N-fertilizer for rice plant. Rhizosphere 2: 5-12.

Etesami H, Beattie GA, 2017. Plant-microbe interactions in adaptation of agricultural crops to abiotic stress conditions. In: Probiotics and plant health. Springer, p. 163-200.

Etesami H, Beattie GA, 2018. Mining halophytes for plant growth-promoting halotolerant bacteria to enhance the salinity tolerance of non-halophytic crops. Front Microbiol 9: 148. 1-20. https://doi.org/10.3389/fmicb.2018.00148.

Etesami H, Maheshwari DK, 2018. Use of plant growth promoting rhizobacteria (PGPRs) with multiple plant growth promoting traits in stress agriculture: action mechanisms and future prospects. Ecotoxicol Environ Saf 156: 225-246. https://doi.org/10.1016/j.ecoenv.2018.03.013.

Etesami H, Alikhani HA, Hosseini HM, 2015a. Indole-3-acetic acid (IAA) production trait, a useful screening to select endophytic and rhizosphere competent bacteria for rice growth promoting agents. MethodsX 2: 72-78.

Etesami H, Alikhani HA, Hosseini HM, 2015b. Indole-3-acetic acid and 1-aminocyclopropane-1-carboxylate deaminase: bacterial traits required in rhizosphere, rhizoplane and/or endophytic competence by beneficial bacteria. In: Bacterial metabolites in sustainable agroecosystem. Springer, pp. 183-258.

Etesami H, Emami S, Alikhani HA, 2017. Potassium solubilizing bacteria (KSB): mechanisms, promotion of plant growth, and future prospects - a review. J Soil Sci Plant Nutr 17: 897-911.

Evelin H, Kapoor R, Giri B, 2009. *Arbuscular mycorrhizal* fungi in alleviation of salt stress: a review. Ann Bot 104: 1263-1280.

Flowers TJ, 2004. Improving crop salt tolerance. J Exp Bot 55: 307-319.

Ford CW, 1984. Accumulation of low molecular weight solutes in water-stressed tropical legumes. Phytochemistry 23: 1007-1015.

Garcia C, Hernandez T, 1996. Influence of salinity on the biological and biochemical activity of a calciorthird soil. Plant Soil 178: 255-263.

Gill SS, Tuteja N, 2010. Reactive oxygen species and antioxidant machinery in abiotic stress tolerance in crop plants. Plant Physiol Biochem 48: 909-930.

Gill SS, 2016. Piriformospora indica: potential and significance in plant stress tolerance. Front Microbiol 7: 332.

Glick BR, 2005. Modulation of plant ethylene levels by the bacterial enzyme ACC deaminase. FEMS Microbiol Lett 251: 1-7.

Glick BR, 2010. Using soil bacteria to facilitate phytoremediation. Biotechnol Adv 28: 367-374.

Glick BR, 2014. Bacteria with ACC deaminase can promote plant growth and help to feed the world. Microbiol Res 169: 30-39.

Glick BR, Todorovic B, Czarny J, Cheng Z, Duan J, McConkey B, 2007. Promotion of plant growth by bacterial ACC deaminase. Crit Rev Plant Sci 26: 227-242.

Godfray HCJ, 2010. Food security: the challenge of feeding 9 billion people. Science 327: 812-818.

Groß F, Durner J, Gaupels F, 2013. Nitric oxide, antioxidants and prooxidants in plant defence responses. Front Plant Sci 4: 419.

Grover M, Ali SZ, Sandhya V, Rasul A, Venkateswarlu B, 2011. Role of microorganisms in adaptation of agriculture crops to abiotic stresses. World J Microbiol Biotechnol 27: 1231-1240.

Gupta B, Huang B, 2014. Mechanism of salinity tolerance in plants: physiological, biochemical, and molecular characterization. Int J Genomics 2014: 701596.

Gupta KJ, Stoimenova M, Kaiser WM, 2005. In higher plants, only root mitochondria, but not leaf mitochondria reduce nitrite to NO, in vitro and in situ. J Exp Bot 56: 2601-2609.

Hamdia MAE-S, Shaddad MAK, Doaa MM, 2004. Mechanisms of salt tolerance and interactive effects of *Azospirillum brasilense* inoculation on maize cultivars grown under salt stress conditions. Plant Growth Regul 44: 165-174.

Hamilton CE, Bever JD, Labbé J, Yang X, Yin H, 2016. Mitigating climate change through managing constructed - microbial communities in agriculture. Agric Ecosyst Environ 216: 304-308.

Han HS, Lee KD, 2005. Physiological responses of soybean-inoculation of Bradyrhizobium japonicum with PGPR in saline soil conditions. Res J Agric Biol Sci 1: 216-221.

Han QQ, 2014. Beneficial soil bacterium *Bacillus subtilis* (GB03) augments salt tolerance of white clover. Front Plant Sci 5: 525.

Hashem A, Abd_ Allah EF, Alqarawi AA, Aldubise A, Egamberdieva D, 2015. Arbuscular mycorrhizal fungi enhances salinity tolerance of Panicum turgidum Forssk by altering photosynthetic and antioxidant pathways. J Plant Interact 10: 230-242.

Heidari M, Jamshid P, 2010. Interaction between salinity and potassium on grain yield, carbohydrate content and nutrient uptake in pearl millet. ARPN J Agric Biol Sci 5: 39-46.

Heidari M, Mousavinik SM, Golpayegani A, 2011. Plant growth promoting rhizobacteria (PGPR) effect on physiological parameters and mineral uptake in basil (*Ocimum basilicum* L.) under water stress. ARPN J Agric Biol Sci 6: 6-11.

Horneck DA, Ellsworth JW, Hopkins BG, Sullivan DM, Stevens RG, 2007. Managing salt-affected soils for crop production. Oregon State University Extension Service, Covallis.

Hu Y, Schmidhalter U, 2004. Limitation of salt stress to plant growth. In: Hock E (ed) Plant toxicology, vol 4. Marcel Dekker, New York, pp. 191-224.

James RA, Blake C, Byrt CS, Munns R, 2011. Major genes for Na^+ exclusion, *Nax1* and *Nax2* (wheat *HKT1*; 4 and *HKT1*; 5), decrease Na^+ accumulation in bread wheat leaves under saline and waterlogged conditions. J Exp Bot 62: 2939-2947.

Jamil A, Riaz S, Ashraf M, Foolad MR, 2011. Gene expression profiling of plants under salt stress. Crit Rev Plant Sci 30: 435-458.

Jha Y, Subramanian RB, Patel S, 2011. Combination of endophytic and rhizospheric plant growth promoting rhizobacteria in *Oryza sativa* shows higher accumulation of osmoprotectant against saline stress. Acta Physiol Plant 33: 797-802.

Kandowangko NY, Suryatmana G, Nurlaeny N, Simanungkalit RDM, 2009. Proline and abscisic acid content in droughted corn plant inoculated with *Azospirillum* sp. and *Arbuscular mycorrhizae* fungi. Hayati J Biosci 16: 15-20.

Kang SM, 2014a. Plant growth-promoting rhizobacteria reduce adverse effects of salinity and osmotic stress by regulating phytohormones and antioxidants in *Cucumis sativus*. J Plant Interact 9: 673-682.

Kang SM, 2014b. Gibberellin secreting rhizobacterium, *Pseudomonas putida* H-2-3 modulates the hormonal and stress physiology of soybean to improve the plant growth under saline and drought conditions. Plant Physiol Biochem 84: 115-124.

Kaushal M, Wani SP, 2016. Rhizobacterial-plant interactions: strategies ensuring plant growth promotion under drought and salinity stress. Agric Ecosyst Environ 231: 68-78.

Kechid M, Desbrosses G, Rokhsi W, Varoquaux F, Djekoun A, Touraine B, 2013. The *NRT2. 5* and *NRT2. 6* genes are involved in growth promotion of *Arabidopsis* by the plant growth-promoting rhizobacterium (PGPR) strain *Phyllobacterium brassicacearum* STM196. New Phytol 198: 514-524.

Khodarahmpour Z, Ifar M, Motamedi M, 2012. Effects of NaCl salinity on maize (*Zea mays* L.) at germination and early seedling stage. Afr J Biotechnol 11: 298-304.

Kim K, Jang YJ, Lee SM, Oh BT, Chae JC, Lee KJ, 2014. Alleviation of salt stress by *Enterobacter* sp. EJ01 in tomato and *Arabidopsis* is accompanied by up-regulation of conserved salinity responsive factors in plants. Mol Cell 37: 109.

Kurepin LV, Park JM, Lazarovits G, Bernards MA, 2015. Burkholderia phytofirmans-induced shoot and root growth promotion is associated with endogenous changes in plant growth hormone levels. Plant Growth Regul 75: 199-207.

Ladha JK, Tirol-Padre A, Punzalan GC, Castillo E, Singh U, Reddy CK, 1998. Nondestructive estimation of shoot nitrogen in different rice genotypes. Agron J 90: 33-40.

Lugtenberg B, Kamilova F, 2009. Plant-growth-promoting rhizobacteria. Annu Rev Microbiol 63: 541-556.

Lugtenberg BJJ, Malfanova N, Kamilova F, Berg G, 2013. Plant growth promotion by microbes. Mol Microb Ecol Rhizosphere 1 & 2: 559-573.

Manchanda G, Garg N, 2008. Salinity and its effects on the functional biology of legumes. Acta Physiol Plant 30: 595-618.

Mantelin S, Touraine B, 2004. Plant growth-promoting bacteria and nitrate availability: impacts on root development and nitrate uptake. J Exp Bot 55: 27-34.

Mantri N, Patade V, Penna S, Ford R, Pang E, 2012. Abiotic stress responses in plants: present and future. In: Abiotic stress responses in plants. Springer, p. 1-19.

Mapelli F, 2013. Potential for plant growth promotion of rhizobacteria associated with Salicornia growing in Tunisian hypersaline soils. BioMed Res Int 2013: 248078.

Mayak S, Tirosh T, Glick BR, 2004a. Plant growth-promoting bacteria confer resistance in tomato plants to salt stress. Plant Physiol Biochem 42: 565-572.

Mayak S, Tirosh T, Glick BR, 2004b. Plant growth-promoting bacteria that confer resistance to water stress in tomatoes and peppers. Plant Sci 166: 525-530.

Miller GAD, Suzuki N, Ciftci-Yilmaz S, Mittler RON, 2010. Reactive oxygen species homeostasis and signalling during drought and salinity stresses. Plant Cell Environ 33: 453-467.

Munns R, 2002. Comparative physiology of salt and water stress. Plant Cell Environ 25: 239-250.

Munns R, 2005. Genes and salt tolerance: bringing them together. New Phytol 167: 645-663.

Munns R, Tester M, 2008. Mechanisms of salinity tolerance. Annu Rev Plant Biol 59: 651-681.

Nabti E, 2010. Restoration of growth of durum wheat (*Triticum durum* var. *waha*) under saline conditions due to inoculation with the rhizosphere bacterium *Azospirillum brasilense* NH and extracts of the marine alga *Ulva lactuca*. J Plant Growth Regul 29: 6-22.

Nadeem SM, Zahir ZA, Naveed M, Arshad M, 2007. Preliminary investigations on inducing salt tolerance in maize through inoculation with rhizobacteria containing ACC deaminase activity. Can J Microbiol 53: 1141-1149.

Nadeem SM, Zahir ZA, Naveed M, Arshad M, 2009. Rhizobacteria containing ACC-deaminase confer salt tolerance in maize grown on salt-affected fields. Can J Microbiol 55: 1302-1309.

Nadeem SM, Zahir ZA, Naveed M, Nawaz S, 2013. Mitigation of salinity-induced negative impact on the growth and yield of wheat by plant growth-promoting rhizobacteria in naturally saline conditions. Ann Microbiol 63: 225-232.

Nadeem SM, Ahmad M, Zahir ZA, Javaid A, Ashraf M, 2014. The role of mycorrhizae and plant growth promoting rhizobacteria (PGPR) in improving crop productivity under stressful environments. Biotechnol Adv 32: 429-448.

Nadeem SM, Ahmad M, Naveed M, Imran M, Zahir ZA, Crowley DE, 2016. Relationship between in vitro characterization and comparative efficacy of plant growth-promoting rhizobacteria for improving cucumber salt tolerance. Arch Microbiol 198: 379-387. https://doi.org/10.1007/s00203-016-1197-5.

Naqvi SM, Ansari R, 1974. Estimation of diffusible auxin under saline growth conditions. Cell MolLife Sci 30: 350-350.

Naz I, Bano A, Ul-Hassan T, 2009. Isolation of phytohormones producing plant growth promoting rhizobacteria from weeds growing in Khewra salt range, Pakistan and their implication in providing salt tolerance to *Glycine max* L. Afr J Biotechnol 8: 5762-5768.

Neamatollahi E, Bannayan M, Darban AS, Ghanbari A, 2009. Hydropriming and osmopriming effects on cumin (*Cuminum cyminum* L.) seeds germination. World Acad Sci Eng Technol 57: 526-529.

Netondo GW, Onyango JC, Beck E, 2004. Sorghum and salinity. Crop Sci 44: 797-805.

Nia SH, Zarea MJ, Rejali F, Varma A, 2012. Yield and yield components of wheat as affected by salinity and inoculation with Azospirillum strains from saline or non-saline soil. J Saudi Soc Agric Sci 11: 113-121.

Nunkaew T, Kantachote D, Nitoda T, Kanzaki H, Ritchie RJ, 2015. Characterization of exopolymeric substances from selected *Rhodopseudomonas palustris* strains and their ability to adsorb sodium ions. Carbohydr Polym 115: 334-341.

Ondrasek G, Rengel Z, Romic D, Poljak M, Romic M, 2009. Accumulation of non/essential elements in radish plants grown in salt-affected and cadmium-contaminated environment. Cereal Res Commun 37: 9-12.

Orhan F, 2016. Alleviation of salt stress by halotolerant and halophilic plant growth-promoting bacteria in wheat (*Triticum aestivum*). Braz J Microbiol 47: 621-627. https://doi.org/10.1016/j.bjm.2016.04.001.

Othman Y, Al-Karaki G, Al-Tawaha AR, Al-Horani A, 2006. Variation in germination and ion uptake in barley genotypes under salinity conditions. World J Agric Sci 2: 11-15.

Parida AK, Das AB, 2005. Salt tolerance and salinity effects on plants: a review. Ecotoxicol Environ Saf 60: 324-349.

Paul D, Lade H, 2014. Plant-growth-promoting rhizobacteria to improve crop growth in saline soils: a review. Agron Sustain Dev 34: 737-752.

Paul D, Nair S, 2008. Stress adaptations in a plant growth promoting rhizobacterium (PGPR) with increasing salinity in the coastal agricultural soils. J Basic Microbiol 48:

378-384.

Paul D, Sarma YR, 2006. Plant growth promoting rhizhobacteria (PGPR) -mediated root proliferation in black pepper (*Piper nigrum* L.) as evidenced through GS root software. Arch Phytopathol Plant Protect 39: 311-314.

Paul D, Anandaraj M, Kumar A, Sarma YR, 2005. Antagonistic mechanisms of fluorescent pseudomonads against *Phytophthora capsici* in black pepper (*Piper nigrum* L.). J Spices Aromat Crop 14: 122-129.

Penrose DM, Glick BR, 2003. Methods for isolating and characterizing ACC deaminase-containing plant growth-promoting rhizobacteria. Physiol Plant 118: 10-15.

Penrose DM, Moffatt BA, Glick BR, 2001. Determination of 1 - aminocycopropane - 1 - carboxylic acid (ACC) to assess the effects of ACC deaminase-containing bacteria on roots of canola seedlings. Can J Microbiol 47: 77-80.

Pérez-Alfocea F, Albacete A, Ghanem ME, Dodd IC, 2010. Hormonal regulation of source-sink relations to maintain crop productivity under salinity: a case study of root-to-shoot signalling in tomato. Funct Plant Biol 37: 592-603.

Pitzschke A, Forzani C, Hirt H, 2006. Reactive oxygen species signaling in plants. Antioxid RedoxSignal 8: 1757-1764.

Postma JA, Lynch JP, 2011. Root cortical aerenchyma enhances the growth of maize on soils with suboptimal availability of nitrogen, phosphorus, and potassium. Plant Physiol 156: 1190-1201.

Prakash L, Prathapasenan G, 1990. Interactive effect of NaCl salinity and gibberellic acid on shoot growth, content of abscisic acid and gibberellin-like substances and yield of rice (*Oryza sativa* L. var. GR-3). Proc Plant Sci 100: 173-181.

Qadir M, Ghafoor A, Murtaza G, 2000. Amelioration strategies for saline soils: a review. Land-Degrad Dev 11: 501-521.

Quispel A, 1988. Bacteria-plant interactions in symbiotic nitrogen fixation. Physiol Plant 74: 783-790.

Rabie GH, Almadini AM, 2005. Role of bioinoculants in development of salt-tolerance of *Vicia faba* plants under salinity stress. Afr J Biotechnol 4: 210.

Rabie GH, Aboul-Nasr MB, Al-Humiany A, 2005. Increased salinity tolerance of cowpea plants by dual inoculation of an *Arbuscular mycorrhizal* fungus Glomus clarum and a nitrogen-fixer *Azospirillum brasilense*. Mycobiology 33: 51-60.

Rahman S, Matsumuro T, Miyake H, Takeoka Y, 2000. Salinity-induced ultrastructural alterationsin leaf cells of rice (*Oryza sativa* L.). Plant Prod Sci 3: 422-429.

Rahnama A, James RA, Poustini K, Munns R, 2010. Stomatal conductance as a screen for osmotic stress tolerance in durum wheat growing in saline soil. Funct Plant Biol 37: 255-263.

Rajput L, Imran A, Mubeen F, Hafeez FY, 2013. Salt-tolerant PGPR strain *Planococcus rifietoensis* promotes the growth and yield of wheat (*Triticum aestivum* L.) cultivated in saline soil. Pak J Bot 45: 1955-1962.

Ramadoss D, Lakkineni VK, Bose P, Ali S, Annapurna K, 2013. Mitigation of salt stress in wheat seedlings by halotolerant bacteria isolated from saline habitats. SpringerPlus 2: 6.

Rengasamy P, 2006. World salinization with emphasis on Australia. J Exp Bot 57: 1017-1023. https://doi.org/10.1093/jxb/erj108.

Rincón A, Valladares F, Gimeno TE, Pueyo JJ, 2008. Water stress responses of two Mediterranean tree species influenced by native soil microorganisms and inoculation with a plant growth promoting rhizobacterium. Tree Physiol 28: 1693-1701.

Rozema J, Flowers T, 2008. Crops for a salinized world. Science 322: 1478-1480.

Sairam RK, Srivastava GC, 2002. Changes in antioxidant activity in sub-cellular fractions of tolerant and susceptible wheat genotypes in response to long term salt stress. Plant Sci 162: 897-904.

Sakhabutdinova AR, Fatkhutdinova DR, Bezrukova MV, Shakirova FM, 2003. Salicylic acid prevents the damaging action of stress factors on wheat plants. Bulg J Plant Physiol 29: 314-319.

Sandhya V, Ali SZ, Grover M, Reddy G, Venkateswarlu B, 2010. Effect of plant growth promoting *Pseudomonas* spp. on compatible solutes, antioxidant status and plant growth of maize under drought stress. Plant Growth Regul 62: 21-30.

Saravanakumar D, Samiyappan R, 2007. ACC deaminase from *Pseudomonas fluorescens* mediated saline resistance in groundnut (*Arachis hypogea*) plants. J Appl Microbiol 102: 1283-1292.

Sarma RK, Saikia R, 2014. Alleviation of drought stress in mung bean by strain *Pseudomonas aeruginosa* GGRJ21. Plant Soil 377: 111-126.

Saxena SC, Kaur H, Verma P, Petla BP, Andugula VR, Majee M, 2013. Osmoprotectants: potential for crop improvement under adverse conditions. In: Plant acclimation to environmental stress. Springer, p. 197-232.

Serraj R, Sinclair TR, 2002. Osmolyte accumulation: can it really help increase crop yield underdrought conditions? Plant Cell Environ 25: 333-341.

Shaharoona B, Arshad M, Zahir ZA, 2006. Effect of plant growth promoting rhizobacteria containing ACC-deaminase on maize (*Zea mays* L.) growth under axenic conditions and on nodulation in mung bean (*Vigna radiata* L.). Lett Appl Microbiol 42: 155-159.

Shahbaz M, Ashraf M, 2013. Improving salinity tolerance in cereals. Crit Rev Plant Sci 32: 237-249.

Shahzad SM, Khalid A, Arshad M, 2010. Screening rhizobacteria containing ACC-deaminase for growth promotion of chickpea seedlings under axenic conditions. Soil Environ 29: 38-46.

Shannon MC, Grieve CM, 1998. Tolerance of vegetable crops to salinity. Sci Horticult 78: 5-38.

Shaterian J, Waterer D, De Jong H, Tanino KK, 2005a. Differential stress responses to NaCl salt application in early-and late-maturing diploid potato (*Solanum* sp.) clones. Environ Exp Bot 54: 202-212.

Shaterian J, Waterer D, Jong HD, Tanino KK, 2005b. Differential stress responses to NaCl salt application in early- and late-maturing diploid potato (*Solanum* sp.) clones. Environ Exp Bot 54: 202-212. https://doi.org/10.1016/j.envexpbot.2004.07.005.

Shintu PV, Jayaram KM, 2015. Phosphate solubilising bacteria (Bacillus polymyxa) -an effective approach to mitigate drought in tomato (*Lycopersicon esculentum* Mill.). Trop Plant Res 2: 17-22.

Shirokova Y, Forkutsa I, Sharafutdinova N, 2000. Use of electrical conductivity instead of soluble salts for soil salinity monitoring in Central Asia. Irrig Drain Syst 14: 199-206.

Shrivastava UP, Kumar A, 2013. Characterization and optimization of 1-aminocyclopropane-1-

carboxylate deaminase (ACCD) activity in different rhizospheric PGPR along with *Microbacterium sp. strain ECI*-12A. Int J Appl Sci Biotechnol 1: 11-15.

Shrivastava P, Kumar R, 2015. Soil salinity: a serious environmental issue and plant growth promoting bacteria as one of the tools for its alleviation. Saudi J Biol Sci 22: 123-131.

Shukla PS, Agarwal PK, Jha B, 2012. Improved salinity tolerance of *Arachis hypogaea* (L.) by the interaction of halotolerant plant-growth-promoting rhizobacteria. J Plant Growth Regul 31: 195-206.

Siddikee MA, Chauhan PS, Anandham R, Han G H, Sa T, 2010. Isolation, characterization, and use for plant growth promotion under salt stress, of ACC deaminase-producing halotolerant bacteria derived from coastal soil. J Microbiol Biotechnol 20: 1577-1584.

Siddikee MA, Glick BR, Chauhan PS, Jong Yim W, Sa T, 2011. Enhancement of growth and salt tolerance of red pepper seedlings (*Capsicum annuum* L.) by regulating stress ethylene synthesis with halotolerant bacteria containing 1-aminocyclopropane-1-carboxylic acid deaminase activity. Plant Physiol Biochem 49: 427-434.

Singh JS, Pandey VC, Singh DP, 2011. Efficient soil microorganisms: a new dimension for sustainable agriculture and environmental development. Agric Ecosyst Environ 140: 339-353.

Singleton PW, Bohlool BB, 1984. Effect of salinity on nodule formation by soybean. Plant Physiol 74: 72-76.

Štajner D, Kevrešan S, Gašić O, Mimica-Dukić N, Zongli H, 1997. Nitrogen and *Azotobacter chroococcum* enhance oxidative stress tolerance in sugar beet. Biol Plant 39: 441.

Sziderics AH, Rasche F, Trognitz F, Sessitsch A, Wilhelm E, 2007. Bacterial endophytes contribute to abiotic stress adaptation in pepper plants (*Capsicum annuum* L.). Can J Microbiol 53: 1195-1202.

Tank N, Saraf M, 2010. Salinity-resistant plant growth promoting rhizobacteria ameliorates sodiumchloride stress on tomato plants. J Plant Interact 5: 51-58.

Tardieu F, Parent B, Simonneau T, 2010. Control of leaf growth by abscisic acid: hydraulic or non-hydraulic processes? Plant Cell Environ 33: 636-647.

Tavakkoli E, Fatehi F, Coventry S, Rengasamy P, McDonald GK, 2011. Additive effects of Na^+ and Cl^- ions on barley growth under salinity stress. J Exp Bot 62: 2189-2203.

Tejada M, Garcia C, Gonzalez JL, Hernandez MT, 2006. Use of organic amendment as a strategy for saline soil remediation: influence on the physical, chemical and biological properties of soil. Soil Biol Biochem 38: 1413-1421. https://doi.org/10.1016/j.soilbio.2005.10.017.

Tester M, Davenport R, 2003. Na^+ tolerance and Na^+ transport in higher plants. Ann Bot 91: 503-527.

Tewari S, Arora NK, 2014. Multifunctional exopolysaccharides from *Pseudomonas aeruginosa* PF23 involved in plant growth stimulation, biocontrol and stress amelioration in sunflower under saline conditions. Curr Microbiol 69: 484-494.

Tripathi AK, Mishra BM, Tripathi P, 1998. Salinity stress responses in the plant growth promoting rhizobacteria, *Azospirillum* spp. J Biosci 23: 463-471.

Upadhyay SK, Singh DP, 2015. Effect of salt-tolerant plant growth-promoting rhizobacteria on wheat plants and soil health in a saline environment. Plant Biol 17: 288-293.

Upadhyay SK, Singh JS, Saxena AK, Singh DP, 2012. Impact of PGPR inoculation on growth and antioxidant status of wheat under saline conditions. Plant Biol 14: 605-611.

Van Loon LC, Bakker P, Pieterse CMJ, 1998. Systemic resistance induced by rhizosphere bacteria. Annu Rev Phytopathol 36: 453-483.

Vardharajula S, Zulfikar Ali S, Grover M, Reddy G, Bandi V, 2011. Drought-tolerant plant growth promoting *Bacillus* spp. : effect on growth, osmolytes, and antioxidant status of maize under drought stress. J Plant Interact 6: 1-14.

Venkateswarlu B, Shanker AK, 2009. Climate change and agriculture: adaptation and mitigation strategies. Indian J Agron 54: 226.

Vimal SR, Singh JS, Karora N, Singh DP, 2016. PGPR: an effective bio-agent in stress agricultural management.

Wang CJ, 2012. Induction of drought tolerance in cucumber plants by a consortium of three plant growth-promoting rhizobacterium strains. PLoS One 7: e52565.

Wang L, Sun X, Li S, Zhang T, Zhang W, Zhai P, 2014. Application of organic amendments to a coastal saline soil in north China: effects on soil physical and chemical properties and tree growth. PLoS One 9: e89185.

Werner JE, Finkelstein RR, 1995. *Arabidopsis* mutants with reduced response to NaCl and osmotic stress. Physiol Plant 93: 659-666.

Wood NT, 2001. Nodulation by numbers: the role of ethylene in symbiotic nitrogen fixation. Trends Plant Sci 6: 501-502.

Xu GY, 2011. A novel rice calmodulin-like gene, *OsMSR2*, enhances drought and salt tolerance and increases ABA sensitivity in *Arabidopsis*. Planta 234: 47-59.

Yamaguchi T, Blumwald E, 2005. Developing salt-tolerant crop plants: challenges and opportunities. Trends Plant Sci 10: 615-620.

Yang J, Kloepper JW, Ryu CM, 2009. Rhizosphere bacteria help plants tolerate abiotic stress. Trends Plant Sci 14: 1-4.

Yao L, Wu Z, Zheng Y, Kaleem I, Li C, 2010. Growth promotion and protection against salt stressby *Pseudomonas putida* Rs-198 on cotton. Eur J Soil Biol 46: 49-54.

Yasmin F, Othman R, Saad MS, Sijam K, 2007. Screening for beneficial properties of Rhizobacteria isolated from sweet potato rhizosphere. Biotechnology 6: 49-52.

Yildirim E, Taylor AG, 2005. Effect of biological treatments on growth of bean plants under salt stress. Science 123: 1.

Yildirim E, Turan M, Ekinci M, Dursun A, Cakmakci R, 2011. Plant growth promoting rhizobacteria ameliorate deleterious effect of salt stress on lettuce. Sci Res Essays 6: 4389-4396.

Zahir ZA, Munir A, Asghar HN, Shaharoona B, Arshad M, 2008. Effectiveness of rhizobacteria containing ACC deaminase for growth promotion of peas (*Pisum sativum*) under drought conditions. J Microbiol Biotechnol 18: 958-963.

Zaki NM, Ahmed MA, Hassanein MS, 2004. Growth and yield of some wheat cultivars irrigated with saline water in newly cultivated land as affected by nitrogen fertilization. Ann Agric Sci Moshtohor 42: 515-525.

Zhang H, Kim MS, Sun Y, Dowd SE, Shi H, Paré PW, 2008. Soil bacteria confer plant salt tolerance by tissue-specific regulation of the sodium transporter HKT1. Mol Plant-Microbe Interact 21: 737-744.

Zhang M, Smith JAC, Harberd NP, Jiang C, 2016. The regulatory roles of ethylene and reactive oxygen species (ROS) in plant salt stress responses. Plant Mol Biol 91: 651-659.

Zhou W, Qin S, Lyu D, Zhang P, 2015. Soil sterilisation and plant growth-promoting rhizobacteria promote root respiration and growth of sweet cherry rootstocks. Arch Agron Soil Sci 61: 361-370.

2 盐生植物生长促生菌在缓解植物受到的盐胁迫危害中的作用

Leila Bensidhoum, Sylia Ait Bessai, Elhafid Nabti

[摘要]

土壤盐渍化是干旱和半干旱地区农业生产活动的主要制约因素之一。在世界范围内的灌溉土壤中，许多耕地都受到严重的土壤盐渍化影响，严重威胁着农业生产系统的可持续发展。由于对生长效率高的植物在生物量和生物能源生产方面具有强烈的需求，因此对于具有很高经济效益的植物而言，生长的可持续性显得更加重要，特别是在用于粮食生产之外的低质量土壤中。利用耐盐植物根际生长促生菌潜在的渗透调节机制，能够有效地促进植物适应盐分含量的土壤，并在这种土壤环境中良好地生长。植物生长促生菌通过自身生命活动产生的代谢物质能够促使植物激活多种生理活动，以应对因盐胁迫而产生的渗透胁迫。在盐胁迫条件下，这些植物生长促生菌通过刺激植物产生 ACC 脱氨酶的合成等调控机制来降低植物体内乙烯的含量，解除乙烯对植物生长的抑制作用，并通过合成产生各种植物激素来刺激植物的生长和发育。耐盐植物生长促生菌还可以通过促使海藻中合成渗透保护剂来协助植物抵抗盐胁迫。必须说明的是，耐盐植物生长促生菌对植物生长的调节作用与盐胁迫的浓度有关。在高盐浓度下，耐盐植物生长促生菌对植物的促生作用并不明显，类似于在低盐浓度下盐敏感细菌对植物生长的刺激作用。

[关键词]

耐盐根际菌；植物生长；盐胁迫；接种；修复

2.1 前言

目前，世界上快速增长的人口与地球维持生命的能力之间的矛盾正在变得不

L. Bensidhoum, S. Ait Bessai, E. Nabti (✉) 阿尔及利亚贝贾亚贝贾亚大学，E-mail：E. Nabti：nabtielhafid1977@yahoo.com

可调和（Ashraf 等，2012）。2004 年，联合国预计至 2025 年底全球的人口将超过 80 亿（亚洲和非洲将达到 68 亿），而到 2050 年底全球的人口规模将会超过 100 亿。因此，为了满足不断增长的人口对粮食的需求，到 2025 年全球粮食的产量需要较 2003 年增加 38%，而到 2050 年全球的粮食产量需要增加 57% 才能将全球增长人口的粮食供应维持在 2003 年的水平（Wild，2003）。

为了能够满足世界上不断增长的人口对口粮的需求，就需要增加粮食产量，特别是在那些由于不受控制的人类活动和环境退化而造成资源破坏的国家。农业生产是人类食物的主要来源，蔬菜在人类的饮食中扮演着非常重要的角色，能够给人体提供多种重要成分，比如矿物质、维生素、复杂的碳水化合物、高膳食纤维、低脂肪和充足的水分等，在人类的膳食结构中具有不可替代的位置。在未来很长的一段时间内，农业生产需要急剧增长 70%~100% 才能够满足全球不断增长的人口对粮食的需求（Lafe，2013）。然而，目前在世界范围内，各个地方的农业生产都受到各种生物和非生物胁迫的威胁，这些逆境胁迫会造成粮食生产的巨大损失（Singh 等，2014）。

众所周知，非生物胁迫造成的农业生产损失往往比生物胁迫造成的损失更为严重（Ashraf 等，2012）。比如，全球每年由于非生物胁迫而造成的粮食减产甚至能够达到 82%，更糟糕的是，目前世界范围内可耕作土地的有效生产面积正在不断地减少（Mona 等，2016）。农业耕作土地受到各种环境胁迫因素的威胁，比如盐渍化、干旱、极端温度、pH、矿物缺乏或毒性、重金属、养分消耗、土壤侵蚀等，生长在这些极端环境土地中的作物产量受到严重的影响（Chedlly 等，2008；Anonymous，2015）。而这些制约因素往往最终都会导致农用土壤的盐渍化，农业用地的盐渍化严重制约粮食作物的生长，给科学家和农民都带来了巨大的挑战。因此，需要投入更多的资源，以寻找治理盐渍土壤恢复农业生产的有效解决方案（Ashraf 等，2012）。

为了达到这一目的，包括物理化学改良、灌溉方法改进、植物繁殖技术和植物遗传改良等在内的多种针对盐渍化土地和粮食作物在内的方法技术都在生产中得到应用。然而这些方法不但费时、费力、成本高昂，其效果往往也很一般（Saghafi 等，2013；Selvakumar 等，2014；Shrivastava 和 Kumar，2015）。微生物产品的使用，特别是盐渍化环境下发现的植物生长促生菌对受到盐胁迫危害的作物具有缓解盐胁迫危害的作用，这说明植物生长促生菌很可能是一种提高在盐渍化土地上粮食作物生产能力的有效物质。不仅如此，植物生长促生菌还具有低成本、生态环境友好的特点，可以在正常土壤条件和逆境环境胁迫条件下提高作物的生产能力，并改善土壤环境。由于它们对植物根际的微生态起着非常重要的作用，因此对农作物的生长、产量和食品安全等具有重要的科学和农艺价值。

目前鉴定到的植物生长促生菌主要属于节杆菌属（*Arthrobacter*）、固氮菌属（*Azotobacter*）、固氮螺菌属（*Azospirillum*）、芽孢杆菌属（*Bacillus*）、肠杆菌属（*Enterobacter*）、假单胞菌属（*Pseudomonas*）、沙雷氏菌属（*Serratia*）（Gray和Smith，2005）和链霉菌属（*Streptomyces* spp.）（Tokala等，2002）等。植物生长促生菌通过改善植物生长和（或）保护其免受非生物和生物逆境胁迫，能够对植物的正常生长和发育产生积极的影响。这些微生物能够在植物根际定殖生长，并且在与其他微生物竞争过程中占据优势，能够占据植物根际最适生存的区域。植物生长促生菌可以利用共生植物根系的分泌物作为营养物质，进行生长并定殖于植物根际，通过多种不同的促生长机制来促进植物的生长和发育（Vacheron等，2013）。最近的研究热点则是通过利用植物生长促生菌来缓解植物由于盐胁迫遭受的危害。Kasim等（2016）研究表明在盐胁迫条件下，植物生长促生菌对植物的一些生长指标（比如幼苗长度、鲜重和干重以及相对含水量等）具有积极的影响，通过接种植物生长促生菌，植物的这些生长指标能够得到一定程度的缓解。

2.2 盐渍化

盐渍化是继干旱之后对粮食生产具有制约作用的第二大不利因素。盐渍化主要发生在干旱和半干旱地区，能够导致作物产量下降，并对世界范围内的植物和作物的正常生长产生不利影响（Pessarakli，1991；Saranga等，2001；Yadavet等，2011）。

土壤盐渍化是指土壤中可溶性盐分的积累达到一定程度，形成盐渍土（Ciseau，2006）。土壤盐渍化对农业生产、环境生态和社会的健康发展均产生负面影响（Rengasamy，2006）。全球每年因土壤盐渍化问题导致的经济损失高达120亿美元（Ghassemi等，1995）。根据联合国粮农组织在2005年对全球农业生产环境的预估，目前全世界受盐分影响的土地面积占整个大陆面积的6%以上，大约为8亿hm^2，包括4.34亿hm^2碱化土和3.97亿hm^2盐渍土。土壤中存在高浓度的Na^+时，这种土壤称之为"碱性土"；而当土壤中氯化物或者其他盐分含量较高时，称之为"盐渍化土"。据估计，目前全世界有20%的可耕地和33%的灌溉农业用地均受到土壤盐渍化的影响（Shrivastava和Kumar，2015）。

2.2.1 土壤盐渍化

土壤盐渍化被认为是限制粮食作物生产的最恶劣环境因素之一，因为目前世界上种植的大多数农作物都对土壤中高浓度的盐分十分敏感，然而遭受盐分影响的可耕作土地的面积却在日益增加（Shrivastava 和 Kumar，2015）。目前，全球每分钟约有 10 hm² 的可耕作土地遭到破坏（Griggs 等，2013），其中盐渍化土壤约占 30%（Buringh，1978）。因此，世界上可用于农业生产的土地面积每年都会减少 1%~2%，这对干旱和半干旱地区的农业生产影响更大（FAO，2002；Shilpi 等，2008）。预计到 2025 年，世界上约有 30% 的可耕作土地可能都会变成盐渍化土地（Munns，2002）；而到 2050 年，将有超过 50% 的农业土壤因盐分增加而变得不适合大多数粮食作物的生长（Jamil 等，2011；Dikilitas 和 Karakas，2012；Stanković 等，2015）。

土地盐渍化面积高速增长是由多种原因造成的，比如低降水量、地表高蒸发量（通过水蒸气蒸发的形式降低土壤含水量，从而提高盐分的相对含量）、原生岩石风化、盐水灌溉、化肥的过度使用、高温以及过度灌溉和其他的一些不良耕作习惯导致土壤下层潜盐通过毛细作用向土壤上层转移（Yao 等，2010；Shrivastava 和 Kumar，2015）。根据 Patel 等（2011）的研究报道，全球可耕作土地盐渍化面积的持续增加是由于在未盐渍化地区（特别是干旱和半干旱地区）不平衡的水分输入和输出。干旱和半干旱地区降水量较少，蒸发作用使土壤中的水分逸散到空气中，加之灌溉条件不足，没有足够的水分补充到土壤中，因此导致土壤表面相对含盐量逐渐升高。

2.2.2 盐化作用

土壤盐渍化是指土壤中水溶性盐分离子的过量积累，这些离子包括 Na^+、K^+、Ca^{2+}、Mg^{2+}、Cl^-、SO_4^{2-}、NO_3^-、HCO_3^- 和 CO_3^{2-} 等（Diby 和 Harshad，2014）。盐分以离子的形式存在于土壤中，这些离子可能通过土壤的风化作用释放出来，也可能来自灌溉用水或施用的肥料，或者经由浅层地下水向上迁移到土壤中。当降水不足以从土壤中浸出这些离子，并通过地表径流或者向下渗透的方式将这些盐分离子带走时，盐分就会在土壤中不断积累，最终导致土壤盐度的升高（Blaylock，1994）。

农业生产中所指的土壤盐渍化是指作物根区土壤水分中可溶性盐的高浓度（Diby 和 Harshad，2014）。据 Ghassemi 等（1995）报道，根据形成原因不同，土壤盐渍化可以分为原生盐渍化和次生盐渍化两种。土壤的原生盐渍化是指在正常的自然过程中形成的盐渍化土壤；而次生盐渍化多为通过人为因素诱发所导致的。

通过自然过程导致的土壤盐渍化因素主要包括沿海（大洋和内海）或大陆（含水盐化石层）地区高含盐量水的侵入以及内陆海洋、湖泊等盐地表水风化沉积物质以及母岩物质的溶解。相比之下，次生盐渍化主要是由一些人为的农业操作措施（比如微咸水灌溉、施肥、无机/有机土壤改良剂的应用、不适当的种植和轮作模式、森林砍伐、水库建设、盐田过度耕作等）造成的（Mitsuchi 等，1986；Rengasamy，2006；Biggs 等，2010）。

Rengasamy（2006）根据世界各地的盐渍化土壤形成过程和地下水状态将土壤盐渍化分为 3 种主要类型。不同于 Ghassemi 等（1995）将土壤盐渍化定义为"初生"或"次生"两种类型，根据这一分类，盐渍化土壤可以分成如下 3 类。

（1）地下水伴生盐渍化（groundwater-associated salinity，GAS）。当地下水由土壤中向土壤表层做向上移动时，会溶解土壤中的盐分，然而含盐水分到达土壤表面时水分会因蒸发作用逸散，而盐分则会逐渐积累在土壤表面。根据 Talsma（1963）报道，当地下水的水位低于土壤表面以下 1.5 m 时，盐分的累积量会增加。

（2）非地下水伴生盐渍化（non-groundwater-associated salinity，NAS）。在这种情况下，盐渍化土壤的盐分是通过雨水、风化和风积物等作用引入的，并在土壤表层中积累。

（3）灌溉水伴生盐渍化（irrigation-associated salinity，IAS）。这种盐渍化土成因是由于灌溉过程中淋洗作用不充分导致的，土壤中的盐分在植物根区积累，造成土壤盐渍化。

此外，目前全球约有 1/3 的灌溉土地（占全球农业产量的 30%以上）受到了盐渍化的影响（Munns 和 Tester，2008）。这是因为人为灌溉增加了额外的水分进入土壤系统，使土壤的水分平衡遭到破坏，并且通过灌溉方式增加水分，还会额外投入盐分，从而导致土壤含盐量逐渐增加，形成盐渍化土壤，不适合农业生产（Wu 等，2008）。

土壤盐分是通过测量土壤样本浸提的饱和土壤溶液的电流大小来定义的，这种"饱和溶液"通过电流的能力称为电导率（electrical conductivity，EC）（Shin 等，2016）。一般来说，如果土壤的含盐量不适合大多数农作物的生长，那么这种土壤就被称为盐渍土（Bui，2013）。但是美国农业部在 1954 年通过对土壤饱和溶液电导率的大小来定义盐渍化土壤：即在 25℃ 下，植物根际区域土壤电导率超过 4 dS/m（约为 40 mmol/L NaCl），可交换性 Na^+ 的百分比超过 15 或者更多，这种土壤才称为盐渍化土壤。

2.3 盐渍化对土壤、植物和微生物的影响

2.3.1 盐渍化对土壤质量的影响

土壤盐渍化不是最近才出现的问题，它在人类历史上一直是限制全球农业发

展的主要问题之一。土壤盐渍化是导致农作物产量降低的主要因素之一,也是农业用地逐渐被抛弃的一个主要原因(Dodd 和 Perez-Alfocea,2012)。

此外,土壤盐渍化不仅会降低大多数农作物的产量,还会影响土壤的理化性质和区域生态平衡(Benlloch-Gonzalez 等,2005; Shrivastava 和 Kumar,2015; Arshadullah 等,2017)。不仅如此,通过利用盐渍化土壤进行生产的经济回报也不容乐观(Hu 和 Schmidhalter,2002)。

大量的研究表明,过量的盐分积累会对土壤的结构产生不利影响,导致土壤团粒结构中渗透压升高,水势降低,植物和微生物细胞膨胀势下降。土壤孔隙是由土壤颗粒相互堆叠形成的大小不等、弯弯曲曲、形状各异的各种孔洞,是土壤中生物活动和土壤保育等土壤过程发生的主要场所。因此,土壤孔隙结构决定了土壤的主要理化特征(Kay,1990; Rengasamy 和 Olsson,1991)。但是,土壤中 Na^+ 浓度的升高会导致黏土微粒聚集并破坏土壤团粒之间的结合力导致土壤团粒结构分散。这些变化会导致黏土微粒堵塞土壤孔隙,特别是在纹理较细的土壤层中(Burrow 等,2002),并因此导致土壤通气量减少(土壤中厌氧条件普遍存在)、水分流失(土壤水分特性恶化)以及土壤压实度降低等严重后果(Abu-Sharar 等,1987)。此外,过量的 Na^+ 浓度还会提高土壤的 pH,增加地表径流带来的土壤侵蚀和地表水流失(Ondrasek 等,2010; Diby 和 Harshad,2014)。同时,多价阳离子的存在可以增加土壤颗粒对土层中有机物质的吸附(Mikutta 等,2007; Mavi 等,2012),从而减少土壤中有机物的分解量(Oades,1988; Six 等,2000),使土壤逐渐变得不适合植物根系的生长和发育。

2.3.2 盐渍化对植物生长发育的影响

土壤的盐渍化是导致全球许多地区农作物产量大幅下降的主要因素之一(Tank 和 Saraf,2010)。对于许多类型的植物尤其是非盐生植物,盐胁迫是一种生长限制性因子(Saghafi 等,2013),土壤中过量的盐分严重制约了农作物的生长。大多数农作物,比如谷类、牧草或园艺作物,特别容易受到土壤盐渍化的危害。尤其是溶解在灌溉水中的过量盐分或自然存在于根际土壤中的过量盐分对这些农作物生长发育的影响普遍存在(Ondrasek 等,2010)。已经有研究指出,蔬菜比牧草和谷物对盐胁迫更加敏感(Waller 和 Yitayew,2016)。

目前所有的土壤类型都含有一些水溶性盐类,这些盐类成分有很多可以作为植物生长发育所需的营养元素,但是这些营养元素的过量积累也不利于植物的正常生长,并最终抑制植物的生长(Paul,2012; Shrivastava,2015)。对植物的耐盐性进行定量量化并不容易,因为植物的耐盐性与许多环境因素直接关联,比如与土壤肥力、土壤理化条件、土壤中盐分的分布、灌溉习惯、气候以及植物自身因素（比如不同的生长阶段、根茎组织和种属多样性等）有关（Juan 等，2005；Niu 等，2012）。盐胁迫对植物生长和发育的影响包括植物种子发芽、营养生长和生殖生长等（Bartels 和 Sunkar，2005）。Rahman 等（2000）和 Jamil 等（2006）研究指出，种子萌发和幼苗生长早期是植物对盐胁迫最敏感的生长阶段，因为此时幼苗的根系与盐分过量积累的土壤直接接触。然而有一些农作物，比如大麦、水稻、小麦和玉米等，它们在生长阶段的早期则对盐胁迫表现出一定的抗性（Baniaghil 等，2013）。

植物根际高浓度的盐分能够通过复杂的相互作用严重抑制植物的生长（Arbona 等，2005）。盐胁迫对植物生长的影响主要通过两种途径实现，即引起植物缺水造成的"生理干旱"和过多摄入低需求元素（主要是 Cl^- 和 Na^+）而导致的离子毒性作用（Setia 等，2013；Bharti 等，2016；Hingole 和 Pathak，2016）。Na^+ 在植物细胞壁中的过度积累会迅速引发渗透胁迫，并导致细胞死亡（Munns，2002）。Cl^- 的过量吸收和积累对植物生长的影响主要体现在其可通过抑制硝酸还原酶的活性而破坏植物的光合作用（Xu 等，2000）。另外，在植物根际过量的 Na^+ 和氯化物的积累会导致植物根系细胞中与其他营养元素离子（如 K^+、NO_3^- 和 $H_2PO_4^-$ 等）的结合位点和转运蛋白被挤占，导致其他营养元素在植物细胞内的分配因竞争不过过量积累的 Na^+ 和氯化物而缺乏（Grattan 和 Grieve，1999；Tester 和 Davenport，2003）。植物细胞储存盐分的能力一旦耗尽，盐分离子就会在细胞间隙积累，导致细胞间隙渗透势迅速升高，从而引起细胞脱水和死亡（White 和 Broadley，2001；Sheldon 等，2004）。土壤环境中盐浓度的增加还会提高植物根际区域土壤溶液的渗透压，并且随着土壤溶液中盐浓度的增加，植物细胞的吸收作用会变得非常困难，最终还会发生失水作用（Baniaghil 等，2013）。如上所述，许多盐分离子也是植物所必需的营养元素，但土壤中的高盐水平会扰乱植物对营养元素吸收和分布的平衡，导致植物对 N、Ca、K、P、Fe 和 Zn 等营养元素摄入的减少，还会对需要在低 Na^+、高 K^+ 或 Ca^{2+} 条件下才能达到最佳功能的某些代谢过程造成紊乱（Blaylock，1994；Diby 和 Harshad，2014）。Shaheen 等（2013）已经证实，盐胁迫能够干扰植物叶片正常的气体交换过程，比如影响植物的光合作用，从而改变气孔下二氧化碳的浓度、影响净二氧化碳同化和植物的蒸腾速率。盐胁迫对植物光合作用的影响主要体现在通过完全或部分关闭气孔以及渗透胁迫作用降低光合作用（Meloni 等，2003）。此外，植物遭受到渗透胁迫的表现是叶面积和叶绿素含量的减少、光合作用中碳同化的减少以及落叶的发生（Shannon 和 Grive，1999）。另外，土壤盐渍化能够显著降低植物对磷元素的吸收，因为磷酸盐能够结合钙离子形成沉淀，不能被植物所吸收（Bano 和 Fatima，2009）。与其他非生物逆境胁迫一样，盐胁迫也会导致植物产生氧化应激反应，这是由植物细胞中单线态氧、超氧阴离子、过氧化氢和羟基自由基等活性氧的产生和积累导致的。活性氧的过量产生对植物的正常生长有很多负面作用，植物细胞内过量积累的活性氧会破坏植物细胞的生物膜、蛋白质、核酸和酶（Azevedo Neto 等，2008；Mishra 等，2009；Shahbazi 等，2011）。盐胁迫还能够改变植物的很多代谢过程，比如膜系统的紊乱、细胞分裂过程紊乱、细胞生长受到抑制、遗传毒性、生长缓慢以及细胞程序性死亡的过早激活等（Flowers，2004；Carillo 等，2011）。盐胁迫能够通过抑制植物小孢子的发生和雄蕊伸长、促进某些类型组织的细胞程序性死亡发生、导致胚珠败育和受精胚胎衰老从而对植物双受精之后的生殖发育产生不利影响（Shrivastava 和 Kumar，2015）。盐胁迫对植物细胞分化的周期也能够产生诸多不利影响。盐胁迫能够通过减少分生组织细胞中的细胞周期蛋白和细胞周期蛋白依赖激酶的表达和活性来抑制细胞分裂周期，从而限制植物细胞的分裂和生长（Javid 等，2011）。另外，盐胁迫还可以促使植物体内产生更多的 1-氨基环丙烷-1-羧酸（ACC），从而增加乙烯的生物合成速率和产量，这可能是导致植物组织发生各种生理变化的一个重要的原因（Zapata 等，2004；Tank 和 Saraf，2010）。对植物而言，乙烯是一种非常重要的气源性植物激素，在种子萌发、根系伸长、果实成熟和器官衰老等生理过程中都有非常重要的作用（Bleecker 和 Kende，2000）。但在植物体内，乙烯浓度超过一定的水平则会成为植物生长的负性调节剂（Holguin 和 Glick，2001；Huang 等，2003）。例如，过量的乙烯会加剧叶片和花瓣的脱落以及器官的衰老，并导致植物在发育早期死亡（Mayak 等，2004；Cheng 等，2007）。此外，乙烯还是植物响应外界刺激的一种应激信号分子（Hahn 和 March，2009）。据报道，在很多豆科植物中，乙烯是一种根系结瘤的负调节信号分子（Schaller，2012）。众所周知，过度的盐分胁迫还能够影响植物细胞内植物激素和其他植物生长调节因子的合成（Xiong 和 Zhu，2002）、蛋白质的合成、脂质代谢（Parida 和 Das，2005），以及维系植物正常生长所需的能量的产生（Larcher，1980）。而由盐胁迫引起的营养代谢紊乱的植物则更容易受到各种病原微生物的侵染（Romic 等，2008）。

目前有许多研究表明，盐胁迫能够抑制多种农作物种子的萌发，这些农作物包括水稻（Xu 等，2011）、小麦（Egamberdieva，2009）、玉米（Khodarahmpour 等，2012 年）、蚕豆（Rabie 和 Almadini，2005）、大豆（Essa，2002）、卷心菜和番茄（Bojovic 等，2010）等。据报道，在盐胁迫条件下，花生（Mensah 等，2006）和鹰嘴豆（Al-Mutawa，2003）的根长会受到盐分的抑制而缩短。这些研究结果与 Egamberdieva（2011）所观察到的结果一致：在无菌沙培条件下，随着盐分含量的增加，豆子幼苗茎和根的长度分别减少了 50% 和 70%。目前已经证实，盐胁迫条件能够抑制豇豆（Taffouo 等，2010）和甜菜（Dadkhah，2011）的光合作用。此外，一项类似的研究报告还指出，盐胁迫的盐分浓度与豆类植物叶绿素 a、叶绿素 b 以及总叶绿素含量之间呈反比关系（Qados，2011）。Golpayegani 和 Tilebeni（2011）研究证实，盐胁迫能够降低罗勒的光合作用、气孔导度、叶绿素含量和对矿物质的吸收。在茄子中，叶片中氯离子的积累往往伴随着硝酸根离子浓度的降低（Savvas 和 Lanz，2000），这说明盐胁迫能够抑制茄子对氮素营养的吸收。此外，Kapoor 和 Srivastava（2010）还研究证实，经不同盐浓度处理的黑革兰植物中蛋白质的含量明显下降；而在番茄中，Takagi 等（2009）研究指出，盐胁迫能够降低植物叶片光合作用并抑制碳同化物的运输，从而导致植物生物量的减少。

2.3.3　盐渍化对根际细菌的影响

尽管土壤微生物占土壤质量的比例不到 0.5%（w/w），但它们在保持土壤性质和肥力方面发挥着关键作用（Tate，2000）。微生物的生物量是土壤有机质组成中的一个重要组分，但是土壤中微生物的生物量对外界环境的各种因素都极为敏感，因此也十分容易发生变化。土壤微生物是土壤中有机质和土壤养分转化和循环的媒介。土壤微生物通过硝化、氨化、固氮以及其他能够导致土壤有机质分解的多种生理过程进行物质养分转化。因此，这些微生物能够直接或间接影响植物的生长和发育（Diby 和 Harshad，2014）。所以影响植物根际微生物及其功能的任何因素都会影响土壤养分的利用率和植物生长（Diby 和 Harshad，2014）。

因此，盐分不仅能够抑制植物的生长和发育，而且还对微生物尤其是植物根际微生物的功能和生物量产生多种影响（Ofek 等，2006）。盐胁迫能够通过改变土壤微生物的群落结构来降低土壤肥力，从而降低农作物的生产能力（Tripathi 等，2007；Andronov 等，2012）。微生物的这种作用被认为是盐胁迫条件下对植物生长发挥的间接调控作用（Chowdhury 等，2011）。

土壤中可溶盐的增加能够增强土壤溶液的渗透势，在这种环境下微生物细胞容易缺水。而对于植物根系而言，土壤溶液的高渗透势也使植物很难从土壤中吸取足够的水分，并不能给微生物提供额外的水分（Oren，1999）。因此，在高渗环境中的微生物会因失水死亡或保持休眠状态。因此，微生物释放的各种降解酶就相应减少，并且随着时间的延长已经释放的降解酶活性也会逐渐降低，最终导致土壤肥力降低（Tripathi 等，2007；Egamberdieva 等，2010）。生物体的细胞质膜对水有渗透性，但对其他代谢物没有渗透性。因此，植物细胞周围突然施加的高渗或低渗环境会导致胞质体积的急剧减少或增加，从而导致胞质分离以致细胞死亡（Nabti 等，2010）。然而，土壤微生物能够适应低渗透势的环境，但这个过程需要消耗更多的能量，因为它们必须消耗能量排出细胞内部的 Na^+ 以保持细胞质和周围介质之间的渗透平衡。但是微生物的生长能力和活性则会发生一定程度的降低（Jiang 等，2007；Ibekwe 等，2010）。此外，盐胁迫条件下，Na^+ 和其他伴生离子（如 Cl^- 和 CO_3^{2-}）的离子毒性作用以及土壤溶液的高 pH 也能够抑制微生物的生长和活性（Zahran，1997）。

Nelson 和 Mele（2007）研究表明，NaCl 能够通过影响植物根系分泌物的数量和/或质量来影响根际微生物的群落结构。Nabti 等（2010）报告称，盐胁迫通过引起参与细菌与植物根系共生相互作用最初附着步骤（吸附和锚定）有关的蛋白质的改变来抑制细菌的结瘤能力和固氮活性。除了可以改变组成细菌细胞表面的胞外多糖（exopolysaccharide，EPS）和脂多糖（lipopolysaccharide，LPS）组分，细菌和植物寄主之间膜葡聚糖含量变化导致的分子信号交换受损，也能够抑制细菌的趋根性迁移。Singleton 和 Bohlool（1984）也证明了土壤盐分能够对豆科植物与根瘤菌（*Rhizobium*）之间的互作产生干扰效应。Rabie 等（2005）报道，在盐碱条件下大豆、普通大豆和蚕豆等豆类植物的结瘤和固氮能力下降，固氮酶活性降低。在固氮螺菌中，高浓度盐胁迫能够抑制 *nifH* 基因的表达水平和固氮酶的活性水平，从而抑制该菌属微生物的固氮作用（Tripathi 等，2002）。此外，还有一些研究表明，随着土壤盐分的增加，土壤的呼吸作用有所降低（Yuan 等，2007；Wong 等，2009；Setia 等，2010）。土壤中的许多酶（比如脲酶、淀粉酶、转化酶、过氧化氢酶、碱性磷酸酶和 β-葡萄糖苷酶等）的活性也受到盐胁迫的强烈抑制（Frankenberger 和 Bingham，1982；Ghollaratta 和 Raiesi，2007；Pan 等，2013）。Omar 等（1994）和 Wichern 等（2006）研究指出，土壤中盐浓度水平增加 5% 以上，土壤中细菌和放线菌的总量将会急剧减少。

2.4 耐盐植物生长促生菌是缓解植物盐胁迫的新手段

值得注意的是，盐胁迫是影响干旱和半干旱地区植物生长和生产能力的主要非生物胁迫因素之一（Yadav 等，2011）。对大多数非盐生植物来说，盐胁迫是一个非常重要的生长限制因子。根据联合国粮农组织报告，全世界有 8.31 亿 hm^2 以上的土地受到盐胁迫的影响（Martinez-Beltran 和 Manzur，2005）。

此外，有许多研究者都得出类似的研究结果，即使用耐盐植物生长促生菌可以有效缓解盐胁迫对辣椒、番茄、大豆、小麦、油菜和莴苣等不同作物生长所造成的危害。有报道称，在盐胁迫下植物生长促生菌能够增加大豆（Naz 等，

2009）和小麦（Bharti 等，2016）的茎长、根长和干物质重量（图 2.1）。此外，有研究报道指出，在盐胁迫条件下极端东方化假单胞菌 TSAU20（*Pseudomonas extremorientalis* TSAU20）和绿假单胞菌 TSAU13（*Pseudomonas chlororaphis* TSAU13）能够促进各种作物的生长（Egamberdieva 和 Kucharova，2009）。Jha 和 Subramanian（2013）报道了植物生长促生菌在盐胁迫条件下能够影响植物细胞中离子浓度、养分吸收和抗氧化酶活性。此外，Sapsirisopa 等（2009）研究表明，使用巨大芽孢杆菌 A12ag（*Bacillus megaterium* A12ag）处理水稻能够显著增加植株茎中 N 和 P 的含量。各种植物生长促生菌还可以提高不同种类植物中叶绿素、类胡萝卜素、氮和磷的含量（Jha 和 Subramanian，2013；Saghafi 等，2013）。此外，有研究报道称，生脂固氮螺菌（*Azospirillum lipoferum*）和荧光假单胞菌（*Pseudomonas fluorescens*）同时接种能够显著增加小麦中叶绿素 a、叶绿素 b 和类胡萝卜素的含量（Saghafi 等，2013）。Nadeem 等（2006）也报道，在 10 dS/m 的盐渍化条件下，玉米种子接种植物生长促生菌后，植株新鲜叶片中的叶绿素 a、叶绿素 b 和类胡萝卜素的含量都有所增加。另一方面，各种植物生长促生菌还可以诱导受到盐胁迫危害的植物产生抗氧化酶，这是消除植物体内活性氧（ROS）的一个重要机制。Kumar 等（2003）、Jha 和 Subramanian（2013）报道，当植物生长促生菌存在时，植物体内的各种抗氧化酶则具有更强的活性。另外，施用植物生长促生菌还可以提高农作物种子的发芽率。在盐渍条件下，接种类产碱假单胞菌（*Pseudomonas pseudoalcaligenes*）和短小芽孢杆菌（*Bacillus*

| 对照 | 对照 | PGPR | PGPR | 对照 | 对照 | PGPR | PGPR |
| -盐 | +盐 | -盐 | +盐 | -盐 | +盐 | -盐 | +盐 |

图 2.1　接种植物生长促生菌对非盐和盐胁迫下小麦植株生长的影响

植物种在土中置于温室条件下生长，灌溉 150 mmol/L 的 NaCl 溶液，萌发后 60d 收获。

（Bharti 等，2016）

pumilus)的水稻种子发芽率较未接种的对照组明显增加（Jha 和 Subramanian，2013）。同时他们还发现，在 10 dS/m 的盐渍化条件下巨大芽孢杆菌 A12ag（*Bacillus megaterium* A12ag）也能够显著提高水稻种子的发芽率。这株菌株在 NaCl 浓度为 30~90 mmol/L 范围内对番茄种子的发芽率也有相似的促进作用，而在 120 mmol/L NaCl 浓度下，巨大芽孢杆菌 A12ag（*Bacillus megaterium* A12ag）还能够增加幼苗的干物质重量（Chookietwattana 和 Maneewan，2012）。

由于植物生长促生菌在盐胁迫下对植物的生长具有良好的促进作用，因此在世界范围内得到了广泛应用。植物生长促生菌除了能够缓解盐胁迫对植物造成的各种危害（表 2.1）外，一些研究还表明，植物生长促生菌还可以增强植物对其他非生物逆境胁迫的耐受性，比如干旱（Sandhya 等，2009；Timmusk 和 Wagner，1999）、冷害（Barka 等，2006）、金属毒性（Dell'Amico 等，2008；Bensidhoum 等，2016）以及高温（Ali 等，2009）等。

表 2.1　植物生长促生菌对不同作物盐胁迫的缓解作用

植物生长促生菌	作物种类	缓解机制	参考文献
表皮短杆菌 RS15（*Brevibacterium epidermidis* RS15）、云南微球菌 RS222（*Micrococcus yunnanensis* RS222）和阿氏芽孢杆菌 RS341（*Bacillus aryabhattai* RS341）	油菜	降低植株内乙烯含量，促进植物生长	Siddikee 等（2010）
根瘤菌（*Rhizobium*）和假单胞菌（*Pseudomonas*）	绿豆	促进植物生长，提高植物耐盐指数	Ahmad 等（2013）
恶臭假单胞菌 R4（*Pseudomonas putida* R4）和绿假单胞菌 R5（*Pseudomonas chlororaphis* R5）	棉花	提高棉花种子萌发率和幼苗生长，缓解盐胁迫	Egamberdieva 等（2015）
碱湖迪茨氏菌 STR1（*Dietzia natronolimnaea* STR1）	小麦	提高植物耐盐性，并增强抗氧化酶基因表达	Bharti 等（2016）
巨大芽孢杆菌（*Bacillus megaterium*）和肠杆菌（*Enterobacter* sp.）	黄秋葵	提高植物生长、增加抗氧化酶活性和上调活性氧途径相关基因	Habib 等（2016）
芽孢杆菌（*Bacillus*）	小麦	增加茎和根的长度	Orhan（2016）
莫哈韦芽孢杆菌 K78（*Bacillus mojavensis* K78）	小麦	改善植物生长，增加叶绿素含量和对营养元素的吸收，降低植物体内乙烯水平	Pourbabaee 等（2016）
根癌土壤杆菌（*Agrobacterium tumefaciens*）和假单胞菌（*Pseudomonas* sp.）	花生	促进植物生长，维持离子平衡，降低活性氧水平	Sharma 等（2016）
蜡状芽孢杆菌 Y5（*Bacillus cereus* Y5）、芽孢杆菌 Y14（*Bacillus* sp. Y14）和枯草芽孢杆菌 Y16（*Bacillus subtilis* Y16）	小麦	促进植物生长	Khan 等（2017）

各种耐盐植物生长促生菌在逆境胁迫环境下提高作物产量的现象涉及多种机制。这些机制总结见图2.2。

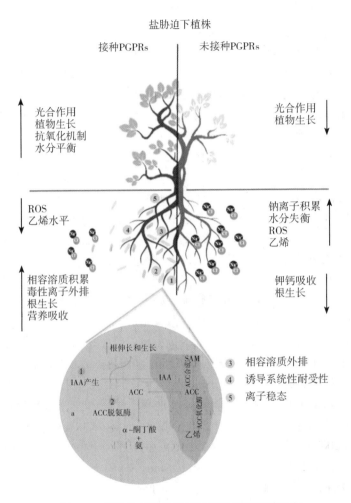

图2.2 植物生长促生菌缓解植物盐胁迫的机制

植物生长促生菌产生的1-氨基环丙烷-1-羧酸脱氨酶降低了植物体内乙烯的含量。ACC—1-氨基环丙烷-1-羧酸, IAA—吲哚-3-乙酸, SAM—腺苷蛋氨酸, ROS—活性氧。

2.4.1.1 1-氨基环丙烷-1-羧酸脱氨酶

乙烯是一种气源性植物激素，参与植物的生长和发育以及植物对各种非生物和生物胁迫的响应过程（Abeles 等，1992）。乙烯存在于植物体内，其对植物的生长发挥促进性还是抑制性作用取决于乙烯的浓度、植物所经历生理过程的性质以及植物处在的生长阶段（Shaharoona 等，2006）。在盐胁迫条件下，植物激素乙烯能够调节植物生长内在稳态，过量的乙烯会导致根和芽生长减少（Shrivastava 和 Kumar，2015；Vurukonda 等，2016）。一些植物生长促生菌通过 ACC 脱氨酶的作用能够降低胁迫条件下植物体内乙烯的浓度，从而促进植物的正常生长（Glick，1995；Penrose 等，2001 年；Glick，2014）。ACC 脱氨酶能将乙烯生物合成的直接前体 1-氨基环丙烷-1-羧酸（ACC）水解成氨和 α-酮丁酸，从而降低植物体内乙烯的水平（Hontzeas 等，2004；Glick 等，2007；Nascimento 等，2014；Santoyo 等，2016；Khan 等，2017）。因此，这些能够产生 ACC 脱氨酶的植物生长促生菌可以保护植物免受盐胁迫和干旱胁迫的有害影响（Glick 等，1998；Saleem 等，2007；Zahir 等，2009 年）。通过对产 ACC 脱氨酶的植物生长促生菌的研究，许多研究者报道了 ACC 脱氨酶在盐渍化土壤中促进植物正常生长的作用。El-Tarabily 等（2008）研究表明，产 ACC 脱氨酶的菲律宾链霉菌 15（*Streptomyces filipinensis* 15）和暗黑微绿链霉菌 26（*Streptomyces atrovirens* 26）可以通过降低番茄根和芽中内源 ACC 的水平进而促进植株生长。同样，正如 Mayak 等（2004）所报道的，番茄幼苗中产生的乙烯也能够在皮氏无色杆菌（*Achromobacter piechaudii*）的干预下减少。研究者发现，即使在高达 172 mmol/L NaCl 浓度的盐胁迫下，接种皮氏无色杆菌（*Achromobacter piechaudii*）番茄的鲜重和干重也得到了显著增加。Nadeem 等（2006）研究还指出，用产 ACC 脱氨酶的丁香假单胞菌（*Pseudomonas syringae*）和假单胞菌（*Pseudomonas* sp.）接种油菜，可促进油菜在盐胁迫下（12 dS/m）的生长，并能够提高油菜的产量。产 ACC 脱氨酶的植物生长促生菌还能够提高油菜、莴苣和玉米等作物的叶绿素含量（Glick 等，1997；Han 和 Lee，2005；Nadeem 等，2006）。Saravanakumar 和 Samiyappan（2007）研究表明，与缺乏 ACC 脱氨酶活性的假单胞菌相比，含有 ACC 脱氨酶活性的荧光假单胞菌（*Pseudomonas fluorescens*）能够增强花生植株对盐胁迫的耐受性，从而提高花生的产量。Shaharoona 等（2006）证实，能够生成 ACC 脱氨酶的植物生长促生菌可以改善豆科植物在盐胁迫条件下的结瘤效果。接种产 ACC 脱氨酶的植物生长促生菌能够与本土中华费氏根瘤菌（*Bradyrhizobium*）竞争促进植物结瘤。因此，对这些细菌的利用可能是促进豆科植物成功结瘤和提高结瘤效率的一个新途径（Zafar-ul-Hye 等，2013）。

2.4.1.2　植物激素的生物合成

植物激素在控制植物生长和发育方面起着至关重要的作用（Davies，1995）。生长素、赤霉素、细胞分裂素、脱落酸和乙烯是目前公认的 5 种经典植物激素

（Baca 和 Elmerich，2007）。研究表明，植物生长促生菌促进植物生长的最重要的机制是促进植物合成相应的植物激素。在正常和逆境条件下，植物激素可以调节植物多种生长发育状态。许多研究表明，盐胁迫会对植物内源激素的合成水平产生负面影响。事实上，外施植物生长调节剂能够缓解一些植物的盐胁迫危害，如小麦（Nabti，2007；Egamberdieva 和 Kucharova，2009）、玉米（Khalid 等，2013；Kaya 等，2013）、棉花（Egamberdieva 等，2015）和绿豆（Ahmad 等，2013）等。吲哚-3-乙酸（IAA）是调节植物生长发育的重要激素之一，它能够调节植物生长发育的许多方面，包括细胞分裂、种子萌发、组织分化、叶片扩展、根伸长和对各种逆境胁迫条件的抵御等（Egamberdieva，2008；Maleki 等，2010；Martínez-Viveros 等，2010）。Egamberdieva 等（2015）报道，在 NaCl 胁迫下接种能够合成 IAA 的植物生长促生菌恶臭假单胞菌 R4（*Pseudomonas putida* R4）和绿假单胞菌 R5（*Pseudomonas chlororaphis* R5）能够将种子的发芽率分别提高 64% 和 73%。在 100 mmol/L NaCl 胁迫下接种这两个菌株株系的试验组棉花的根和茎的长度比未进行接种的棉花显著增加。Khalid 等（2013）还报道，外施 IAA 能够减轻盐胁迫对玉米植株的危害。研究发现，能够过量产生 IAA 的植物生长促生菌菌株可以增加苜蓿属植物中脯氨酸的含量，以缓解各种逆境胁迫因素对植物生长造成的危害（Saghafi 等，2013）。已有报道从阿根廷沿海土壤（Sgroy 等，2009；Siddake 等，2010）、荒漠地区生长的辣椒根际土中（Marasco 等，2012）和盐碱地盐生植物根际（Mapelli 等，2013）中分离出的耐盐细菌可以产生 IAA，以调节植物对逆境胁迫的反应。植物生长促生菌产生的 IAA 除了具有刺激植物生长的作用外，还可以被植物吸收，与植物体内的内源 IAA 一起刺激 ACC 合成酶的合成（Penrose 和 Glick，2001；Gerhardt 等，2006）。ACC 是合成乙烯的直接前体，ACC 合成酶能够将 S-腺苷蛋氨酸（S-adenosylmethionine，SAM）转化为 ACC（Kende，1993）。植物合成的 ACC 大部分可以从种子和根中渗出，被植物生长促生菌吸收后，在 ACC 脱氨酶的作用下将 ACC 水解为氨和 α-酮丁酸酯（Glick 等，1998；Penrose，2000；Holguin 和 Glick，2001；Gerhardt 等，2006）。因此，植物体内 ACC 含量的降低能够导致乙烯含量降低，从而减轻高浓度乙烯对植物生长发育的抑制作用（图 2.2）。高浓度的 IAA 可以提高 ACC 的合成速率，因此可能会对植物根的生长和种子的萌发产生不利影响（Siddake 等，2011；Nakbanpote 等，2014）。Siddake 等（2010）研究表明，与只能够产生 ACC 脱氨酶的菌株相比，能够同时产生 ACC 脱氨酶和合成 IAA 的耐盐细菌能够在很大程度上提高油菜的根长和干物质重量。Glick 等（2007）研究表明，IAA 和 ACC 脱氨酶可以发生协同作用促进植物根系的生长。由于 IAA 可以诱导 ACC 的合成，并进一步转化为乙烯，但是 ACC 也能够在 ACC 脱氨酶的作用下水解，从而降低乙烯的水平，因此植物体内乙烯对植物根系生长等生理过程的影响就取决于 IAA 和 ACC 脱氨酶之间的平衡（Arora 等，2012）。正如 Glick（2014）所指出的，在盐胁迫条件下植物生长促生菌 ACC 脱氨酶和 IAA 之间的协同作用对于促进植物的生长和发育是必不可少的。

2.4.1.3 诱导系统性抗性：抗氧化酶的合成

作为固着生物，植物不能随意移动生长位置，因此经常受到多种生物和非生物逆境因素的胁迫。植物生长促生菌有助于缓解这两类胁迫对植物正常生长发育的危害。不同于植物生长促生菌通过直接机制（如植物激素合成、磷酸盐增溶和固氮）促进植物生长，进化过程中产生的间接机制能够保护植物，同时缓解甚至免受生物和非生物胁迫带来的危害。Jha 和 Saraf（2012）指出，植物生长促生菌可以诱导植物产生防御反应，产生获得性系统性抗性（systemic acquired resistance，SAR）和诱导性系统性抗性（induced systemic resistance，ISR），从而减少植物根际周围对植物有毒性作用的微生物群落。近年来，已经发表过许多关于植物生长促生菌在诱导植物获得耐受干旱、盐害、冷害、高温等非生物逆境胁迫能力的报道。"诱导系统性耐受性"一词是植物生长促生菌在植物防御反应中发挥的作用而提出的，植物生长促生菌能够诱导植物的多个生理过程产生物理和（或）化学变化，从而增强植物对非生物逆境胁迫的耐受性（Mantelin 和 Toureine，2004）。在盐胁迫条件下，植物生长促生菌诱导植物产生对盐胁迫耐受能力的机制中，抗氧化酶的产生非常重要。

盐胁迫等非生物逆境胁迫能够诱导植物产生大量的活性氧物质，包括超氧自由基、羟基自由基和过氧化氢等，这些物质会对生物大分子（如核酸、脂质和蛋白质等）造成氧化损伤，对植物正常的生理过程造成破坏，并最终导致植物的死亡（Del Rio 等，2003）。植物细胞氧化还原状态在逆境胁迫下的扰动需要通过植物的抗氧化系统来调节（Gill 和 Tutjja，2010；Jha 和 Subramanian，2013）。因此，这就需要植物产生各种抗氧化酶以及各种抗氧化剂，例如超氧化物歧化酶（SOD）、过氧化物酶（POX）、过氧化氢酶（CAT），以及非酶抗氧化剂如抗坏血酸、谷胱甘肽和 α-生育酚等（Del Rio 等，2003）。最近有很多研究报道了植物生长促生菌能够通过激活植物抗氧化系统而减少盐胁迫对植物的危害。Han 和 Lee（2005）提到了植物生长促生菌具有能够诱导生菜产生抗氧化酶的能力，他们认为，将两个植物生长促生菌菌株黏质沙雷氏菌（*Serratia marcescens*）和根瘤菌（*Rhizobium*）接种到盐渍土中，可以明显提高生菜中各种抗氧化酶、抗坏血酸、过氧化物酶以及谷胱甘肽还原酶等的含量和活性，减轻盐胁迫对生菜生长的影响。Kohler 等（2009）研究也指出，植物生长促生菌菌株诱导植物产生抗氧化酶提高了生菜对盐胁迫的耐受性。此外，有报道称植物生长促生菌菌株蜡状芽孢杆菌 Pb25（*Bacillus cereus* Pb25）能够显著增加豇豆中抗氧化酶（POD、SOD 和 CAT）的活性（Islam 等，2015）。为了检测盐胁迫下编码抗氧化酶基因表达的变化，Gururani 等（2013）将接种植物生长促生菌和未接种植物生长促生菌的马铃薯植株进行了不同的盐胁迫处理。结果表明，在逆境胁迫条件下，接种细菌的植株中编码 SOD 和 APX 蛋白的基因，其 mRNA 表达水平明显高于未接种的植株。类似地，在接种植物生长促生菌的马铃薯中，编码其他抗氧化酶蛋白如 CAT 和 GR 的基因 mRNA 表达水平也有所增加。因此，接种合适的植物生长促生菌是减轻植物氧化应激危害的有用工具。根据 Bharti 等（2016）研究，在盐胁迫条件下，与未接种植物生长促生菌的植物相比，接种耐盐植物生长促生菌的植物耐受逆境胁迫相关的基因表达水平有所提高。同时，研究人员还注意到，盐胁迫下接种植物生长促生菌的植株中编码 APX、CAT 和 POD 等蛋白的基因表达水平比未接种植物生长促生菌的植株高 3 倍。

2.4.1.4 离子稳态和养分吸收

盐胁迫能够导致植物细胞被水分平衡或离子稳态失衡，进而对植物的生长和发育产生不利的影响。大量的研究表明，盐胁迫条件增加了植物细胞内 Na^+ 的积累，但是 Ca^{2+} 和 K^+ 的吸收受到了抑制（Yildirim 等，2006；Zhang 等，2008；Tank 和 Saraf，2010；Shukla 等，2011；Rojas-Tapias 等，2012；Aamir 等，2013；Jha 和 Subramanian，2013；Younesi 等，2013；Sharma 等，2016）。Na^+ 和 Cl^- 在植物细胞内的过度积累会产生离子毒性，扰乱细胞正常的水分平衡和离子稳态，最终干扰植物的代谢过程和激素状态，并对植物细胞中酶的产生和活性、光合作用，以及营养运输等过程造成不利影响（Munn，2002）。维持植物细胞内较低的 Na^+/K^+ 比例是植物生长促生菌在高盐浓度下能够促进植物生长的主要机制之一（Munns 和 Tester，2008）。植物生长促生菌能够确保植物对矿物质养分的吸收和交换，缓解因盐胁迫条件下 Na^+ 和 Cl^- 的大量涌入而造成的植株养分失衡。根据 Hamdia 等（2004）、Ilangumaran 和 Smith（2017）的研究结果，植物生长促生菌可以通过诱导植物改变对阳离子（Na^+、K^+、Ca^+）吸收的选择性，减少叶片中 Na^+ 和 Cl^- 的积累，从而保持植物细胞内较高的 K^+/Na^+ 比值；并且增加植物根中 Na^+ 的外排，提高高亲和力 K^+ 转运蛋白的活性以促进植株对 K^+ 的吸收，从而促进遭受盐胁迫的植物的生长。此外，还有一些研究指出在盐胁迫条件下植物生长促生菌在降低植物 Na^+/K^+ 比值中的作用。研究发现，在盐胁迫条件下，经植物生长促生菌菌株（MBE01、MBE02 和 MBE05）处理的花生幼苗的 Na^+/K^+ 比值远远低于未接种细菌的花生幼苗（Sharma 等，2016）。Han 和 Lee（2005）的研究表明，植物生长促生菌菌株能够通过降低 Na^+/K^+ 的比值提高玉米植株对盐胁迫的耐受性。接种植物生长促生菌菌株苏拉氏短杆菌（*Brachybacterium saurashtrense*）和乳酪短杆菌（*Brevibacterium casei*）能够显著降低植物细胞中 Na^+ 的浓度。在 100 mmol/L NaCl 处理下，植物细胞中 Na^+/K^+ 的比率要明显低于对照处理（0 mmol/L NaCl）（SukLa 等，2011）。盐胁迫也影响植物离子稳态相关基因的表达。Pinedo 等（2015）阐明了盐胁迫对植物离子稳态相关基因（拟南芥 K^+ 转运蛋白 1、高亲和力 K^+ 转运蛋白 1、钠氢交换蛋白 2 和拟南芥 SOS1 蛋白）表达的影响。研究结果表明，接种植物生长促生菌菌株伯克霍尔德氏菌 PsJN（*Burkholderia phytofirmans* PsJN）之后，无论植物受到盐胁迫时间的长短，植物生长促生菌都能够诱导植物调节重要的离子稳态相关基因的表达。土壤枯草芽孢杆菌 GB03（*Bacillus subtilis* GB03）可以通过对植物 *HKT1* 基因的表达进行组织特异性调控，从而赋予拟南芥对盐胁迫的耐受性。在盐胁迫条件下，该菌株能够下调 *HKT1* 基因在植物根中的表达，但上调其在芽中的表达，从而导致整个植株中 Na^+ 的积累低于未接种的对照植株（Zhang 等，2008）。Niu 等（2015）报道了枯草芽孢杆菌 GB03（*Bacillus subtilis* GB03）菌株在 200 mmol/L NaCl 胁迫下能够诱导植物上调 *PtSOS1* 基因的表达，这与植株的芽、根和全株的 Na^+ 积累减少以及植株耐盐性的提高相一致。

盐胁迫降低了植物对养分的吸收能力，导致植物的生长发育迟缓。并且，盐胁迫还能够导致有毒离子和营养离子在根细胞中的转运蛋白结合位点之间的竞争性结合作用，导致养分在植物内部的转移、沉积和分配异常（Tester 和 Davenport，2003）。在这种竞争性相互作用下，整个植物所吸收到的营养元素就会减少。同时，大量研究还表明，在盐胁迫条件下植物生长促生菌还能促进植物对营养元素的吸收（Yildirim 等，2006；Zhang 等，2008；Shukla 等，2011；Aamir 等，2013；Jha 和 Subramanian，2013；Niu 等，2015；Sharma 等，2016）。Jha 和 Subramanian（2013）发现，接种植物生长促生菌菌株的植物体内 N、P 和 K 元素的浓度增加，而 Na 和 Ca 元素的浓度降低。Aamir 等（2013）研究指出，通过对植物同时接种根瘤菌和植物生长促生菌，谷物作物中的 N、P、K 元素浓度和蛋白质的含量都相应增加。

2.4.1.5 渗透相容保护物质

在盐胁迫条件下，为了适应盐胁迫带来的渗透压的波动，细菌主要采取了两种策略来减少外界环境渗透势的升高对自身的影响，即"盐入"机制和渗透调节物质的产生（Korcan 等，2015；Mohammadipanah 等，2015）。产生渗透调节物质是原核生物和真核生物缓解盐胁迫危害的主要机制。嗜盐和耐盐细菌适应渗透胁迫的策略主要包括 K^+ 和谷氨酸（胁迫应激第一反应的产物）的积累（Sleator 和 Hill，2001；Le Rudulier 等，2002）。如果渗透应激持续存在或强度增加，细菌会选择积累其他的相容性溶质（胁迫应激第二反应）（Lucht

和Bremer，1994；Bartlett和Roberts，2004）。之所以将这类物质命名为相容性溶质，是因为在细胞质中即使这些物质的浓度很高，也不会干扰重要的细胞生理过程（Kempf和Bremer，1998）。这些物质可以由嗜盐生物合成，也可以通过从周围环境中吸收并在细胞内积累（Mohammadipanah等，2015）。这些有机渗透调节物质包含不同种类的分子，其中有氨基酸及其衍生物（脯氨酸、谷氨酸、谷氨酰胺、四氢嘧啶和牛磺酸）、糖（蔗糖、海藻糖）、甲胺（甜菜碱）和多元醇（甘露醇、赤藓糖醇和甘油）（Oren，2002；Yacey，2005）。盐渍化土壤中耐盐细菌的存在可能与植物发生相互作用，在植物抵御盐胁迫危害的过程中发挥重要的作用（Nabti等，2015）。在培养基中，微生物可以通过释放这些渗透调节物质分子来促进植物的生长。而对于植物而言，它们吸收和积累的能够在高渗透压下促进植物生长的这些有机化合物称为渗透保护剂。因此，渗透保护剂的概念假定了一个生态循环，在这个循环中，相容性渗透保护溶质从生产者运输到受渗透胁迫危害的需求者（Arora等，2002）。大量对嗜盐和耐盐植物生长促生菌的研究表明，植物生长促生菌在逆境条件下能够促使植物体内的相容性溶质增加。Sandhya等（2010）指出，在水分胁迫条件下假单胞菌菌株GAP45（*Pseudomonas* strain GAP 45）的存在增加了玉米植株中相容性溶质的含量，并且植株的抗氧化状态也有所提升。在细胞中积累脯氨酸是非生物胁迫条件下植物的一种常见反应。Saghafi等（2013）研究表明，植物生长促生菌菌株能够提高胁迫条件下小麦植株中脯氨酸的含量。Naz等（2009）报道了植物生长促生菌菌株在盐胁迫条件下显著提高了大豆植株茎、根中脯氨酸的含量。值得注意的是，接种高产IAA的植物生长促生菌菌株的苜蓿属植物能够适应多种不同的逆境胁迫条件，植株内脯氨酸的水平明显升高（Saghafi等，2013）。另外，有研究报道指出，经植物生长促生菌处理的小麦植株中脯氨酸和可溶性总糖含量的增加对其耐渗性有显著贡献（Upadhyay等，2012）。根瘤菌的海藻糖代谢对于提高豆科植物的生长、产量和适应非生物逆境胁迫也很重要（Suarez等，2008）。因此，耐盐植物生长促生菌能够诱导植物增加相容性溶质，缓解盐胁迫条件下植株受到的危害。

2.5 耐盐植物生长促生菌——藻类：一个很有前途的缓解植物盐胁迫危害的联合体

如果这些植物生长促生菌本身不能抵御盐胁迫，在盐胁迫条件下植物生长促生菌改善植物生长的机制就可能被抑制或削弱。如前所述，外源相容性溶质的积累能够保护和刺激高渗环境中细菌的生长。对于微生物而言，从培养基中直接吸收相容性物质并在细胞内积累要优于从头生物合成，因为从能量角度讲，直接吸收比从头生物合成更划算（Imhoff，1986）。Nabti 等（2010）强调，偶氮螺菌属（*Azospirillum*）的细菌能够通过积累谷氨酸、脯氨酸、甜菜碱和海藻糖等相容性溶质来适应盐胁迫条件下带来的高渗透胁迫。

藻类植物富含植物激素、多糖、营养元素和维生素，因此在正常和逆境环境中都能够保持正常的生长。此外，它们还是植物和细菌在逆境环境中所吸收的相容性溶质的主要来源。Ghoul 等（1995）研究表明，海藻提取物含有大量的相容性溶质，如甜菜碱、氨基酸和二甲基亚砜等。他们注意到，在盐胁迫条件下海藻提取物对大肠杆菌（*Escherichia coli*）具有显著的渗透保护作用。甜菜碱和石莼提取物的存在可使巴西固氮螺菌 NH（*Azospirillum brasilense* NH）对 NaCl 的耐受性提高到 600 mmol/L（Nabti 等，2007）。Pichereau 等（1998）研究证明，在逆境胁迫环境中藻类能够通过其作为渗透保护剂的有机渗透调节物质三甲基甘氨酸（trimethylglycine，GB）和 3-二甲基磺酸（3-dimethylsulfonic acid，DMSP）来刺激细菌的生长。然而，目前关于植物生长促生菌-藻类联合体对植物生长的刺激作用的研究还不多。现有的一些研究也主要是在逆境胁迫条件下对目标农作物同时应用植物生长促生菌和藻类提取物，其中，藻类提取物是被当作渗透保护调节剂的来源使用。这样组合的主要作用是恢复细菌生长，从而改善它们与植物之间的相互作用，改善植物在逆境胁迫中的生存和生产能力。正如 Nabti 等（2007）报道，海藻提取物能够提高盐胁迫条件下细菌的各种性能，从而强化其对于植物生长的促进作用。Arif（2016）研究表明，在盐胁迫条件下添加石莼提取物恢复了接种荧光假单胞菌（*Pseudomonas fluorescens*）的小麦发芽率、生物量、叶绿素含量以及蛋白质含量。Nabti 等（2007）研究结果表明，在盐胁迫条件下添加石莼提取物的植物生长促生菌菌株巴西固氮螺菌（*Azospirillum brasilense*）能够更好地恢复小麦的生长。

2.6 结论

接种耐盐植物生长促生菌来改善盐渍化土壤环境下植物的生长环境，以期作为在这种恶劣环境下提高作物生长和产量的一种合适方法，在未来的农业应用中很可能会取得非常理想的效果。此外，这些植物生长促生菌在盐碱条件下通过诱导植物产生多种渗透调节物质来抵御盐害胁迫带来的渗透胁迫压力，从而表现出很强的促进植物生长的能力。

参考文献

Abeles FB, Morgan PW, Saltveit ME Jr, 1992. Ethylene in plant biology, 2nd edn. Academic, New York, p. 414.

Abu-Sharar TM, Bingham FT, Rhoades JD, 1987. Stability of soil aggregates as affected by electrolyte concentration and composition. Soil Sci Soc Am J 51: 309-314.

Ahmad M, Zahir AZ, Nazli F, Akram F, Arshad M, Khalid M, 2013. Effectiveness of halotolerant, auxin producing *Pseudomonas* and *Rhizobium* strains to improve osmotic stress tolerance in mung bean (*Vigna radiata* L). Braz J Microbiol 44: 1341-1348.

Ali SZ, Sandhya V, Grover M, Rao LV, Kishore VN, Venkateswarlu B, 2009. *Pseudomonas* sp. strain AKM-P6 enhances tolerance of sorghum seedlings to elevated temperatures. Biol Fertil Soils 46: 45-55.

Al-Mutawa MM, 2003. Effect of salinity on germination and seedling growth of chick pea (*Cicer arietinum* L.) genotypes. Int J Agron Biol 5: 227-229.

Amir M, Aslem A, Khan MY, Jamshaid MU, Ahmed M, Asghar HN, Zahir ZA, 2013. Co-inoculation with *Rhizobium* and plant growth promoting Rhizobacteria (PGPR) for inducing salinity tolerance in mung bean under field condition of semi-arid climate. Asian J Biol 1: 17-22.

Andronov EE, Petrova SN, Pinaev AG, Pershina EV, Rakhimgalieva SZ, Akhmedenov KM, Gorobets AV, Sergaliev NK, 2012. Analysis of the structure of microbial community in soils with different degrees of salinization using T-RFLP and real time PCR techniques. Eurasian Soil Sci 45: 147-156.

Anonymous, 2015. Soils are endangered, but the degradation can be rolled back. Food and Agriculture Organization of United Nations, Rome.

Arbona V, Macro AJ, Iglessias DJ, Lopez-Climent MF, Talon M, Gomez-Cadnas A, 2005. Carbohydrate depletion in roots and leaves of salt-affected potted *Citrus clementina* L. Plant Growth Regul 46: 153-160.

Arif F, 2016. Effet du stress salin et d'osmoprotecteurs naturels sur la germination de blé dur (*Triticumdurum*) inoculé par *Pseudomonas fluorescens*. Thèse de Doctorat, Université Ferhat Abbas, Sétif 1. Faculté des Sciences de la Nature et de la Vie, Sétif, 156p.

AroraNK TS, Singh S, Lal N, Maheshwari DK, 2012. PGPR for protection of plant health under saline conditions. In: Maheshwari DK (ed) Bacteria in agrobiology-stress management. Springer, Berlin/Heidelberg, pp. 239-258.

Arshadullah M, Hyder SI, Mahmood IA, Sultan T, Naveed S, 2017. Mitigation of salt stress in wheat plant (*Triticum aestivum*) by plant growth promoting rhizobacteria for ACC deaminase. Int J Adv Res Biol Sci 4: 41-46. ISSN: 2348-8069.

Ashraf M, Ahmad MSA, Öztürk M, Aksoy A, 2012. Crop improvement through different means: challenges and prospects. In: Ashraf M et al (eds) Crop production for agricultural improvement. Springer, Dordrecht, pp. 1-15.

Azevedo Neto AD, Gomes-Filho E, Prisco JT, 2008. Salinity and oxidative stress. In: Khan NA, Sarvajeet S (eds) Abiotic stress and plant responses. IK International, New Delhi, pp.

58-82.

Baca BE, Elmerich C, 2007. Microbial production of plant hormones. In: Elmerich C, Newton WE (eds) Associative and endophytic nitrogen-fixing bacteria and cyanobacterial associations. Springer, Dordrecht, pp. 113-143.

Baniaghil N, Arzanesh MH, Ghorbanli M, Shahbazi M, 2013. The effect of plant growth promoting rhizobacteria on growth parameters, antioxidant enzymes and microelements of canola under salt stress. J Appl Environ Biol Sci 3: 17-27.

Bano A, Fatima M, 2009. Salt tolerance in *Zea mays* (L) following inoculation with *Rhizobium* and *Pseudomonas*. Biol Fertil Soils 45: 405-413.

Barka EA, Nowak J, Clement C, 2006. Enhancement of chilling resistance of inoculated grapevine plantlets with a plant growth-promoting rhizobacterium *Burkholderia phytofirmans* strain PsJN. Appl Environ Microbiol 72: 7246-7252.

Bartels D, Sunkar R, 2005. Drought and salt tolerance in plants. Crit Rev Plant Sci 24: 23-58.

Bartlett DH, Roberts M, 2004. Osmotic stress. In: Schaechter M (ed) The desk encyclopedia of microbiology. Elsevier/Academic Press, Amsterdam/Boston, pp. 754-766.

Benlloch-Gonzalez M, Fournier JM, Ramos J, Benlloch M, 2005. Strategies underlying salt tolerance in halophytes are present in *Cynara cardunculus*. Plant Sci 168: 653-659.

Bensidhoum L, Nabti E, Tabli N, Kupferschmied P, Weiss A, Rothballer M, Schmid M, Keel C, Hartmann A, 2016. Heavy metal tolerant *Pseudomonas protegens* isolates from agricultural well water in northeastern Algeria with plant growth promoting, insecticidal and antifungal activities. Eur J Soil Biol 75: 38-46.

Bharti N, Pandey SS, Barnawal D, Patel VK, Kalra A, 2016. Plant growth promoting rhizobacteria *Dietzia natronolimnaea* modulates the expression of stress responsive genes providing protection of wheat from salinity stress. Sci Rep 6: 1-16. https://doi.org/10.1038/srep34768.

Biggs AJW, Watling KM, Cupples N, Minehan K, 2010. Salinity risk assessment for the Queensland Murray-Darling Region. Queensland Department of Environment and Resource Management, Toowoomba.

Blaylock AD, 1994. Soil salinity, salt tolerance and growth potential of horticultural and landscape plants. Co-operative Extension Service, University of Wyoming, Department of Plant, Soil and Insect Science, College of Agriculture, Laramie.

Bleecker AB, Kende H, 2000. Ethylene: a gaseous signal molecule in plants. Annu Rev Cell Dev Biol 16: 1-18.

Bojovic B, Delic G, Topuzovic M, Stankovic M, 2010. Effects of NaCl on seed germination in some species from families *Brassicaceae* and *Solanaceae*. Kragujevac J Sci 32: 83-87.

Bui EN, 2013. Soil salinity: a neglected factor in plant ecology and biogeography. J Arid Environ 92: 14-25.

Buringh P, 1978. Food production potential of the world. In: Sinha R (ed) The world food problem: consensus and conflict. Pergamon Press, Oxford, p. 477e485.

Burrow DP, Surapaneni A, Rogers ME, Olsson KA, 2002. Groundwater use in forage production: the effect of saline-sodic irrigation and subsequent leaching on soil sodicity. Aust J Exp Agric 42: 237-247.

Carillo P, Annunziata MG, Pontecorvo G, Fuggi A, Woodrow P, 2011. Salinity stress and salt tolerance. In: Arun S (ed) Abiotic stress in plants—mechanisms and adaptations. InTech, Croatia, pp. 22-38.

Chedlly A, Ozturk M, Ashraf M, Grignon C, 2008. Biosaline agriculture and high salinity tolerance. Birkhauser Verlag-AG (Springer Science), Basel, p. 367.

Cheng Z, Park E, Glick BR, 2007. 1-Aminocyclopropane-1-carboxylic acid deaminase from *Pseudomonas putida* UW4 facilitates the growth of canola in the presence of salt. Can J Microbiol 53: 912-918.

Chookietwattana K, Maneewan K, 2012. Screening of efficient halotolerant solubilizing bacterium and its effect on promoting plant growth under saline conditions. World Appl Sci J 16: 1110-1117.

Chowdhury N, Marschner P, Burns RG, 2011. Soil microbial activity and community composition: impact of changes in matric and osmotic potential. Soil Biol Biochem 43: 1229-1236.

CISEAU, 2006. Extent of salinization and strategies for salt-affected land prevention and rehabilitation. Background Paper, In Electronic Conference on Salinization Organized and Coordinated by IPTRID 2006.

Dadkhah A, 2011. Effect of salinity on growth and leaf photosynthesis of two sugar beet (*Beta vulgaris* L.) cultivars. J Agric Sci Technol 13: 1001-1012.

Davies PJ, 1995. Plant hormones physiology, biochemistry and molecular biol, 2nd edn. Kluwer Academic Publishers, Dordrecht, p. 833.

Del Rio LA, Corpas FJ, Sandalio LM, Palma JM, Barroso JB, 2003. Plant peroxisomes, reactive oxygen metabolism and nitric oxide. IUBMB Life 55: 71-81.

Dell'Amico E, Cavalca L, Andreoni V, 2008. Improvement of *Brassica napus* growth under cadmium stress by cadmium resistant rhizobacteria. Soil Biol Biochem 40: 74-84.

Diby P, Harshad L, 2014. Plant-growth-promoting rhizobacteria to improve crop growth in saline soils: a review. Agron Sustain Dev 34: 737-752.

Dikilitas M, Karakas S, 2012. Behavior of plant pathogens for crops under stress during the determination of physiological, biochemical, and molecular approaches for salt stress tolerance. In: Ashraf M et al (eds) Crop production for agricultural improvement. Springer, Dordrecht, pp. 417-441.

Dodd IC, Perez-Alfocea F, 2012. Microbial alleviation of crop salinity. J Exp Bot 63: 3415-3428.

Egamberdieva D, 2008. Plant growth promoting properties of rhizobacteria isolated from wheat and pea grown in loamy sand soil. Turk J Biol 32: 9-15.

Egamberdieva D, 2009. Alleviation of salt stress by plant growth regulators and IAA producing bacteria in wheat. Acta Phys Plant 31: 861-864.

Egamberdieva D, 2011. Survival of *Pseudomonas extremorientalis* TSAU20 and *P. chlororaphis* TSAU13 in the rhizosphere of common bean (*Phaseolus vulgaris*) under saline conditions. Plant Soil Environ 57: 122-127.

Egamberdieva D, Kucharova Z, 2009. Selection for root colonizing bacteria stimulating wheat growth in saline soils. BiolFertil Soils 45: 563-571.

Egamberdieva D, Renella G, Wirth S, Islam R, 2010. Secondary salinity effects on soil microbial biomass. Biol Fertil Soils 46: 445-449.

Egamberdieva D, Jabborova D, Hashem A, 2015. *Pseudomonas* induces salinity tolerance in cotton (*Gossypium hirsutum*) and resistance to *Fusarium* root rot through the modulation of indole-3-acetic acid. Saudi J Biol Sci 22: 773-779.

El-Tarabily KA, Nassar AH, Sivasithamparam K, 2008. Promotion of growth of bean (*Phaseolus vulgaris* L.) in a calcareous soil by a phosphate–solubilizing, rhizosphere–competent isolate of *Micromonospora endolithica*. Appl Soil Ecol 39: 161-171.

Essa TA, 2002. Effect of salinity stress on growth and nutrient composition of three soybean *Glycine max* (L.) Merrill cultivars. J Agron Crop Sci 188: 86-93.

FAO, 2005. Global network on integrated soil management for sustainable use of salt-affected soils. FAO, Rome. ftp://ftp.fao.org/agl/agll/docs/misc23.pdf.

FAO: The Food and Agriculture Organization of the United Nations, 2002. Crops and drops: making the best use of water for agriculture. FAO, Rome, http://www.fao.org/docrep/w5146e/w5146e0a.htm.

Flowers TJ, 2004. Improving crop salt tolerance. J Exp Bot 55: 307-319.

Frankenberger WT, Bingham FT, 1982. Influence of salinity on soilless enzyme activities. Soil Sci Soc Am J 46: 1173-1177.

Gerhardt KE, Greenberg BM, Glick BR, 2006. The role of ACC deaminase in facilitating the phytoremediation of organics, metals and salt. Curr Trends Microbiol 2: 61-73.

Ghassemi F, Jakeman AJ, Nix HA, 1995. Salinization of land and water resources: human causes, extent, management and case studies. Center for resource and environmental studies. The Australian National University, Canberra.

Ghollaratta M, Raiesi F, 2007. The adverse effects of soil salinization on the growth of *Trifolium alexandrinum* L. and associated microbial and biochemical properties in a soil from Iran. Soil Biol Biochem 39: 1699-1702.

Ghoul M, Minet J, Bernard T, Dupray E, Cornier M, 1995. Marine macroalgae as a source for osmoprotection for *Escherichia coli*. Microb Ecol 30: 171-181.

Gill SS, Tuteja N, 2010. Reactive oxygen species and antioxidant machinery in abiotic stress. Plant Physiol Biochem 48: 909-930.

Glick BR, 1995. The enhancement of plant growth by free-living bacteria. Can J Microbiol 41: 109-117.

Glick BR, 2014. Bacteria with ACC deaminase can promote plant growth and help to feed the world. Microbiol Res 169: 30-39.

Glick BR, Liu C, Ghosh S, Dumbrof EB, 1997. Early development of canola seedlings in the presence of the plant growth-promoting rhizobacterium *Pseudomonas putida* GR12-2. Soil Biol Biochem 29: 1233-1239.

Glick BR, Penrose DM, Li J, 1998. A model for the lowering of plant ethylene concentrations by plant growth promoting bacteria. J Theor Biol 190: 63-68.

Glick BR, Chenz Z, Czarny J, Duan J, 2007. Promotion of plant growth by ACC-deaminase producing soil bacteria. Eur J Plant Pathol 119: 329-339.

Golpayegani A, Tilebeni HG, 2011. Effect of biological fertilizers on biochemical and physiological parameters of Basil (*Ociumum basilicm* L.). Med Plant Am-Eur J Agric Environ Sci 11: 411-416.

Grattan SR, Grieve CM, 1999. Salinity-mineral nutrient relations in horticultural crops. Sci Hortic 78: 127-157.

Gray EJ, Smith DL, 2005. Intracellular and extracellular PGPR: commonalities and distinctions in the plant-bacterium signaling processes. Soil Biol Biochem 37: 395-412.

Griggs D, Stafford-Smith M, Gaffney O, Rockström J, Ohman MC, Shyamsundar P, Steffen W, Glaser G, Kanie N, Noble I, 2013. Policy: sustainable development goals for people and planet. Nature 495: 305-307.

Gururani MA, Upadhyaya CP, Baskar V, Venkatesh J, Nookaraju A, 2013. Plant growth promoting rhizobacteria enhance abiotic stress tolerance in *Solanum tuberosum* through inducing changes in the expression of ROS - scavenging enzymes and improved photosynthetic performance. J Plant Growth Regul 32: 245-258.

Habib SH, Kausar H, Saud HM, 2016. Plant growth-promoting rhizobacteria enhance salinity stress tolerance in Okra through ROS-scavenging enzymes. Biomed Res. https://doi.org/10.1155/2016/6284547.

Hahn A, March HK, 2009. Mitogen activated protein kinase cascades and ethylene: signaling, biosynthesis, or both? Plant Physiol 149: 1207-1210.

Hamdia MA, Shaddad MAK, Doaa MM, 2004. Mechanisms of salt tolerance and interactive effects of *Azospirillum brasilense* inoculation on maize cultivars grown under salt stress conditions. Plant Growth Regul 44: 165-174.

Han HS, Lee KD, 2005. Plant growth promoting rhizobacteria effect on antioxidant status, photosynthesis, mineral uptake and growth of lettuce under soil salinity. Res J Agric Biol Sci 1: 210-215.

Hingole HS, Pathak AP, 2016. Saline soil microbiome: a rich source of halotolerant PGPR. J Crop Sci Biotechnol 19: 231-239.

Holguin G, Glick BR, 2001. Expression of ACC deaminase gene fromEnterobacter cloacae UW4 in *Azospirillum brasilense*. Microb Ecol 41: 81-288.

Hontzeas N, Saleh SS, Glick BR, 2004. Changes in gene expression in canola roots by ACC deaminase - containing plant growth - promoting bacteria. Mol Plant - Microbe Interact 17: 865-871.

Hu Y, Schmidhalter U, 2002. Limitation of salt stress to plant growth. In: Hock B, Elstner CF (eds) Plant toxicology. Marcel Dekker, New York, pp. 91-224.

Huang Y, Hutchison LH, Laskey CE, Kieber JJ, 2003. Biochemical and functional analysis of CTR1, a protein kinase that negatively regulates ethylene signaling in *Arabidopsis*. Plant J 33: 221-233 44.

Ibekwe AM, Poss JA, Grattan SA, Grieve CM, Suarez D, 2010. Bacterial diversity in cucumber (*Cucumis sativus*) rhizosphere in response to salinity, soil pH, and boron. Soil Biol Biochem 42: 567-575.

Ilangumaran G, Smith DL, 2017. Plant growth promoting rhizobacteria in amelioration of salinity stress: a systems biology perspective. Front Plant Sci 8: 1-14. https://doi.org/10.3389/fpls.2017.01768.

Imhoff JF, 1986. Osmoregulation and compatible solutes in eubacteria. FEMS Microbiol Rev 39: 57-66.

Islam F, Yasmeen T, Arif MS, Ali S, Ali B, Hameed S, Zhou W, 2015. Plant growth promoting bacteria confer salt tolerance in Vigna radiata by up-regulating antioxidant defense and biological soil fertility. Plant Growth Regul. https://doi.org/10.1007/s10725-015-0142-y.

Jamil M, Lee DB, Jung KY, Ashraf M, Lee SC, Rhal ES, 2006. Effect of salt (NaCl) stress on germination and early seedling growth of four vegetables species. J Cent Eur Agric 7 (273): 282.

Jamil A, Riaz S, Ashraf M, Foolad MR, 2011. Gene expression profiling of plants under salt stress. Crit Rev Plant Sci 30: 435-458.

Javid MG, Sorooshzadeh A, Moradi F, Sanavy Seyed AMM, Allahdadi I, 2011. The role of phytohormones in alleviating salt stress in crop plants. AJCS 5: 726-734.

Jha CK, Saraf M, 2012. Hormonal signaling by PGPR improves plant health under stress conditions. In: Maheshwar DK (ed) Bacteria in agrobiology: stress management. Springer, Berlin/Heidelberg, pp. 119-140.

Jha Y, Subramanian RB, 2013. Paddy plants inoculated with PGPR show better growth physiology and nutrient content under saline conditions. Chil J Agric Res 73: 213-219.

Jiang HC, Dong HL, Yu BS, Liu XQ, Li YL, Ji SS, Zhang CL, 2007. Microbial response to salinity change in Lake Chaka, a hypersaline lake on Tibetan plateau. Environ Microbiol 9: 2603-2621. https://doi.org/10.1111/j.1462-2920.2007.01377.x.

Juan M, Rivero RM, Romero L, Ruíz JM, 2005. Evaluation of some nutritional and biochemical indicators in selecting salt-resistant tomato cultivars. Environ Exp Bot 54: 193-201.

Kapoor K, Srivastava A, 2010. Assessment of salinity tolerance of *Vigna mungo* var. Pu-19 using ex vitro and in vitro methods. Asian J Biotechnol 2: 73-85.

Kasim WA, Gaafar RM, Abou-Ali RM, Omar MN, Hewait HM, 2016. Effect of biofilm forming plant growth promoting rhizobacteria on salinity tolerance in barley. AOAS 61: 217-227.

Kay BD, 1990. Rates of change of soil structure under different cropping systems. Adv Soil Sci 12: 1-52.

Kaya C, Ashraf M, Dikilitas M, Tuna AL, 2013. Alleviation of salt stress induced adverse effects on maize plants by exogenous application of indole acetic acid (IAA) and inorganic nutrients-a field trial. Aust J Crop Sci 7: 249-254.

Kempf B, Bremer E, 1998. Uptake and synthesis of compatible solutes as microbial stress responses to high-osmolality environments. Arch Microbiol 170: 319-330.

Kende H, 1993. Ethylene biosynthesis. Annu Rev Plant Physiol Plant Mol Biol 44: 283-307.

Khalid S, Parvaiz M, Nawaz K, Hussain H, Arshad A, Shawaka S, Sarfaraz ZN, Waheed T, 2013. Effect of indole acetic acid (IAA) on morphological, biochemical and chemical attributes of two varieties of maize (*Zea mays* L.) under salt stress. World Appl Sci J 26: 1150-1159.

Khan MY, Zahir ZA, Asghar HN, Waraich EA, 2017. Preliminary investigations on selection of synergistic halotolerant plant growth promoting rhizobacteria for inducing salinity tolerance in wheat. Pak J Bot 49: 1541-1551.

Khodarahmpour Z, Ifar M, Motamedi M, 2012. Effects of NaCl salinity on maize (*Zea mays* L.) at germination and early seedling stage. Afr J Biotechnol 11: 298-304.

Kohler J, Hernandezb A, Caravacaa F, Roldan A, 2009. Induction of antioxidant enzymes is involved in the greater effectiveness of a PGPR versus AM fungi with respect to increasing the tolerance of lettuce to severe salt stress. Environ Exp Bot 65: 245-252.

Korcan SE, Konuk M, Erdoğmuş SF, 2015. Beneficial usages of halophilic microorganisms. In:

Maheshwari DK, Saraf M (eds) Halophile: biodiversity and sustainable exploitation. Springer, Cham, pp. 261-276.

Kumar SG, Reddy AM, Sudhakar C, 2003. NaCl effects on proline metabolism in two high yielding genotypes of mulberry (Morus alba L.) with contrasting salt tolerance. Plant Sci 165: 1245-1251.

Lafe O, 2013. Agriculture. In: Lafe O (ed) Abulecentrism, rapid development of society catalyzed at the local community level. Springer, Cham, pp. 127-132.

Larcher W, 1980. Physiological plant ecology: ecophysiology and stress physiology of functional groups, 2nd edn. Springer, Berlin.

Le Rudulier D, Mandon K, Dupont L, Trinchant JC, 2002. Salinity effects on physiology of soil microorganisms. In: Bitton G (ed) Encyclopedia of environmental microbiology. Wiley Interscience, Canada, pp. 2774-2789.

Lucht JM, Bremer E, 1994. Adaptation of Escherichia coli to high osmolarity environments: osmoregulation of the high-affinity glycine betaine transport system ProU. FEMS Microbiol Rev 14: 3-20.

Maleki M, Mostafaee S, Mokhtarnejad L, Farzaneh M, 2010. Characterization of *Pseudomonas fluorescens* strain CV6 isolated from cucumber rhizosphere in Varamin as a potential biocontrol agent. AJCS 4: 676-683.

Mantelin S, Touraine B, 2004. Plant growth-promoting bacteria and nitrate availability: impacts on root development and nitrate uptake. J Exp Bot 55: 27-34.

Mapelli F, Marasco R, Rolli E, Barbato M, Cherif H, Guesmi A, Ouzari I, Daffonchio D, Borin S, 2013. Potential for plant growth promotion of rhizobacteria associated with *Salicornia* growing in Tunisian hypersaline soils. Biomed Res Int. https://doi.org/10.1155/2013/248078.

Marasco R, Rolli E, Ettoumi B, Vigani G, Mapelli F, Borin S, Abou-Hadid AF, El-Behairy UA, Sorlini C, Cherif A, Zocchi G, Daffonchio D, 2012. A drought resistance-promoting microbiome is selected by root system under desert farming. PLoS ONE 7: e48479. https://doi.org/10.1371/journal.pone.0048479.

Martinez-Beltran J, Manzur CL, 2005. Overview of salinity problems in the world and FAO strategies to address the problem In: Proceedings of the international salinity forum, Riverside, California. USDA-ARS Salinity Lab Riverside, p. 311-313.

Martinez-Viveros O, Jorquera MA, Crowley DE, Gajardo G, Mora ML, 2010. Mechanisms and practical considerations involved in plant growth promotion by rhizobacteria. J Soil Sci Plant Nutr 10: 293-319.

Mavi MS, Sanderman J, Chittleborough DJ, Cox JW, Marschner P, 2012. Sorption of dissolved organic matter in salt-affected soils: effect of salinity, sodicity and texture. Sci Total Environ 435: 337-344.

Mayak S, Tirosh T, Glick BR, 2004. Plant growth-promoting bacteria that confer resistance in tomato and pepper to salt stress. Plant Physiol Biochem 42: 565-572.

Meloni DA, Oliva MA, Martinez CA, Cambraia J, 2003. Photosynthesis and activity of superoxide dismutase, peroxidase and glutathione reductase in cotton under salt stress. Environ Exp Bot 49: 69-76. https://doi.org/10.1016/S0098-8472(02)00058-8.

Mensah JK, Akomeah PA, Ikhajiagbe B, Ekpekurede EO, 2006. Effects of salinity on germination, growth and yield offive groundnut genotypes. Afr J Biotechnol 5: 1973-1979.

Mikutta R, Mikutta C, Kalbitz K, Scheel T, Kaiser K, Jahn R, 2007. Biodegradation of forestfloor organic matter bound to minerals via different binding mechanisms. Geochim Cosmochim Acta 71: 2569-2590.

Mishra M, Mishra PK, Kumar U, Prakash V, 2009. NaCl phytotoxicity induces oxidative stress and response of antioxidant systems in Cicer arietinum L. CV. Abrodhi. Bot Res Int 2: 74-82.

Mitsuchi M, Wichaidit P, Jeungnijnirund S, 1986. Outline of soils of the North-East plateau, Thailand: their characteristics and constraints. Technical p. 131, paper No. 1. Khon-Kaen: Agricultural Development Research Center in the North-East.

Mohammadipanah F, Hamedi J, Dehhaghi M, 2015. Halophilic bacteria: potentials and applications in biotechnology. In: Maheshwari DK, Saraf M (eds) Halophile: biodiversity and sustainable exploitation. Springer, Cham, pp. 277-321.

Mona BH, Soha SM, Eman AES, Mostafa AA, Fayez M, 2016. Cereal-PGPR interweave in saltaffected environments: towards plant persistence and growth promotion. IJSER 7: 1774-1814.

Munns R, 2002. Comparative physiology of salt and water stress. Plant Cell Environ 25: 239-250.

Munns R, Tester M, 2008. Mechanisms of salinity tolerance. Annu Rev Plant Biol 59: 651-681.

Nabti E, 2007. Restauration de la croissance de Azospirillum brasilense et de Blé dur et leur osmoprotection par Ulva lactuca en Milieux Salés. Thèse de Doctorat en Science Biologique. Université Abderrahmane Mira, Faculté des sciences de la nature et de la vie, Bejaia, algérie 147p.

Nabti E, Sahnoune M, Adjrad S, Van Dommelen A, Ghoul M, Schmid M, Hartmann A, 2007. A halophilic and osmotolerant Azospirillum brasilense strain from Algerian soil restores wheat growth under saline conditions. Eng Life Sci 7: 354-360.

Nabti E, Sahnoune M, Ghoul M, Fischer D, Hofmann A, Rothballer M, Schmid M, Hartmann A, 2010. Restoration of growth of durum wheat (Triticum durum var. Waha) under saline conditions due to inoculation with the rhizosphere bacterium Azospirillum brasilense NH and extracts of the marine alga Ulva lactuca. J Plant Growth Regul 29: 6-22.

Nabti E, Schmid M, Hartmann A, 2015. Application of halotolerant bacteria to restore plant growth under salt stress. In: Maheshwari DK, Saraf M (eds) Halophile: biodiversity and sustainable exploitation. Springer, Cham, pp. 235-359.

Nadeem SM, Hussain I, Naveed M, Asghar HN, Zahir ZA, Arshad M, 2006. Performance of plant growth promoting rhizobacteria containing ACC-deaminase activity for improving growth of maize under salt-stressed conditions. Pak J Sci 43: 114-120.

Nakbanpote W, Panitlurtumpai N, Sangdee A, Sakulpone N, Sirisom P, Pimthong A, 2014. Salt-tolerant and plant growth-promoting bacteria isolated from Zn/Cd contaminated soil: identification and effect on rice under saline conditions. J Plant Interact 9: 379-387.

Nascimento FX, Rossi MJ, Soares CRFS, McConkey BJ, Glick BR, 2014. New insights into 1-aminocyclopropane-1-carboxylic acid (ACC) deaminase phylogeny, evolution and ecological significance. PLoS ONE 9: e99168. https://doi.org/10.1371/journal.pone.0099168.

Naz I, Bano A, Ul-Hassan T, 2009. Isolation of phytohormones producing plant growth promoting rhizobacteria from weeds growing in Khewra salt range, Pakistan, and their implication in providing salt tolerance to *Glycine max* (L.). Afr J Biotechnol 8: 5762-5766.

Nelson DR, Mele PM, 2007. Subtle changes in the rhizosphere microbial community structure in response to increased boron and sodium chloride concentrations. Soil Biol Biochem 39: 340-351.

Niu G, Xu W, Rodriguez D, Sun Y, 2012. Growth and physiological responses of maize and sorghum genotypes to salt stress. ISRN Agron 2012: 145072., 12 pages. https://doi.org/10.5402/2012/145072.

Niu SQ, Li HR, Paré PW, Aziz M, Wang SM, Shi H, Li J, Han QQ, Guo SQ, Li J, Guo Q, Ma Q, Zhang JL, 2015. Induced growth promotion and higher salt tolerance in the halophyte grass *Puccinellia tenuiflora* by beneficial rhizobacteria. Plant Soil. https://doi.org/10.1007/s11104-015-2767-z.

Oades JM, 1988. The retention of organic-matter in soils. Biogeochemistry 5: 35-70.

Ofek M, Ruppel S, Waisel Y, 2006. Effects of salinity on rhizosphere bacterial communities associated with different root types of *Vicia faba* L. In: Ozturk M, Waisel Y, Khan A, Gork G (eds) Biosaline agriculture and salinity tolerance in plants. Birkhauser, Basel, pp. 1-21.

Omar SA, Abdel-Sater MA, Khallil AM, Abdalla MH, 1994. Growth and enzyme activities of fungi and bacteria in soil salinized with sodium chloride. Folia Microbiol 39: 23-28. https://doi.org/10.1007/BF02814524.

Ondrasek G, Rengel Z, Romic D, Savic R, 2010. Environmental salinisation processes inagroecosystem of Neretva river estuary. Novenytermeles 59: 223-226.

Oren A, 1999. Bioenergetic aspects of halophilism. Microbiol Mol Biol Rev 63: 334-340.

Oren A, 2002. Diversity of halophilic microorganisms: environments, phylogeny, physiology, and applications. J Ind Microbiol Biotechnol 28: 56-63.

Orhan F, 2016. Alleviation of salt stress by halotolerant and halophilic plant growth-promoting bacteria in wheat (*Triticum aestivum*). BJM 4 (7): 621-627.

Pan CC, Liu CA, Zhao HL, Wang Y, 2013. Changes of soil physicochemical properties and enzyme activities in relation to grassland salinization. Eur J Soil Biol 55: 13-19.

Parida AK, Das AB, 2005. Salt tolerance and salinity effects on plants: a review. Ecotoxicol Environ Safe 60: 324-349. https://doi.org/10.1016/j.ecoenv.2004.06.010.

Patel BB, Patel Bharat B, Dave RS, 2011. Studies on infiltration of saline-alkali soils of several parts of Mehsana and Patan districts of north Gujarat. J Appl Technol Environ Sanitation 1: 87-92.

Paul D, 2012. Osmotic stress adaptations in rhizobacteria. J Basic Microbial 52: 1-10.

Penrose DM, 2000. The role of ACC deaminase in plant growth promotion. PhD Thesis, University of Water Loo, Canada..

Penrose DM, Glick BR, 2001. Levels of 1-aminocyclopropane-1-carboxylic acid (ACC) in exudates and extracts of canola seeds treated with plant growth promoting bacteria. Can J Microbiol 41: 368-372.

Penrose DM, Moffatt BA, Glick BR, 2001. Determination of 1-aminocyclopropane-1-carboxylic acid (ACC) to assess the effects of ACC deaminase-containing bacteria on roots of canola seedlings. Can J Microbiol 47: 77-80.

Pessarakli M, 1991. Dry matter yield nitrogen-15 absorption and water uptake by green bean under sodium chloride stress. Crop Sci 31: 1633-1640.

Pichereau V, Pocard JA, Hamelin J, Blanco C, Bernard T, 1998. Differential effects of dimethylsulfoniopropionate, dimethylsulfonioacetate, and other S-methylated compounds on the growth of *Sinorhizobium meliloti* at low and high osmolarities. Appl Environ Microbiol 64: 1420-1429.

Pinedo I, Ledger T, Greve M, Poupin MJ, 2015. *Burkholderia phytofirmans* PsJN induces longterm metabolic and transcriptional changes involved in *Arabidopsis thaliana* salt tolerance. Front Plant Sci. https://doi.org/10.3389/fpls.2015.00466.

Pourbabaee AA, Bahmani E, Alikhani HA, Emami S, 2016. Promotion ofwheat growth under salt stress by halotolerant bacteria containing ACC deaminase. J Agric Sci Technol 18: 855-864.

Qados AMSA, 2011. Effect of salt stress on plant growth and metabolism of bean plant *Vicia faba* (L.). J Saudi Soc Agric Sci 10: 7-15.

Rabie GH, Almadini AM, 2005. Role of bioinoculants in development of salt-tolerance of *Vicia faba* plants under salinity stress. Afr J Biotechnol 4: 210-222.

Rabie GH, Aboul-Nasr MB, Al-Humiany A, 2005. Increase salinity tolerance of cowpea plants by dual inoculation of Am fungus *Glomus clarum* and nitrogen-fixer *Azospirillum brasilense*. Mycobiology 33: 51-61.

Rahman MS, Matsumuro T, Miyake H, Takeoka Y, 2000. Salinity-induced ultrastructural alternations in leaf cells of rice (*Oryza sativa* L.). Plant Prod Sci 3: 422-429.

Rengasamy P, 2006. World salinization with emphasis on Australia. J Exp Bot 57: 1017-1023.

Rengasamy P, Olsson KA, 1991. Sodicity and soil structure. Aust J Soil Res 29: 935-952.

Rojas-Tapias D, Moreno-Galvan A, Pardo-Diaz S, Obando M, Rivera D, Bonilla R, 2012. Effect of inoculation with plant growth promoting bacteria (PGPB) on amelioration of saline stress in maize (*Zea mays*). Appl Soil Ecol 61: 264-272.

Romic D, Ondrasek G, Romic M, Josip B, Vranjes M, Petosic D, 2008. Salinity and irrigation method affect crop yield and soil quality in watermelon (*Citrullus lanatus* L.) growing. Irrig Drain 57: 463-469. https://doi.org/10.1002/ird.358.

Saghafi K, Ahmadi J, Asgharzadeh A, Bakhtiari S, 2013. The effect of microbial inoculants on physiological responses of two wheat cultivars under salt stress. Int J Adv Biol Biomed Res 4: 421-431.

Saleem M, Arshad M, Hussain S, Bhatti AS, 2007. Perspective of plant growth promoting rhizobacteria (PGPR) containing ACC deaminase in stress agriculture. J Ind Microbiol Biotechnol 34: 635-648.

Sandhya V, Ali S, Grover M, Reddy G, Venkateswarlu B, 2009. Alleviation of drought stress effects in sunflower seedlings by the exopolysaccharides producing *Pseudomonas putida* strain GAP-P45. Biol Fertil Soils 46: 17-26.

Sandhya V, Ali SZ, Grover M, Reddy G, Venkateswarulu B, 2010. Effect of plant growth promoting *Pseudomonas* spp. on compatible solutes antioxidant status and plant growth of maize under drought stress. Plant Growth Regul. https://doi.org/10.1007/s10725-010-9479-4.

Santoyo G, Moreno-Hagelsieb G, Orozco-Mosqueda MC, Glick BR, 2016. Plant growthpromoting bacterial endophytes. Microbiol Res 183: 92-99.

Sapsirisopa S, Chookietwattana K, Maneewan K, Khaengkhan P, 2009. Effect of salt-tolerant

Bacillus inoculum on rice KDML 105 cultivated in saline soil. Asian J Food Ag-Ind Special Issue S69-S74.

Saranga Y, Menz M, Jiang CX, Robert JW, Yakir D, Andrew HP, 2001. Genomic dissection of genotype X environment interactions conferring adaptation of cotton to arid conditions. Genome Res 11: 1988-1995.

Saravanakumar D, Samiyappan R, 2007. ACC deaminase from *Pseudomonas fluorescens* mediated saline resistance in groundnut (*Arachis hypogea*) plants. J Appl Microbiol 102: 1283-1292.

Savvas D, Lanz F, 2000. Effects of NaCl or nutrient-induced salinity on growth, yield, and composition of eggplant grown in rockwool. Sci Hortic 84: 37-47.

Schaller GE, 2012. Ethylene and the regulation of plant development. BioMed Cent Biol 10: 9.

Selvakumar G, Kim K, Hu S, Sa T, 2014. Effect of salinity on plants and the role of Arbuscular mycorrhizal fungi and plant growth-promoting rhizobacteria in alleviation of salt stress. In: Ahmad P, Wani MR (eds) Physiological mechanisms and adaptation strategies in plants under changing environment. Springer, New York, pp. 115-144.

Setia R, Marschner P, Baldock J, Chittleborough D, 2010. Is CO_2 evolution in saline soils affected by an osmotic effect and calcium carbonate? Biol Fertil Soils 46: 781-792.

Setia R, Gottschalk P, Smith P, Marschner P, Baldock J, Setia D, Smith J, 2013. Soil salinity decreases global soil organic carbon stocks. Sci Total Environ 465: 267-272.

Sgroy V, Cassán F, Masciarelli O, Papa MFD, Lagares A, Luna V, 2009. Isolation and characterization of endophytic plant growth-promoting (PGPB) or stress homeostasis-regulating (PSHB) bacteria associated to the halophyte *Prosopis strombulifera*. Appl Microbiol Biotechnol 85: 371-381.

Shaharoona B, Arshad M, Zahir ZA, Khalid A, 2006. Performance of *Pseudomonas* spp. containing ACC-deaminase for improving growth and yield of maize (*Zea mays* L.) in the presence of nitrogenous fertilizer. Soil Biol Biochem 38: 2971-2975.

Shahbazi M, Arzani A, Saeidi G, 2011. Effect of NaCl treatments of seed germination and antioxidant activity of Canola (*Brassica Napus*) Cultivars. Bangladesh J Bot 41: 67-73.

Shaheen S, Naseer S, Ashraf M, Akram NA, 2013. Salt stress affects water relations, photosynthesis, and oxidative defense mechanisms in *Solanum melongena* L. J. Plant Interact 8: 85-96.

Shannon MC, Grieve CM, 1999. Tolerance of vegetable crops to salinity. Sci Hortic 78 (5): 38.

Sharma S, Kulkarni J, Jha B, 2016. Halotolerant rhizobacteria promote growth and enhance salinity tolerance in peanut. Front Microbiol 7: 1600. https://doi.org/10.3389/fmicb.2016.01600.

Sheldon A, Menzies NW, Bing SH, Dalal RC, 2004. The effect of salinity on plant available water. Supersoil 2004: 3rd Australian New Zealand soil conference.

Shilpi M, Girdhar K, Tuteja PN, 2008. Calcium and salt-stress signaling in plants: shedding light on SOS pathway. Arch Biochem Biophys 471: 146-158.

Shin W, Siddikee MA, Joe MM, Benson A, Kim K, Selvakumar G, Kang Y, Jeon S, Samaddar S, Chatterjee P, Walitang D, Chanratana M, Sa T, 2016. Halotolerant plant growth promoting bacteria mediated salinity stress amelioration in plants. Korean J Soil Sci Fertil 49:

355-367.

Shrivastava P, Kumar R, 2015. Soil salinity: a serious environmental issue and plant growth promoting bacteria as one of the tools for its alleviation. Saudi J Biol Sci 22: 123-131.

Shukla PS, Agarwal PK, Jha B, 2011. Improved salinity tolerance of *Arachis hypogaea* (L.) by the interaction of halotolerant plant-growth-promoting rhizobacteria. J Plant Growth Regul. https://doi.org/10.1007/s00344-011-9231-y.

Siddikee MA, Chauhan PS, Anandham R, Gwang-Hyun H, Tongmin S, 2010. Isolation, characterization, and use for plant growth promotion under salt stress of ACC deaminase-producing halotolerant bacteria derived from coastal soil. J Microbiol Biotechnol 20: 1577-1584.

Siddikee MA, Glick BR, Chauhan PS, Sa T, 2011. Enhancement of growth and salt tolerance of red pepper seedlings (*Capsicum annuum* L.) by regulating stress ethylene synthesis with halotolerant bacteria containing 1-aminocyclopropane-1-carboxylic acid deaminase activity. Plant Physiol Biochem 49: 427-434.

Singh PK, Kumar V, Maurya N, Choudhary H, Kumar V, Gupta AK, 2014. Grassroots solutions to overcome abiotic and biotic environmental stress in agriculture. In: Singh SB et al (eds) Translational research in environmental and occupational stress. Springer, New Delhi, pp. 11-16.

Singleton PW, Bohlool BB, 1984. Effect of salinity on nodule formation by soybean. Plant Physiol 74: 72-76.

Six J, Paustian K, Elliott ET, Combrink C, 2000. Soil structure and organic matter: I. Distribution of aggregate-size classes and aggregate-associated carbon. Soil Sci Soc Am J 64: 681-689.

Sleator RD, Hill C, 2001. Bacterial osmoadaptation: the role of osmolytes in bacterial stress and virulence. FEMS Microbiol Rev 26: 49-71.

Stanković MS, Petrović M, Godjevac D, Stevanović ZD, 2015. Screening inland halophytes from the central Balkan for their antioxidant activity in relation to total phenolic compounds and flavonoids: are there any prospective medicinal plants? J Arid Environ 120: 26-32.

Suarez R, Wong A, Ramirez M, Barraza A, Orozco MD, Cevallos MA, Lara M, Hernandez G, Iturriaga G, 2008. Improvement of drought tolerance and grain yield in common bean by overexpressing trehalose-6-phosphate synthase in rhizobia. Mol Plant-Microbe Interact 21: 958-966.

Taffouo VD, Wamba OF, Yombi E, Nono GV, Akoe A, 2010. Growth, yield, water status and ionic distribution response of three bambara groundnut (*Vigna subterranean*) (L.) verdc. landraces grown under saline conditions. Int J Bot 6: 53-58.

Takagi M, El-Shemy HA, Sasaki S, Toyama S, Kanai S, Saneoka H, Fujita K, 2009. Elevated CO_2 concentration alleviates salinity stress in tomato plant. Acta Agric Scand B—S P 59: 87-96.

Talsma T, 1963. The control of saline groundwater. Thesis for the degree of doctor in land technology, University of Wageningen. Repr Bull Univ Wagening 63: 1-68.

Tank N, Saraf M, 2010. Salinity-resistant plant growth promoting rhizobacteria ameliorates sodium chloride stress on tomato plants. J Plant Interact 5: 51-58.

Tate RL III, 2000. Soil microbiology. Wiley, New York.

Tester M, Davenport R, 2003. Na^+ tolerance and Na^+ transport in higher plants. Ann Bot 91:

503-527.

Timmusk S, Wagner GH, 1999. The plant-growth-promoting rhizobacterium *Paenibacillus polymyxa* induces changes in *Arabidopsis thaliana* gene expression: a possible connection between biotic and abiotic stress responses. Mol Plant-Microbe Interact 12: 951-959.

Tokala RK, Strap JL, Jung CM, Crawford DL, Salove H, Deobald LA, Bailey FJ, Morra MJ, 2002. Novel plant microbe rhizosphere interaction involving *S. lydicus* WYEC108 and the pea plant (*Pisum sativum*). Appl Environ Microbiol 68: 2161-2171.

Tripathi AK, Verma SC, Ron EZ, 2002. Molecular characterization of a salt-tolerant bacterial community in the rice rhizosphere. Res Microbiol 153: 579-584.

Tripathi S, Chakraborty A, Chakrabartia K, Bandyopadhyayc BK, 2007. Enzyme activities and microbial biomass in coastal soils of India. Soil Biol Biochem 39: 2840-2848.

United Nations, Department of Economic and Social Affairs, Population Division, 2004. World population prospects: world population to 2300. Working Paper No. ST/ESA/SER. A/236. New York.

United States Department of Agriculture, 1954. Diagnosis and improvement of saline and alkali soils. In: Richards LA (ed) Agriculture hand book no. 60. Oxford & IBH, New Delhi, pp. 1-160.

Upadhyay SK, Maurya SK, Singh DP, 2012. Salinity tolerance in free-living plant growth promoting rhizobacteria. Indian J Res 3: 73-78.

Vacheron J, Desbrosses G, Bouffaud ML, Touraine B, Moënne-Loccoz Y, Muller D, Legendre L, Wisniewski-Dyé F, Prigent-Combaret C, 2013. Plant growth promoting rhizobacteria and root system functioning. Front Plant Sci 4: 1-19.

Vurukonda SSKP, Vardharajula S, Shrivastava M, SkZ A, 2016. Enhancement of drought stress tolerance in crops by plant growth promoting rhizobacteria. Microbiol Res 184: 13-24.

Waller P, Yitayew M, 2016. Water and salinity stress. In: Waller P, Yitayew M (eds) Irrigation and drainage engineering. Springer, Cham, pp. 51-65.

White PJ, Broadley MR, 2001. Chloride in soils and its uptake and movement within the plant: a review. Ann Bot 88: 967-988. https://doi.org/10.1006/anbo.2001.1540.

Wichern J, Wichern F, Joergensen RG, 2006. Impact of salinity on soil microbial communities and the decomposition of maize in acidic soils. Geoderma 137: 100-108.

Wild A, 2003. Soils, land and food: managing the land during the twenty-first century. Cambridge University Press, Cambridge.

Wong VNL, Dalal RC, Greene RSB, 2009. Carbon dynamics of sodic and saline soils following gypsum and organic material additions: a laboratory incubation. Appl Soil Ecol 41: 29-40.

Wu J, Vincent B, Yang J, Bouarfa S, Vidal A, 2008. Remote sensing monitoring of changes in soil salinity: a case study in Inner Mongolia, China. Sensors 8: 7035-7049.

Xiong L, Zhu JK, 2002. Salt-stress signal transduction. In: Scheel D, Wasternack C (eds) Plant signal transduction. Frontiers in molecular biology series. Oxford University Press, Oxford, pp. 165-197.

Xu ZH, Saffigna PG, Farquhar GD, Simpson JA, Haines RJ, Walker S, Osborne DO, Guinto D, 2000. Carbon isotope discrimination and oxygen isotope composition in clones of the F (1) hybrid between slash pine and Caribbean pine in relation to tree growth, water-use efficiency and foliar nutrient concentration. Tree Physiol 20: 1209-1217.

Xu GY, Rocha PS, Wang ML, Xu ML, Cui YC, Li LY, Zhu YX, Xia X, 2011. A novel rice calmodulin-like gene, *OsMSR2*, enhances drought and salt tolerance and increases ABA sensitivity in *Arabidopsis*. Planta 234: 47-59.

Yadav S, Irfan M, Ahmad A, Hayat S, 2011. Causes of salinity and plant manifestations to salt stress: a review. J Environ Biol 32: 667-685.

Yancey PH, 2005. Organic osmolytes as compatible, metabolic and counteracting cytoprotectants in high osmolarity and other stresses. J Exp Biol 208: 2819-2830.

Yao LX, Wu ZS, Zheng YY, Kaleem I, Li C, 2010. Growth promotion and protection against salt stress by *Pseudomonas putida* Rs-198 on cotton. Eur J Soil Biol 46: 49-54. https://doi.org/10.1016/j.ejsobi.2009.11.002.

Yildirim E, Taylor AG, Spittler TD, 2006. Ameliorative effects of biological treatments on growth of squash plants under salt stress. Sci Hortic 111: 1-6.

Younesi O, Chaichi MR, Postini K, 2013. Salt tolerance in alfalfa following inoculation with Pseudomonas. Middle-East J Sci Res 16: 101-107.

Yuan BC, Xu XG, Li ZZ, Gao TP, Gao M, Fan XW, Deng HM, 2007. Microbial biomass and activity in alkalized magnesic soils under arid conditions. Soil Biol Biochem 39: 3004-3013.

Zafar-ul-Hye M, Ahmed M, Shahzad SM, 2013. Synergistic effect of Rhizobia and plant growth promoting rhizobacteria on the growth and nodulation of lentil seedlings under axenic conditions. Soil Environ 32: 79-86.

Zahir ZA, Ghani U, Naveed M, Nadeem SM, Asghar HN, 2009. Comparative effectiveness of *Pseudomonas* and *Serratia* sp. containing ACC-deaminase for improving growth and yield of wheat (*Triticum aestivum* L.) under salt-stressed conditions. Arch Microbiol 191: 415-424.

Zahran HH, 1997. Diversity, adaptation and activity of the bacterialflora in saline environments. Biol Fertil Soils 25: 211-223.

Zapata PJ, Serrano M, Pretel MT, Amoros A, Botella MA, 2004. Polyamines and ethylene changes during germination of different plant species under salinity. Plant Sci 167: 781-788.

Zhang H, Kim MS, Dowd SE, Shi H, Pare PW, 2008. Soil bacterial confer plant tolerance by tissue-specific regulation of the sodium transporter HKT1. Mol Plant-Microbe Interact 21: 737-744.

3 耐盐根际菌：一种很有前途的盐渍土农业益生菌

Ankita Alexander，Avinash Mishra，Bhavanath Jha

[摘要]

土壤盐渍化严重威胁着现代农业的可持续发展。目前利用多种育种方法和基因工程手段来提高作物抗盐能力的研究正在如火如荼地进行，但是这些育种方法不仅费时费力，而且常常面临产量与质量目标的不协调，以及许多道德伦理和社会问题。为了农业的可持续发展，探索其他更加稳定和环保的土壤保育以及农作物生产方法势在必行。对盐生植物根际的植物生长促生菌的挖掘及在盐渍化土壤中作为一种农用益生菌的应用，是替代传统方法来提高农作物耐盐能力的有效途径之一。盐胁迫是干旱和半干旱地区农业耕作中遇到的主要非生物胁迫因素之一，给农业生产造成了重大的损失。盐生植物因其遗传组成和与相关微生物群落的共生作用，能够在盐渍化土壤环境中生长。这些与盐生植物共生的微生物能够在盐渍化土壤环境中存活下来，但它们的耐盐机制以及对植物的促生作用还没有被研究透彻。目前，有一些研究表明，与盐生植物共生的细菌能够直接或间接地提高盐生植物在盐渍化土地中的生长和生物产量。因此，这些细菌可以作为提高盐敏感植物（甜土植物）生产能力的益生菌，提高农作物在盐渍化土地上的生产能力。在盐胁迫条件下，植物生长促生菌会引起植物发生很多形态、生理和遗传层面的变化，以抵消因盐胁迫对农作物正常生长的抑制作用。植物生长促生菌引起的植物在遗传水平上产生的变化称为诱导性系统性抗性（ISR）。植物生长促生菌能够分泌一些有益物质，包括有机溶质、铁载体蛋白等，这些物质能够增强农作物在恶劣条件下的生存能力。植物生长促生菌还能够诱导植物产生相容性溶质，以维持渗透压平衡。植物生长促生菌能够影响盐胁迫条件下植物的生长和发育，还能在胁迫条件下调控激素

A. Alexander，A. Mishra（✉），B. Jha（✉），印度古吉拉特邦巴夫纳格尔 CSIR 中央盐和海洋化学品研究所，印度加齐阿巴德科学与工业研究理事会科学与创新研究院，E-mail：A. Mishra：avinash@csmcri.res.in，B. Jha：jha.bhavanath@gmail.com

合成，提高植物的耐盐能力。

[关键词]

盐生植物；盐敏感植物；耐盐植物生长促生菌；盐胁迫；作物

3.1 前言

相对于人类益生菌，植物益生菌（plant-probiotic microorganisms，PPM）是一种对植物生长有益的微生物，可以作为植物生长保护剂以减少病原体对植物的侵染，作为生物肥料以提高农作物的产量，以及作为植物刺激剂来调控植物和农作物的生长发育。此外，植物益生菌还能够降低与其共生的植物在遭受各种生物和非生物胁迫时受到的危害（De Souza Vandenberghe 等，2017）。据联合国粮农组织/世卫组织专家的咨询报告称，植物益生菌是一种"当施用足够数量时，能够给宿主植物带来益处的活体微生物"（Hill 等，2014）。

植物的生长环境易受重金属、盐害、寒冷、干旱、高温以及洪水等多种非生物胁迫因素的影响。其中，土壤盐渍化是对农作物的生长和发育以及粮食产量产生不利影响的主要非生物胁迫因素之一（Príncipe 等，2007）。当农作物根际土壤（根区）的饱和水溶液的电导率（electroconductibility，EC）超过 4 dS/m（相当于在 25℃条件下浓度约为 40 mmol/L 的 NaCl 溶液）时，就称为盐渍化土壤。一般来说，土壤溶液的电导率越高，表示土壤的含盐量越高，因此对农作物产量的负面影响也就越大（Munn 和 Tunter，2008）。土壤的初生盐渍化是由土壤中自然存在的盐分积累引起的，而次生盐渍化则主要是由于后期的人类活动，比如过度使用化肥、土壤改良剂以及灌溉管理不当等造成的（Carillo 等，2011）。据联合国粮农组织在 2016 年的报告称，目前世界上有 20%农业用地的土壤盐分含量过高。

土壤盐渍化会致使植物在生理、生化以及分子水平上发生变化，进而对自身的生长发育产生不利影响，从而降低农作物的生长和粮食产量（Roy 等，2014）。转基因耐盐农作物的培育是改善农作物在盐渍化土壤上生长，并提高农作物产量的一种有效的方法。然而在现阶段，转基因农作物在生产上的规模化应用还存在不少分歧，可持续农业的发展需要环境友好、符合伦理道德规范以及自然规律的农作物（Ashraf，2009）。盐生植物是能够自然适应盐渍化土地的植物种类，这类植物依靠其特殊的遗传组成以及与其共生的多种根际细菌组成的根际微生物群能够在盐胁迫条件下保证盐生植物的正常生长（Mishra 和 Tanna，2017；Mukhtar 等，2017）。从与盐生植物共生的微生物群中分离到的根际细菌可以提高盐生植物在盐渍土环境中的生长，并提高盐生植物的生物量。此外，这些根际细菌也可用于改善土壤质量。

3.2 盐胁迫对植物生长的影响

盐胁迫可在多个方面对植物的正常生长造成影响。土壤盐渍化对农作物生长发育和产量的影响主要包括以下几个方面。

3.2.1 渗透胁迫

土壤盐渍化导致土壤溶液中离子浓度升高，因此农作物根系周围渗透势升高产生渗透胁迫，导致农作物根系吸水发生困难，老叶的蒸腾速率和叶片的膨胀率都降低，叶片气孔关闭，水分保持和利用效率也都显著降低，导致农作物的衰老加速（Munns 和 Tester，2008）。此外，渗透胁迫还能够导致农作物体内活性氧（ROS）物质的产生，从而发生氧化损伤（Han 等，2011）。

3.2.2 营养失衡

盐渍化土壤中的可溶性盐分在很大程度上影响着植物对营养元素的吸收和分配，高含量的可溶性盐分会导致植物养分的缺乏，抑制植物生长（Blaylock，1994）。在盐渍化土壤中，由于磷酸根离子可以与钙离子一起形成难溶性沉淀，因此磷元素很难被植物吸收利用（Bano 和 Fatima，2009）。

3.2.3 光合作用减少

盐胁迫能够破坏植物的光合能力。在盐胁迫条件下，植物正常的生长和发育受到影响，导致叶面积减少、叶绿素和胡萝卜素含量降低以及气孔导度降低，对光系统Ⅰ和光系统Ⅱ产生有害影响，从而导致植物的光合作用能力下降（Netondo 等，2004）。

3.2.4 对生殖能力的影响

盐胁迫能够严重扰乱植物的繁殖机制。在植物正常的生殖过程中，盐胁迫会抑制小孢子发育，并导致胚珠败育以及受精卵衰老和发育停止。不仅如此，盐胁迫还在一些生殖组织中诱导细胞凋亡（细胞程序性死亡）（Shrivastava 和 Kumar，2015）。

3.2.5 离子毒性

离子毒性是土壤盐渍化衍生出来的主要问题。在盐胁迫条件下，土壤溶液中积累的过量的 Na⁺ 会导致叶片枯萎和落叶（Podmore，2009）。氯化物离子毒性能够引起植物叶片发生变色反应，并且产生坏死斑点（RaMaNa 等，2010）。高浓度的 Cl⁻ 能够导致植物细胞内硝酸还原酶（nitrate reductase，NR）活性降低，干扰植物对 NO_3^- 的吸收，进而影响植物的氮素同化作用（Baki 等，2000）。此外，土壤溶液中过量的可溶性盐会导致植物细胞中 Na⁺ 的过量积累，使细胞内 Na⁺/K⁺ 的比例变高，从而使细胞中许多发挥生理功能的蛋白质和酶的构象发生变化，影响植物的各种代谢途径（Zhu，2002；Chinnusamy 等，2006）。

3.2.6 氧化应激胁迫

盐胁迫能够诱导植物细胞产生活性氧（ROS），引发氧化应激胁迫，同时降低植物细胞中抗氧化酶的含量，比如过氧化氢酶（CAT）、超氧化物歧化酶（SOD）、过氧化物酶（POX）、谷胱甘肽（GSH）、抗坏血酸过氧化物酶（APX）、谷胱甘肽还原酶（GR）、谷胱甘肽 S-转移酶（GST）等，以及植物细胞中的单脱氢抗坏血酸还原酶（AbdElgawad 等，2016）。活性氧能够对生物大分子（如蛋白质和核酸等）造成氧化损伤，并损害脂质分子形成膜系统，对植物细胞造成伤害。此外，长期暴露在活性氧环境中的植物宿主会逐渐衰老死亡（Del Río 等，2003）。

3.2.7 过量乙烯的产生

植物在盐胁迫条件下会产生过量的乙烯，而高浓度的乙烯会严重阻碍植物根系的生长发育（Mahajan 和 Tuteja，2005）。在生物和非生物胁迫条件下，乙烯能够导致作物根系生长减缓，并促使作物早衰（Ma 等，2003）。一般来说，盐胁迫能够通过降低编码细胞周期素和细胞周期素依赖性激酶基因的表达对细胞分裂和分化产生不利影响。有研究表明，盐胁迫也会降低种子的发芽率，对植物幼苗的生长和植物体内相关酶的活性产生抑制作用（Seckin 等，2009），最终影响农作物的生物量和产量。

3.3 对植物益生菌的需求

在应对生物和非生物胁迫方面，转基因农作物的培育和应用是农业可持续发展中的常用手段。但目前转基因农作物的培育和应用存在许多缺点和局限性，比如转基因操作过程耗费的时间长、从实验室验证到大田检验应用的过程需要大量人力，并且获得理想性状的转基因农作物成功率较低（Coleman-Derr 和 Tringe，2014）。另外，由于参与生物胁迫和非生物胁迫等逆境胁迫响应的基因并不只有一个，以及转基因农作物下一代基因遗传具有不确定性，这导致通过转基因方法获得的转基因农作物的后代面临性状分离的风险（Jewell 等，2010）。此外，转基因农作物还面临着许多社会问题，比如民众接受程度低以及一些潜在的道德伦理等问题（Fedoroff，2010）。同时，利用转基因技术培育转基因农作物品种还受到农作物本身遗传背景的限制。对于四倍体和六倍体的农作物来说，服务于转基因技术所开发的各种分子技术在大多数情况下都不适用或效率低下。土壤的盐胁迫和碱胁迫在自然条件下通常是伴生的（Saslis Lagoudakis 等，2014）。而通过转基因技术获得的转基因农作物在通常情况下并不能同时耐受高土壤碱度（pH）和/或盐碱混合的土壤环境（Yamaguchi 和 Blumwald，2005）。有研究指出，利用转基因技术改造植物基因培育的耐盐农作物，特别是通过对转录因子进行改造培育的农作物会导致产量的降低（Roy 等，2014）。当农作物面临盐胁迫危害时，细胞内许多基因、蛋白质和代谢产物同时被激活，共同抵御盐胁迫危害，但转基因农作物对植物的改造往往只集中在一个基因或一个启动子上，因此对盐胁迫的耐受能力并不能达到预期水平（Bhatnagar Mathur 等，2008）。

为了克服这些问题，并能够通过环境友好的方式满足日益增长的粮食需求，探索其他更加有效的、能够满足农业可持续发展的技术势在必行。植物正常的生长发育以及对任何逆境胁迫的应激反应不仅受到本身基因表达的影响，还受到周围微生物群落的影响（Munns 和 Gilliham，2015；Vannier 等，2015）。植物通过根系能够向外界释放出大量的营养物质，这些物质统称为根系渗出物（如根细胞渗出液、边缘细胞和黏液质等）。根系渗出物作为微生物的营养来源能够通过调节植物根系周围的微生物多样性和群落结构反作用于植物的生长（Cook 等，1995）。植物根际是包括植物根系在内的一个非常狭小的土壤区域，但是每克植物根系中却可含有高达 10^{11} 个微生物细胞（Egamberdieva，2008）。这些微生物与共生宿主共同进化，并能够根据所处的环境变化来积极地调整其群落结构，同时促进植物的生长（Lau 和 Lennon，2012）。因此，植物根系、土壤和微生物之间的相互作用所形成的网络调控体系，对在不利条件（比如生物和非生物逆境胁迫）下微生物本身和宿主植物的正常生长和防御机制起到至关重要的作用。这其中有一些微生物则能够在生物和非生物逆境胁迫下促进宿主植物的生长并提高产量（图 3.1）。由于这些细菌具有促进植物生长的特性，因此它们被称为植物生长促生菌（plant growth-promoting rhizobacteria，PGPR）（Kloeper 和 Schroth，1981）。植物生长促生菌可分为胞外植物生长促生菌（extracellular PGPR，ePGPR）和胞内植物生长促生菌（intracellular PGPR，iPGPR）。胞外植物生长促生菌存在于植物根系的周围、根表面细胞以及根皮层细胞之间等；而胞内植物生长促生菌则存在于植物根系细胞内部，一般作为内生菌存在于特异的组织细胞中（Gray 和 Smith，2005）。

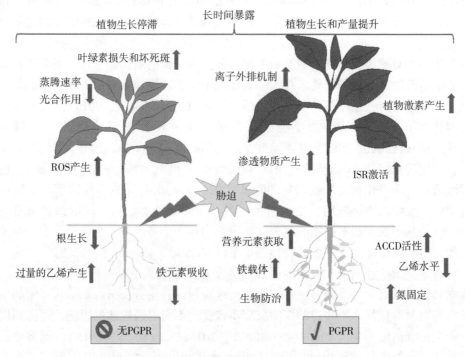

图 3.1 植物生长促生菌在促进植物生长中的作用

3.4 根际细菌可作为植物生长益生菌

系统性诱导抗性（IST）这一术语是源于植物生长促生菌诱导目标植物在细胞水平产生的物理和化学变化而提出的，这些变化能够增强植物对非生物胁迫的耐受性（Shrivastava 和 Kumar，2015）。耐盐植物生长促生菌具有许多特性，能够有利于改善植物面临逆境胁迫时的生长条件（图 3.2），其中包括固氮、溶磷、产生植物激素和胞外多糖、1-氨基环丙烷-1-羧酸脱氨酶（aminocyclopropane-1-carboxylic acid deaminase，ACCD）的产生以及相关生物降解酶的合成和分泌（de Souza 等，2015）。

3.4.1 氨基环丙烷-1-羧酸脱氨酶活性

有研究指出，许多耐盐菌株能够表现出植物健康生长所需要的 1-氨基环丙烷-1-羧酸脱氨酶活性（Santoyo 等，2016）。在逆境条件下，植物自然地选择与产1-氨基环丙烷-1-羧酸脱氨酶的细菌共生。植物生长促生菌产生1-氨基环丙烷-1-羧酸脱氨酶,该酶能够将植物中乙烯激素的前体1-氨基环丙烷-1-羧酸分解为氨和α-酮丁酸,因此植物幼苗和植株体内乙烯的含量就会降低。此外,1-氨基环丙烷-1-羧酸脱氨酶还能够刺激吲哚-3-乙酸的产生,有利于提高植物在非生物逆境胁迫下的存活率(Glick,2014;Mayak等,2004a,2004b)。

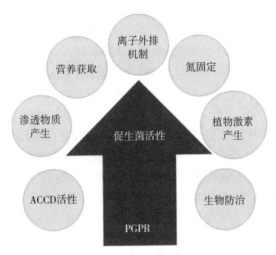

图3.2 植物生长促生菌促进植物生长的促生活性

3.4.2 植物渗透调节物质的产生

植物渗透调节物质是一类高度可溶的物质,在植物细胞正常生理条件下不带电荷,因此它们不与细胞中的蛋白质发生相互作用,从而不会阻碍细胞的正常生理功能。植物渗透调节物质还能够与许多细胞的功能兼容,如DNA-蛋白质之间的相互作用、蛋白质-蛋白质相互作用、细胞代谢和渗透平衡(Lippert和Galinski,1992;Welsh,2000)。目前,研究最多的植物渗透调节物质是由耐盐植物生长促生菌产生和分泌的海藻糖、葡萄糖基甘油、脯氨酸、甘氨酸、甜菜碱和胆碱等(Qurashi和Sabri,2013)。海藻糖是一种非还原性的二糖,由两个葡萄糖分子以1,1-糖苷键构成,在陆生维管植物中并不常见。但海藻糖作为一种植物渗透调节保护物质,能起到稳定细胞和生物膜的脱水作用,从而防止其因脱水而造成功能障碍。海藻糖对豆科植物在非生物逆境胁迫条件下的植物生长和产量具有明显的促进作用(Suarez等,2008)。窄食单胞菌(*Stenotrophomonas rhizophila*)是一种耐盐根际细菌,能在浓度为4%的NaCl胁迫条件下生长,并且能够在这种盐胁迫条件下产生海藻糖(Roder等,2005)。

3.4.3 离子排斥机制

植物根际细菌可以通过对寄主植物的生理状态产生轻微变化（通过对离子转运蛋白进行调节）和改变植物根系周围的物理屏障来改变根对有毒离子的吸收（Siddike 等，2011），或者直接减少植物对 Na^+ 和 Cl^- 等的积累。植物根际细菌还能够通过维持植物体内必需的大量元素（N、P、K）和微量元素（Zn、Fe、Cu、Mn）的水平来改善植物的营养状况。Zahir 等（2009）的研究结果表明，接种假单胞菌（*Pseudomonas*）和沙雷氏菌（*Serratia* sp.）后，在盐渍化环境中生长的作物（比如小麦）植株中的 K^+/Na^+ 比值较高。

3.4.4 磷、铁等可溶性增加

磷（P）在土壤中以有机和无机两种形式大量存在，但土壤中的大多数磷以不溶于水的形式存在，不易被植物吸收。植物根际微生物可以促使植物根系分泌有机酸等物质，引起根际 pH 发生变化，将土壤中不易溶于水的磷转化为可溶态供植物吸收利用（El Tarabily 和 Youssef，2010）。另外，土壤中的铁元素通常以 Fe^{3+} 的形式存在，易与土壤中的碱性阴离子（如 OH^- 等）形成不溶性的氢氧化物，阻碍植物对铁元素的正常吸收。植物生长促生菌能够分泌铁-铁载体蛋白等有机大分子，这些铁-铁载体蛋白能够螯合土壤溶液中的 Fe^{3+}，供植物吸收铁元素（Rajkumar 等，2010）。有研究结果表明，肠杆菌 638（*Enterobacter* sp. 638）和假单胞菌（*Pseudomonas* sp.）产生的铁-铁载体分别促进了杨树和番茄的生长（Taghavi 等，2009；Nishma 等，2014）。

3.4.5 固氮作用

固氮作用是将大气中的 N_2 转化为 NH_3 的过程。与植物共生或寄生的根际细菌都能够参与 N_2 的固定。所有这些微生物都有编码固定 N_2 的特定固氮酶基因，如 *nifH* 基因等，它编码的二硝化酶还原酶可以将空气中的 N_2 转化为能被植物吸收的 NH_3（Raymond 等，2004）。植物根际相关细菌利用植物根系分泌物中的碳化合物作为自身生长的能量来源，通过合成固氮酶来固定空气中的 N_2 供植物利用。

3.4.6 植物激素的合成

植物根际细菌还可以合成多种植物激素，比如细胞分裂素（cytokinin）、吲哚-3-乙酸（Indole-3-acetic acid，IAA）、脱落酸（abscisic acid，ABA）和赤霉素（gibberellin，GA）等（Glick，2012）。由植物生长促生菌产生的植物激素在许多植物中发挥重要的作用，主要集中在增加植物根系表面积和根尖数目两个方面。植物根系表面积和根尖数目的增加能够增强植物对土壤溶液中养分的吸收能力（Narula 等，2006）。类似地，有报道指出，赤霉素能够提高植物茎的生长和总生物量（Boiero 等，2007）。脱落酸可以调节植物叶片的气孔活性，从而影响生长在盐渍化土壤上的植物的光合反应（Dodd，2003）。目前，在各种研究中使用的植物生长促生菌大多都是从盐渍化土壤中分离出来的，这些菌合成的脱落酸能够促进盐渍化条件下植物的生长（Naz 等，2009）。

3.4.7 生物防控物质

植物生长促生菌还可以合成并分泌一些抗生素类化合物，这些化合物在减少植物病原菌方面发挥着主要作用（Stutz 等，1986）。利用植物根际细菌作为生防菌，可以间接地刺激植物生长。根际细菌能够合成可以水解病原体细胞壁的胞外酶，通过降解病原体细胞壁抑制病原菌的增殖和对植物的侵染。并且，植物根际细菌还能够与有害细菌竞争植物根际生态位，从而降低病原菌在植物根际的密度（Zahir 等，2004；Van Loon，2007）。植物生长促生菌还能够激活宿主植物的诱导性系统性抗性，在分子水平上与植物病原体抗争（Chandler 等，2008）。

3.5 耐盐根际细菌

有研究报道指出，盐生植物根际细菌能够耐受高达 4%~30% 的盐浓度胁迫（Yuan 等，2016），并在盐生植物对恶劣环境的适应和生存中发挥重要的作用，因此，这些根际细菌又被称为盐生植物根际相关微生物（halophyte rhizosphere-associated microbes，HRAM）。这类细菌能够激活对植物在盐胁迫下的生存起到至关重要的特定代谢信号（Szymanska 等，2013）。植物和微生物的相互作用能够有效地改善土壤质量和肥力（Yuan 等，2016）。盐生植物根际相关微生物也是一种能够耐受高浓度盐胁迫的植物生长促生菌，因此它们能够以最环保的方式促进甜土植物在盐渍化土壤上的生长，这是诸多能够提高作物产量方法中的最佳选择。

3.6 耐盐细菌是盐渍土农业的植物益生菌

适当的植物渗透物质调节是盐渍化土地上植物正常生长的关键因素。植物细胞中没有适当的渗透调节物质会导致植物细胞膨大或细胞脱水、植物光合作用机能失效、植物营养失调、离子毒性频发、作物产量降低，甚至最终导致植物细胞和植株的死亡（Ashraf，2004；SurravaVa 和 Kumar，2015）。

植物根际耐盐细菌能够在盐胁迫条件下存活，是因为它们能够适应盐渍土环境下的渗透压（Egamberdieva，2011）。植物渗透调节物质可以通过植物细胞合成，也可以从周围环境中吸收，被植物利用后可以充当细胞内的自由基清除剂、光合作用的调节器以及植物亚细胞结构的稳定剂（Yang 等，2009），进而影响植物的正常生长（Miller 和 Wood，1996）。因此，植物渗透调节物质被认为是促进植物生长的有效促进剂。

此前，许多研究表明，耐盐植物生长促生菌能够改善盐胁迫对罗勒、油菜、玉米和番茄等植株生长的不利影响（Kang 等，2014；Rojas Tapias 等，2012）。Karuppasamy 等（2011）的研究结果也表明，耐盐植物根际细菌的施用可以促进豆科植物雨树（*Samanea saman*）的生长。另外，从鳄梨树中分离得到的耐盐微生物菌群也可以改善其他作物遭受的盐胁迫危害，比如小麦（Barra 等，2016）。Upadhyay 等（2009）从盐渍化土壤区域生长的小麦根际分离出了 130 株根际细菌，其中 24 株能够耐受高达 8% 的 NaCl 胁迫。这 24 个耐盐菌株均能够产生吲哚-3-乙酸，其中有 10 个菌株具有溶解磷的能力，有 8 个菌株能够合成分泌铁-铁载体，有 6 个菌株参与赤霉素的合成和分泌，另外还有 2 个菌株含有 *nifH* 基因，能够参与空气中 N_2 的固定，这些结果说明整个菌群都具有促进植物生长的能力。

在另一项研究中，Nadeem 等（2010）评估了在不同盐胁迫条件下植物根际细菌菌株对促进植物生长的潜力。在高盐度水平（15 dS/m）下，植物生长促生菌-菌剂复合物使植株的株高、根长、生物量和产量分别提高了 37%、70%、116% 和 111%。此外，处理后的植株还表现出较高的 K^+/Na^+ 比、相对含水量和叶绿素含量以及较低的脯氨酸含量。表 3.1 列出了在盐渍化土壤中具有应用潜力的耐盐植物根际细菌。

表 3.1 具有不同促生长特性的潜在耐盐根际细菌

细菌名称	分离地点	植物促生长性状观察	参考文献
皮氏无色杆菌 ARV8（*Achromobacter piechaudii* ARV8）	干旱盐渍化环境植物根际	盐胁迫下番茄幼苗生物量增加	Mayak 等（2004a，b）
短小芽孢杆菌（*Bacillus pumilus*）、海水芽孢杆菌（*Bacillus aquimaris*）、亚砷酸芽孢杆菌（*Bacillus arsenicus*）、耐热芽孢杆菌（*Bacillus sporothermodurances*）、节杆菌属（*Arthrobacter* sp.）、蜡状芽孢杆菌（*Bacillus cereus*）、地中海假单胞菌（*Pseudomonas medicona*）和枯草芽孢杆菌（*Bacillus subtilis*）	盐渍土 30 日龄小麦根际	吲哚-3-乙酸的产生、磷的增溶、铁载体的产生、赤霉素的产生和 *nifH* 基因的存在	Upadhyay 等（2009）

(续表)

细菌名称	分离地点	植物促生长性状观察	参考文献
不动杆菌（*Acinetobacter*）、粪产碱杆菌（*Alcaligenes faecalis*）、蜡状芽孢杆菌（*Bacillus cereus*）、霍氏肠杆菌（*Enterobacter hormaechei*）、潘托亚菌（*Pantoae*）和铜绿假单胞菌（*Pseudomonas aeruginosa*）	从生长在壤土中的小麦和豌豆中分离的根际微生物	盐胁迫下促进植物生长和营养的改善	Egamberdieva（2008）
窄食单胞菌 DSM14405T（*Stenotrophomonas rhizophila* strain DSM144-05T）	高盐胁迫土壤	提高小麦、番茄、生菜、甜椒、甜瓜、芹菜和胡萝卜的发芽率，促进芽和根的生长	Egamberdieva（2011）
巨大芽孢杆菌（*Bacillus megaterium*）	退化土壤	玉米根系导水率的增加	Marulanda 等（2010）
门多萨假单胞菌（*Pseudomonas mendocina*）	根际土	促进莴苣（*Lactuca sativa*）生长和养分吸收	Kohler 等（2009）
褐球固氮菌（*Azotobacter chroococcum*）	根际土	提高甜菜抗氧化能力	Stajner 等（1997）
恶臭假单胞菌 Rs-198（*Pseudomonas putida* Rs-198）	碱土	提高 K^+/Na^+ 比率	Yao 等（2010）
促进植物生长的根菌群	盐渍土中各种杂草的根际，如金须茅属植物 *Chrysopogon aucheri*、裂叶莴苣（*Lactuca dissecta*）、牛茄子（*Solanum surattense*）和苣荬菜（*Sonchus arvensis*）	提高大豆（*Glycine max* L.）的耐盐性	Naz 等（2009）
短小芽孢杆菌（*Bacillus pumilus*）和类产碱假单胞菌（*Pseudomonas pseudoalcaligenes*）	水稻根与水稻根际土壤	诱导渗透保护剂诱导	Jha 等（2011）
固氮螺菌属（*Azospirillum*）	盐渍土和非盐渍土	提高小麦茎干重和产量	Nia 等（2012）
嗜盐芽孢杆菌（*Halobacillus* sp.）和盐反硝化枝芽孢杆菌（*Bacillus halodenitrificans*）	盐渍土	改善小麦幼苗盐胁迫和增加根长	Ramadoss 等（2013）
枝芽孢菌（*Virgibacillus marismortui*）和嗜盐土地芽孢杆菌（*Terribacillus halophilus*）	甜土	促进番茄茎的生长	Essghaier 等（2014）
原小单孢菌属 SE188（*Promicromonospora* sp. SE188）	田野土壤样品	促进番茄茎的生长	Kang 等（2012）

（续表）

细菌名称	分离地点	植物促生长性状观察	参考文献
根际细菌群	生长在高盐环境下的盐角草属（Salicornia）植物根	促进植物生长	Mapelli 等（2013）
枯草芽孢杆菌（Bacillus subtilis）、萎缩芽孢杆菌（Bacillus atrophaeus）、球形芽孢杆菌（Bacillus spharicus）、克卢氏葡萄球菌（Staphylococcus kloosii）和玫瑰考克氏菌（Kocuria erythromyxa）	天然高盐土壤植物根际	盐渍条件下提高草莓和莴苣的产量和生长	Karlidag 等（2010）和 Yildirim 等（2011）
耐盐刘志恒氏菌（Zhihengliuella halotolerans）、葡萄球菌（Staphylococcus succinus）、吉氏芽孢杆菌（Bacillus gibsonii）、小鳟鱼大洋芽孢杆菌（Oceanobacillus oncorhynchi）、盐单胞菌属（Halomonas）和食有机物深海芽孢杆菌（Thalassobacillus sp.）	盐渍危害的土壤	提高盐胁迫下小麦（Triticum aestivum L.）的生长	Orhan（2016）
固氮螺菌属（Azospirillum spp.）	红树林根际	促进比吉洛氏海蓬子（Salicornia bigelovii）的生长	Bashan 等（2000）
植物内生菌（Endophytic bacteria）、假单胞菌属 ISE-12（Pseudomonas sp. ISE-12）和黄单胞菌 CSE-34（Xanthomonadales sp. CSE-34）	盐生植物盐角草（Salicornia europaea）根	促进甜菜（Beta vulgaris）在盐渍环境下的生长	Piernik 等（2017）
联合根瘤菌（Consortiaod rhizobacteria）和假单胞菌（Pseudomonas sp.）	海岸沙丘植物根系	促进植物生长	Shin 等（2007）
产酸克雷伯氏菌 Rs-5（Klebsiella oxytoca Rs-5）	受盐渍危害的棉花田	缓解棉花盐胁迫并促进生长	Yue 等（2007）
铜绿假单胞菌 PF23（Pseudomonas aeruginosa PF23）	盐渍条件下不同轮作植物根际（改良向日葵）	植物生长促进，缓解生物控制和胁迫	Tewari 和 Arora（2014）
芽孢杆菌（Bacillus）、假单胞菌（Pseudomonas）、克雷伯氏菌（Klebsiella）、沙雷氏菌（Serratia）、节杆菌（Arthrobacter）、链霉菌属（Streptomyces）、居白蚁菌（Isoptericola）和微杆菌属（Microbacterium sp.）	盐生植物补血草（Limonium sinense）根	盐胁迫下促进补血草种子萌发和幼苗生长的研究	Qin 等（2014）
新鞘氨醇菌（Novosphingobium pokkalii sp. Nov）	耐盐谷子根际	增加谷子品种 VTL-2 的生长	Krishnan 等（2017）

(续表)

细菌名称	分离地点	植物促生长性状观察	参考文献
芽孢杆菌（*Bacillus*）和假单胞菌（*Pseudomonas*）	海滨盐草（*Distichlis spicata*）根际土	促进拟南芥（*Arabidopsis thaliana*）、黄瓜（*Cucumis sativus*）和西瓜（*Citrullus lanatus*）的生长	Palacio Rodríguez 等（2017）
阿氏肠杆菌（*Enterobacter asburiae*）、苏云金芽孢杆菌（*Bacillus thuringiensis*）、莫拉菌（*Moraxella pluranimalium*）和施氏假单胞菌（*Pseudomonas stutzeri*）	盐渍土生长的阿拉伯金合欢树（*Acacia arabia*）	缓解小麦盐胁迫	Raheem 和 Ali（2015）
约氏黄杆菌（*Flavobacterium johnsoniae*）、恶臭假单胞菌（*Pseudomonas putida*）、木糖氧化产碱菌（*Achromobacter xylosoxidans*）和褐球固氮菌（*Azotobacter chroococcum*）	用于小麦种植的农业用地	显著提高小麦种子发芽率	Abdelwahab 等（2017）
巴西固氮螺菌（*Azospirillum brasilense*）	多种禾本科植物根际	缓解白车轴草（*Trifolium repens*）盐胁迫	Khalid 等（2017）
山野壳菌 Jp-root-44（*Montagnulaceae* sp. Jp-root-44）	耐盐植物盐地碱蓬（*Suaeda salsa*）根际	在正常和盐胁迫条件下促进黄瓜（*Cucumis sativus*）和水稻（*Oryza sativa*）的生长	Yuan 等（2016）
丁香假单胞菌（*Pseudomonas syringae*）和荧光假单胞菌（*Pseudomonas fluorescens*）	盐渍地绿豆根际及根瘤	绿豆总干重和耐盐性的提高	Ahmad 等（2013）
肠杆菌属 EJ01（*Enterobacter* sp. EJ01）	复垦地盐生日本石竹（*Dianthus japonicus*）根际土壤	促进番茄幼苗生长提高抗盐性	Kim 等（2014）
表皮短杆菌 RS15（*Brevibacterium epidermidis* RS15）、云南微球菌 RS222（*Micrococcus yunnanensis* RS222）和阿氏芽孢杆菌 RS341（*Bacillus aryabhattai* RS341）	滨海盐碱地6种原生盐生植物根际	促进盐胁迫下油菜幼苗的生长	Siddikee 等（2010）
荧光假单胞菌（*Pseudomonas fluorescens*）、铜绿假单胞菌（*Pseudomonas aeruginosa*）和施氏假单胞菌（*Pseudomonas stutzeri*）	农田番茄根际	盐胁迫下番茄鲜重和干重的提高	Tank 和 Saraf（2010）
纤维菌属 S16（*Cellulosimicrobium* sp. S16）	盐渍农业根际土壤	大麦幼苗生长及生物防治能力的提高	Nabti 等（2014）

(续表)

细菌名称	分离地点	植物促生长性状观察	参考文献
短小芽孢杆菌（*Bacillus pumilus*）、门多萨假单胞菌（*Pseudomonas mendocina*）、节杆菌（*Arthrobacter* sp.）、盐单胞菌属（*Halomonas*）和拉齐斯庞氏硝化菌（*Nitrinicola lacisaponensis*）	高盐土地	提高小麦（*Triticum aestivum* L.）的耐盐能力	Tiwari 等（2011）
重氮营养哈特曼杆菌（*Hartmannibacter diazotrophicus*）	天然盐渍草地冬车前根际	促进盐胁迫下大麦（*Hordeum vulgare*）的生长	Suarez 等（2015）
巴西固氮螺菌（*Azospirillum brasilense*）	盐渍麦田土壤根际	促进盐胁迫条件下硬粒小麦（*Triticum durum* var. *waha*）的生长	Nabti 等（2010）
链霉菌（*Streptomyces isolate*）	小麦根际	促进小麦在盐胁迫条件下的生长发育	Sadeghi 等（2012）
粪产碱杆菌（*Alcaligenes faecalis*）、短小芽孢杆菌（*Bacillus pumilus*）和苍白杆菌属（*Ochrobactrum* sp.）	沿海水稻根区土壤	对盐胁迫下水稻不同生长参数产生积极影响	Bal 等（2013）
枯草芽孢杆菌（*Bacillus subtilis*）和节杆菌（*Arthrobacter* sp.）	盐渍土 30 日龄小麦根际	减轻土壤盐分对小麦生长的不利影响	Upadhyay 等（2012）
固氮螺菌属（*Azospirillum* sp.）	非盐渍土和盐渍土	对生长在盐胁迫条件下的小麦产生积极影响	Zarea 等（2012）
类产碱假单胞菌（*Pseudomonas pseudoalcaligenes*）和恶臭假单胞菌（*Pseudomonas putida*）	盐渍土区域	促进盐胁迫条件下鹰嘴豆的生长	Patel 等（2012）
假单胞菌（*Pseudomonas* sp.）	生长在滨海盐渍土中茄子（*Solanum melongena* L.）根际土壤	促进盐胁迫条件下茄子幼苗的生长	Fu 等（2010）
芽孢杆菌（*Bacillus* sp.）和苍白杆菌属（*Ochrobactrum* sp.）	盐渍土	促进盐胁迫条件下玉米（*Zea mays*）和冰草属植物（*Agropyron elongatum*）的生长	Príncipe 等（2007）
荧光假单胞菌（*Pseudomonas fluorescens*）	沿海和林业生态系统	提高花生植株的抗盐性	Saravanakumar 和 Samiyappan（2007）
固氮螺菌属（*Azospirillum*）	玉米根际	促进两个玉米品种在盐胁迫下的生长	Hamdia 等（2004）
云南微球菌（*Micrococcus yunnanensis*）、莱比托游动球菌（*Planococcus rifietoensis*）和争论贪噬菌（*Variovorax paradoxus*）	沙漠边缘的盐渍地	显著提高甜菜（*Beta vulgaris* L.）的耐盐性	Zhou 等（2017）

(续表)

细菌名称	分离地点	植物促生长性状观察	参考文献
肠杆菌（*Enterobacter*）、沙雷氏菌（*Serratia*）、假单胞菌（*Pseudomonas*）、微杆菌属（*Microbacterium* sp.）和无色杆菌（*Achromobacter*）	海边鳄梨根际土壤	改善盐胁迫对小麦幼苗出苗、生长和生物量的影响	Barra 等（2016）
类产碱假单胞菌（*Pseudomonas pseudoalcaligenes*）、反硝化假单胞杆菌（*Pseudomonas denitrificans*）、多黏黏孢杆菌（*Bacillus polymyxa*）和分枝杆菌（*Mycobacterium phlei*）	甜瓜根区土壤	盆栽试验显著促进棉花和豌豆的生长和养分吸收	Egamberdiyeva 和 Hoflich（2004）
巨大芽孢杆菌（*Bacillus megaterium*）和肠杆菌（*Enterobacter* sp.）	作物田块	提高秋葵的耐盐性	Habib 等（2016）
皮氏无色杆菌 ARV8（*Achromobacter piechaudii* ARV8）	干旱和盐渍环境	促进盐胁迫条件下番茄生物量的提高	Mayak 等（2004b）
恶臭假单胞菌（*Pseudomonas putida*）、阴沟肠杆菌属（*Enterobacter cloacae*）、阴沟肠杆菌属（*Enterobacter cloacae*）、无花果沙雷氏菌（*Serratia ficaria*）和荧光假单胞菌（*Pseudomonas fluorescens*）	盐渍土	提高高盐胁迫下小麦的发芽率	Nadeem 等（2013）

3.7 结论

利用耐盐植物根际细菌提高作物耐盐能力为基础的盐土农业比从分子水平上对农作物品种进行改良后再利用更有效和环保。植物微生物组目前被认为具有增强宿主抗逆能力，是植物的第二基因组。盐生植物的根际细菌群落具备多种能够促进植物生长的特性。目前报道了很多从盐生植物根际分离到耐盐细菌群落，并对它们在盐渍化土壤环境下促进植物生长的潜力进行了研究。虽然单一的植物生长促生菌菌株可以提高植物的生长和对非生物逆境胁迫的耐受性，比如盐胁迫，但也有一些研究报道指出，植物对盐胁迫的耐受能力通常由不止 1 种共生微生物决定，含有两种或两种以上微生物的植物-微生物共生体系对植物在盐胁迫条件下的生长促进效果更加明显。然而，植物生长促生菌在农业中的应用仍需要更加广泛和深入地研究和提高认识。除了对宿主植物的生理和对微生物的研究外，还需要关注植物与微生物之间的相互作用，以了解植物在应对生物和非生物胁迫下的响应机制。鉴定植物-微生物分泌物、植物-微生物互作交流信号分子和根际微生物群中的关键代谢物质，可以阐明植物生长促生菌促进植物生长的分子机制，此外还可以揭示植物是否以及如何招募和刺激有益微生物共同抵御逆境胁迫的过程。

[致谢]

CSIR-CSMCRI 通信号：PRIS-42/2018。

参考文献

AbdElgawad H, Zinta G, Hegab MM, Pandey R, Asard H, Abuelsoud W, 2016. High salinity induces different oxidative stress and antioxidant responses in maize seedlings organs. Front Plant Sci 7: 276.

Abdelwahab RAI, Cherif A, Cristina C, Nabti E, 2017. Extracts from seaweeds and Opuntiaficusindica cladodes enhance diazotrophic-PGPR halotolerance, their enzymatic potential, and their impact on wheat germination under salt stress. Pedosphere.

Ahmad M, Zahir ZA, Nazli F, Akram F, Arshad M, Khalid M, 2013. Effectiveness of halotolerant, auxin producing *Pseudomonas* and *Rhizobium* strains to improve osmotic stress tolerance in mung bean (*Vigna radiata* L.). Braz J Microbiol 44 (4): 1341-1348.

Ashraf M, 2004. Some important physiological selection criteria for salt tolerance in plants. FloraMorphology, Distrib, Funct Ecol Plants 199 (5): 361-376.

Ashraf M, 2009. Biotechnological approach of improving plant salt tolerance using antioxidants as markers. Biotechnol Adv 27 (1): 84-93.

Baki GK, Siefritz F, Man HM, Weiner H, Kaldenhoff R, Kaiser WM, 2000. Nitrate reductase in *Zea mays* L. under salinity. Plant Cell Environ 23 (5): 515-521.

Bal HB, Nayak L, Das S, Adhya TK, 2013. Isolation of ACC deaminase producing PGPR from rice rhizosphere and evaluating their plant growth promoting activity under salt stress. Plant and Soil: 1-13.

Bano A, Fatima M, 2009. Salt tolerance in *Zea mays* (L). following inoculation with Rhizobium and *Pseudomonas*. Biol Fertil Soils 45 (4): 405-413.

Barra PJ, Inostroza NG, Acuña JJ, Mora ML, Crowley DE, Jorquera MA, 2016. Formulation of bacterial consortia from avocado (*Persea americana* Mill.) and their effect on growth, biomass and superoxide dismutase activity of wheat seedlings under salt stress. Appl Soil Ecol 102: 80-91.

Bashan Y, Moreno M, Troyo E, 2000. Growth promotion of the seawater-irrigated oilseed halophyte Salicornia bigelovii inoculated with mangrove rhizosphere bacteria and halotolerant *Azospirillum* spp. Biol Fertil Soils 32 (4): 265-272.

Bhatnagar-Mathur P, Vadez V, Sharma KK, 2008. Transgenic approaches for abiotic stress tolerance in plants: retrospect and prospects. Plant Cell Rep. 27 (3): 411-424.

Blaylock AD, 1994. Soil salinity, salt tolerance, and growth potential of horticultural and landscape plants. University of Wyoming, Cooperative Extension Service, Department of Plant,

Soil, and Insect Sciences, College of Agriculture.

Boiero L, Perrig D, Masciarelli O, Penna C, Cassán F, Luna V, 2007. Phytohormone production by three strains of *Bradyrhizobium japonicum* and possible physiological and technological implications. Appl Microbiol Biotechnol 74 (4): 874-880.

Carillo P, Annunziata MG, Pontecorvo G, Fuggi A, Woodrow P, 2011. Salinity stress and salt tolerance. In: Abiotic stress in plants-mechanisms and adaptations. InTech.

Chandler D, Davidson G, Grant W, Greaves J, Tatchell G, 2008. Microbial biopesticides for integrated crop management: an assessment of environmental and regulatory sustainability. Trends Food Sci Technol 19 (5): 275-283.

Chinnusamy V, Zhu J, Zhu JK, 2006. Gene regulation during cold acclimation in plants. Physiol Plant 126 (1): 52-61.

Coleman-Derr D, Tringe SG, 2014. Building the crops of tomorrow: advantages of symbiont-based approaches to improving abiotic stress tolerance. Front Microbiol 5: 283.

Cook RJ, Thomashow LS, Weller DM, Fujimoto D, Mazzola M, Bangera G, Kim DS, 1995. Molecular mechanisms of defense by rhizobacteria against root disease. Proc Natl Acad Sci 92 (10): 4197-4201.

De Souza R, Ambrosini A, Passaglia LM, 2015. Plant growth-promoting bacteria as inoculants in agricultural soils. Genet Mol Biol 38 (4): 401-419.

De Souza Vandenberghe LP, Garcia LMB, Rodrigues C, Camara MC, de Melo Pereira GV, de Oliveira J, Soccol CR, 2017. Potential applications of plant probiotic microorganisms in agriculture and forestry. AIMS Microbiol 3 (3): 629-648.

Del Río LA, Corpas FJ, Sandalio LM, Palma JM, Barroso JB, 2003. Plant peroxisomes, reactive oxygen metabolism and nitric oxide. IUBMB Life 55 (2): 71-81.

Dodd IC, 2003. Hormonal interactions and stomatal responses. J Plant Growth Regul 22 (1): 32-46.

Egamberdieva D, 2008. Plant growth promoting properties of rhizobacteria isolated from wheat and pea grown in loamy sand soil. Turk J Biol 32 (1): 9-15.

Egamberdieva D, 2011. Survival of *Pseudomonas extremorientalis* TSAU20 and P. chlororaphis TSAU13 in the rhizosphere of common bean (*Phaseolus vulgaris*) under saline conditions. Plant Soil Environ 57 (3): 122-127.

Egamberdiyeva D, Höflich G, 2004. Effect of plant growth-promoting bacteria on growth and nutrient uptake of cotton and pea in a semi-arid region of Uzbekistan. J Arid Environ 56 (2): 293-301.

El-Tarabily KA, Youssef T, 2010. Enhancement of morphological, anatomical and physiological characteristics of seedlings of the mangrove *Avicennia marina* inoculated with a native phosphate-solubilizing isolate of *Oceanobacillus picturae* under greenhouse conditions. Plant Soil 332 (1-2): 147-162.

Essghaier B, Dhieb C, Rebib H, Ayari S, Boudabous ARA, Sadfi-Zouaoui N, 2014. Antimicrobial behavior of intracellular proteins from two moderately halophilic bacteria: strain J31 of *Terribacillus halophilus* and strain M3-23 of *Virgibacillus marismortui*. J Plant Pathol Microbiol 5 (1): 1.

FAO, 2016. FAO Soil Portal. http://www.fao.org/soils-portal/en/.

Fedoroff NV, 2010. The past, present and future of crop genetic modification. New Biotechnol 27

(5): 461-465.

Fu Q, Liu C, Ding N, Lin Y, Guo B, 2010. Ameliorative effects of inoculation with the plant growth-promoting rhizobacterium *Pseudomonas* sp. DW1 on growth of eggplant (*Solanum melongena* L.) seedlings under salt stress. Agric Water Manag 97 (12): 1994-2000.

Glick BR, 2012. Plant growth-promoting bacteria: mechanisms and applications. Scientifica 2012: 963401. https://doi.org/10.6064/2012/963401.

Glick BR, 2014. Bacteria with ACC deaminase can promote plant growth and help to feed the world. Microbiol Res 169 (1): 30-39.

Gray EJ, Smith DL, 2005. Intracellular and extracellular PGPR: commonalities and distinctions in the plant-bacterium signaling processes. Soil Biol Biochem 37 (3): 395-412.

Habib SH, Kausar H, Saud HM, 2016. Plant growth-promoting rhizobacteria enhance salinity stress tolerance in okra through ROS-scavenging enzymes. BioMed Res Int 2016: 6284547.

Hamdia MAE-S, Shaddad M, Doaa MM, 2004. Mechanisms of salt tolerance and interactive effects of *Azospirillum brasilense* inoculation on maize cultivars grown under salt stress conditions. Plant Growth Regul 44 (2): 165-174.

Han S, Yu B, Wang Y, Liu Y, 2011. Role of plant autophagy in stress response. Protein Cell 2 (10): 784-791.

Hill C, Guarner F, Reid G, Gibson GR, Merenstein DJ, Pot B, Morelli L, Canani RB, Flint HJ, Salminen S, 2014. Expert consensus document: The international scientific association for probiotics and prebiotics consensus statement on the scope and appropriate use of the term probiotic. Nat Rev Gastroenterol Hepatol 11 (8): 506.

Jewell MC, Campbell BC, Godwin ID, 2010. Transgenic plants for abiotic stress resistance. In: Transgenic crop plants. Springer, pp. 67-132.

Jha Y, Subramanian R, Patel S, 2011. Combination of endophytic and rhizospheric plant growth promoting rhizobacteria in *Oryza sativa* shows higher accumulation of osmoprotectant against saline stress. Acta Physiol Plant 33 (3): 797-802.

Kang S M, Khan AL, Hamayun M, Hussain J, Joo GJ, You YH, Kim JG, Lee IJ, 2012. Gibberellin-producing *Promicromonospora* sp. SE188 improves *Solanum lycopersicum* plant growth and influences endogenous plant hormones. J Microbiol 50 (6): 902.

Kang S M, Khan AL, Waqas M, You YH, Kim JH, Kim JG, Hamayun M, Lee IJ, 2014. Plant growth-promoting rhizobacteria reduce adverse effects of salinity and osmotic stress by regulating phytohormones and antioxidants in *Cucumis sativus*. J Plant Interact 9 (1): 673-682.

Karlidag H, Esitken A, Yildirim E, Donmez MF, Turan M, 2010. Effects of plant growth promoting bacteria on yield, growth, leaf water content, membrane permeability, and ionic composition of strawberry under saline conditions. J Plant Nutr 34 (1): 34-45.

Karuppasamy K, Nagaraj S, Kathiresan K, 2011. Stress tolerant rhizobium enhances the growth of *Samanea saman* (JECQ) Merr. Afr J Basic Appl Sci 3 (6): 278-284.

Khalid M, Bilal M, Hassani D, Iqbal HM, Wang H, Huang D, 2017. Mitigation of salt stress in white clover (*Trifolium repens*) by *Azospirillum brasilense* and its inoculation effect. Bot Stud 58 (1): 5.

Kim K, Jang YJ, Lee SM, Oh BT, Chae JC, Lee KJ, 2014. Alleviation of salt stress by *Enterobacter* sp. EJ01 in tomato and *Arabidopsis* is accompanied by up-regulation of conserved salinity responsive factors in plants. Mol Cells 37 (2): 109.

Kloepper J, Schroth M, 1981. Relationship of in vitro antibiosis of plant growth-promoting rhizobacteria to plant growth and the displacement of root microflora. Phytopathology 71 (10): 1020-1024.

Kohler J, Hernández JA, Caravaca F, Roldán A, 2009. Induction of antioxidant enzymes is involved in the greater effectiveness of a PGPR versus AM fungi with respect to increasing the tolerance of lettuce to severe salt stress. Environ Exp Bot 65 (2): 245-252.

Krishnan R, Menon RR, Busse HJ, Tanaka N, Krishnamurthi S, Rameshkumar N, 2017. *Novosphingobium pokkalii* sp nov, a novel rhizosphere-associated bacterium with plant beneficial properties isolated from saline-tolerant pokkali rice. Res Microbiol 168 (2): 113-121.

Lau JA, Lennon JT, 2012. Rapid responses of soil microorganisms improve plantfitness in novel environments. Proc Natl Acad Sci 109 (35): 14058-14062.

Lippert K, Galinski EA, 1992. Enzyme stabilization be ectoine-type compatible solutes: protection against heating, freezing and drying. Appl Microbiol Biotechnol 37 (1): 61-65.

Ma W, Guinel FC, Glick BR, 2003. Rhizobium leguminosarum biovar viciae 1-aminocyclopropane-1-carboxylate deaminase promotes nodulation of pea plants. Appl Environ Microbiol 69 (8): 4396-4402.

Mahajan S, Tuteja N, 2005. Cold, salinity and drought stresses: an overview. Arch Biochem Biophys 444 (2): 139-158.

Mapelli F, Marasco R, Rolli E, Barbato M, Cherif H, Guesmi A, Ouzari I, Daffonchio D, Borin S, 2013. Potential for plant growth promotion of rhizobacteria associated with *Salicornia* growing in Tunisian hypersaline soils. BioMed Res Int 2013: 248078.

Marulanda A, Azcón R, Chaumont F, Ruiz-Lozano JM, Aroca R, 2010. Regulation of plasma membrane aquaporins by inoculation with a *Bacillus megaterium* strain in maize (*Zea mays* L.) plants under unstressed and salt-stressed conditions. Planta 232 (2): 533-543.

Mayak S, Tirosh T, Glick BR, 2004a. Plant growth-promoting bacteria confer resistance in tomato plants to salt stress. Plant Physiol Biochem 42 (6): 565-572.

Mayak S, Tirosh T, Glick BR, 2004b. Plant growth-promoting bacteria that confer resistance to water stress in tomatoes and peppers. Plant Sci 166 (2): 525-530.

Miller KJ, Wood JM, 1996. Osmoadaptation by rhizosphere bacteria. Annu Rev Microbiol 50 (1): 101-136.

Mishra A, Tanna B, 2017. Halophytes: potential resources for salt stress tolerance genes and promoters. Front Plant Sci 8: 829.

Mukhtar S, Ishaq A, Hassan S, Mehnaz S, Mirza MS, Malik KA, 2017. Comparison of microbial communities associated with halophyte (*Salsola stocksii*) and non-halophyte (*Triticum aestivum*) using culture-independent approaches. Pol J Microbiol 66 (3): 353-364.

Munns R, Gilliham M, 2015. Salinity tolerance of crops - what is the cost? New Phytol 208 (3): 668-673.

Munns R, Tester M, 2008. Mechanisms of salinity tolerance. Annu Rev Plant Biol 59: 651-681.

Nabti E, Sahnoune M, Ghoul M, Fischer D, Hofmann A, Rothballer M, Schmid M, Hartmann A, 2010. Restoration of growth of durum wheat (*Triticum durum* var. waha) under saline conditions due to inoculation with the rhizosphere bacterium *Azospirillum brasilense* NH and extracts of the marine alga Ulva lactuca. J Plant Growth Regul 29 (1): 6-22.

Nabti E, Bensidhoum L, Tabli N, Dahel D, Weiss A, Rothballer M, Schmid M, Hartmann A, 2014. Growth stimulation of barley and biocontrol effect on plant pathogenic fungi by a *Cellulosimicrobium* sp. strain isolated from salt-affected rhizosphere soil in northwestern Algeria. Eur J Soil Biol 61: 20-26.

Nadeem SM, Zahir ZA, Naveed M, Asghar HN, Arshad M, 2010. Rhizobacteria capable of producing ACC-deaminase may mitigate salt stress in wheat. Soil Sci Soc Am J 74 (2): 533-542.

Nadeem SM, Zahir ZA, Naveed M, Nawaz S, 2013. Mitigation of salinity-induced negative impact on the growth and yield of wheat by plant growth-promoting rhizobacteria in naturally saline conditions. Ann Microbiol 63 (1): 225-232.

Narula N, Deubel A, Gans W, Behl R, Merbach W, 2006. Paranodules and colonization of wheat roots by phytohormone producing bacteria in soil. Plant Soil Environ 52 (3): 119.

Naz I, Bano A, Ul-Hassan T, 2009. Isolation of phytohormones producing plant growth promoting rhizobacteria from weeds growing in Khewra salt range, Pakistan and their implication in providing salt tolerance to *Glycine max* L. Afr J Biotechnol 8 (21): 5762-5768.

Netondo GW, Onyango JC, Beck E, 2004. Sorghum and salinity. Crop Sci 44 (3): 797-805.

Nia SH, Zarea MJ, Rejali F, Varma A, 2012. Yield and yield components of wheat as affected by salinity and inoculation with *Azospirillum* strains from saline or non-saline soil. J Saudi Soc Agric Sci 11 (2): 113-121.

Nishma K, Adrisyanti B, Anusha S, Rupali P, Sneha K, Jayamohan N, Kumudini B, 2014. Induced growth promotion under in vitro salt stress tolerance on solanum lycopersicum by fluorescent pseudomonads associated with rhizosphere. IJASER 3: 422-430.

Orhan F, 2016. Alleviation of salt stress by halotolerant and halophilic plant growth-promoting bacteria in wheat (*Triticum aestivum*). Braz J Microbiol 47 (3): 621-627.

Palacio-Rodríguez R, Coria-Arellano JL, López-Bucio J, Sánchez-Salas J, Muro-Pérez G, Castañeda-Gaytán G, Sáenz-Mata J, 2017. Halophilic rhizobacteria from *Distichlis spicata* promote growth and improve salt tolerance in heterologous plant hosts. Symbiosis 73 (3): 179-189.

Patel D, Jha CK, Tank N, Saraf M, 2012. Growth enhancement of chickpea in saline soils using plant growth-promoting rhizobacteria. J Plant Growth Regul 31 (1): 53-62.

Piernik A, Hrynkiewicz K, Wojciechowska A, Szymańska S, Lis MI, Muscolo A, 2017. Effect of halotolerant endophytic bacteria isolated from *Salicornia europaea* L. on the growth of fodder beet (*Beta vulgaris* L.) under salt stress. Arch Agron Soil Sci 63 (10): 1404-1418.

Podmore C, 2009. Irrigation salinity-causes and impacts. Primefact 937 (1): 1-4.

Príncipe A, Alvarez F, Castro MG, Zachi L, Fischer SE, Mori GB, Jofré E, 2007. Biocontrol and PGPR features in native strains isolated from saline soils of Argentina. Curr Microbiol 55 (4): 314-322.

Qin S, Zhang YJ, Yuan B, Xu PY, Xing K, Wang J, Jiang JH, 2014. Isolation of ACC deaminase-producing habitat-adapted symbiotic bacteria associated with halophyte *Limonium sinense* (Girard) Kuntze and evaluating their plant growth-promoting activity under salt stress. Plant Soil 374 (1-2): 753-766.

Qurashi AW, Sabri AN, 2013. Osmolyte accumulation in moderately halophilic bacteria improves salt tolerance of chickpea. Pak J Bot 45: 1011-1016.

Raheem A, Ali B, 2015. Halotolerant rhizobacteria: beneficial plant metabolites and growth enhancement of *Triticum aestivum* L. in salt–amended soils. Arch Agron Soil Sci 61 (12): 1691-1705.

Rahnama A, James RA, Poustini K, Munns R, 2010. Stomatal conductance as a screen for osmotic stress tolerance in durum wheat growing in saline soil. Funct Plant Biol 37 (3): 255-263.

Rajkumar M, Ae N, Prasad MNV, Freitas H, 2010. Potential of siderophore–producing bacteria for improving heavy metal phytoextraction. Trends Biotechnol 28 (3): 142-149.

Ramadoss D, Lakkineni VK, Bose P, Ali S, Annapurna K, 2013. Mitigation of salt stress in wheat seedlings by halotolerant bacteria isolated from saline habitats. SpringerPlus 2 (1): 6.

Raymond J, Siefert JL, Staples CR, Blankenship RE, 2004. The natural history of nitrogen fixation. Mol Biol Evol 21 (3): 541-554.

Roder A, Hoffmann E, Hagemann M, Berg G, 2005. Synthesis of the compatible solutes glucosylglycerol and trehalose by salt-stressed cells of *Stenotrophomonas* strains. FEMS Microbiol Lett 243 (1): 219-226.

Rojas-Tapias D, Moreno-Galván A, Pardo-Díaz S, Obando M, Rivera D, Bonilla R, 2012. Effect of inoculation with plant growth–promoting bacteria (PGPB) on amelioration of saline stress in maize (*Zea mays*). Appl Soil Ecol 61: 264-272.

Roy SJ, Negrão S, Tester M, 2014. Salt resistant crop plants. Curr Opin Biotechnol 26: 115-124.

Sadeghi A, Karimi E, Dahaji PA, Javid MG, Dalvand Y, Askari H, 2012. Plant growth promoting activity of an auxin and siderophore producing isolate of *Streptomyces* under saline soil conditions. World J Microbiol Biotechnol 28 (4): 1503-1509.

Santoyo G, Moreno-Hagelsieb G, del Carmen Orozco-Mosqueda M, Glick BR, 2016. Plant growth-promoting bacterial endophytes. Microbiol Res 183: 92-99.

Saravanakumar D, Samiyappan R, 2007. ACC deaminase from *Pseudomonas fluorescens* mediated saline resistance in groundnut (*Arachis hypogea*) plants. J Appl Microbiol 102 (5): 1283-1292.

Saslis-Lagoudakis CH, Hua X, Bui E, Moray C, Bromham L, 2014. Predicting species' tolerance to salinity and alkalinity using distribution data and geochemical modelling: a case study using Australian grasses. Ann Bot 115 (3): 343-351.

Seckin B, Sekmen AH, Türkan I, 2009. An enhancing effect of exogenous mannitol on the antioxidant enzyme activities in roots of wheat under salt stress. J Plant Growth Regul 28 (1): 12.

Shin DS, Park MS, Jung S, Lee MS, Lee KH, Bae KS, Kim SB, 2007. Plant growth-promoting potential of endophytic bacteria isolated from roots of coastal sand dune plants. J Microbiol Biotechnol 17 (8): 1361-1368.

Shrivastava P, Kumar R, 2015. Soil salinity: a serious environmental issue and plant growth promoting bacteria as one of the tools for its alleviation. Saudi J Biol Sci 22 (2): 123-131.

Siddikee MA, Chauhan PS, Anandham R, Han GH, Sa T, 2010. Isolation, characterization, and use for plant growth promotion under salt stress, of ACC deaminase-producing halotolerant bacteria derived from coastal soil. J Microbiol Biotechnol 20 (11): 1577-1584.

Siddikee MA, Glick BR, Chauhan PS, Jong Yim W, Sa T, 2011. Enhancement of growth and salt tolerance of red pepper seedlings (*Capsicum annuum* L.) by regulating stress ethylene synthesis with halotolerant bacteria containing 1-aminocyclopropane-1-carboxylic acid deaminase activity. Plant Physiol Biochem 49 (4): 427-434.

Štajner D, Kevrešan S, Gašić O, Mimica-Dukić N, Zongli H, 1997. Nitrogen and Azotobacter chroococcum enhance oxidative stress tolerance in sugar beet. Biol Plant 39 (3): 441.

Stutz EW, Défago G, Kern H, 1986. Naturally occurring fluorescent pseudomonads involved in suppression of black root rot of tobacco. Phytopathology 76 (2): 181-185.

Suárez R, Wong A, Ramírez M, Barraza A, Orozco MC, Cevallos MA, Lara M, Hernández G, Iturriaga G, 2008. Improvement of drought tolerance and grain yield in common bean by overexpressing trehalose-6-phosphate synthase in rhizobia. Mol Plant-Microbe Interact 21 (7): 958-966.

Suarez C, Cardinale M, Ratering S, Steffens D, Jung S, Montoya AMZ, Geissler-Plaum R, Schnell S, 2015. Plant growth-promoting effects of Hartmannibacter diazotrophicus on summer barley (*Hordeum vulgare* L.) under salt stress. Appl Soil Ecol 95: 23-30.

Szymańska S, Piernik A, Hrynkiewicz K, 2013. Metabolic potential of microorganisms associated with the halophyte *Aster tripolium* L. in saline soils. Ecol Quest 18 (1): 9-19.

Taghavi S, Garafola C, Monchy S, Newman L, Hoffman A, Weyens N, Barac T, Vangronsveld J, van der Lelie D, 2009. Genome survey and characterization of endophytic bacteria exhibiting a beneficial effect on growth and development of poplar trees. Appl Environ Microbiol 75 (3): 748-757.

Tank N, Saraf M, 2010. Salinity-resistant plant growth promoting rhizobacteria ameliorates sodium chloride stress on tomato plants. J Plant Interact 5 (1): 51-58.

Tewari S, Arora NK, 2014. Multifunctional exopolysaccharides from *Pseudomonas aeruginosa* PF23 involved in plant growth stimulation, biocontrol and stress amelioration in sunflower under saline conditions. Curr Microbiol 69 (4): 484-494.

Tiwari S, Singh P, Tiwari R, Meena KK, Yandigeri M, Singh DP, Arora DK, 2011. Salt-tolerant rhizobacteria-mediated induced tolerance in wheat (*Triticum aestivum*) and chemical diversity in rhizosphere enhance plant growth. Biol Fertil Soils 47 (8): 907.

Upadhyay SK, Singh DP, Saikia R, 2009. Genetic diversity of plant growth promoting rhizobacteria isolated from rhizospheric soil of wheat under saline condition. Curr Microbiol 59 (5): 489-496.

Upadhyay SK, Singh JS, Saxena AK, Singh DP, 2012. Impact of PGPR inoculation on growth and antioxidant status of wheat under saline conditions. Plant Biol 14 (4): 605-611.

Van Loon L, 2007. Plant responses to plant growth-promoting rhizobacteria. Eur J Plant Pathol 119 (3): 243-254.

Vannier N, Mony C, Bittebière AK, Vandenkoornhuyse P, 2015. Epigenetic mechanisms and microbiota as a toolbox for plant phenotypic adjustment to environment. Front Plant Sci 6: 1159.

Welsh DT, 2000. Ecological significance of compatible solute accumulation by micro-organisms: from single cells to global climate. FEMS Microbiol Rev 24 (3): 263-290.

Yamaguchi T, Blumwald E, 2005. Developing salt-tolerant crop plants: challenges and opportunities. Trends Plant Sci 10 (12): 615-620.

Yang J, Kloepper JW, Ryu CM, 2009. Rhizosphere bacteria help plants tolerate abiotic stress. Trends Plant Sci 14 (1): 1-4.

Yao L, Wu Z, Zheng Y, Kaleem I, Li C, 2010. Growth promotion and protection against salt stress by *Pseudomonas putida* Rs-198 on cotton. Eur J Soil Biol 46 (1): 49-54.

Yildirim E, Turan M, Ekinci M, Dursun A, Cakmakci R, 2011. Plant growth promoting rhizobacteria ameliorate deleterious effect of salt stress on lettuce. Sci Res Essays 6 (20): 4389-4396.

Yuan Z, Druzhinina IS, Labbé J, Redman R, Qin Y, Rodriguez R, Zhang C, Tuskan GA, Lin F, 2016. Specialized microbiome of a halophyte and its role in helping non-host plants to withstand salinity. Sci Rep 6: 32467.

Yue H, Mo W, Li C, Zheng Y, Li H, 2007. The salt stress relief and growth promotion effect of Rs-5 on cotton. Plant Soil 297 (1-2): 139-145.

Zahir ZA, Arshad M, Frankenberger WT, 2004. Plant growth promoting rhizobacteria: applications and perspectives in agriculture. Adv Agron 81: 98-169.

Zahir ZA, Ghani U, Naveed M, Nadeem SM, Asghar HN, 2009. Comparative effectiveness of *Pseudomonas* and *Serratia* sp. containing ACC-deaminase for improving growth and yield of wheat (*Triticum aestivum* L.) under salt-stressed conditions. Arch Microbiol 191 (5): 415-424.

Zarea M, Hajinia S, Karimi N, Goltapeh EM, Rejali F, Varma A, 2012. Effect of *Piriformospora indica* and *Azospirillum* strains from saline or non-saline soil on mitigation of the effects of NaCl. Soil Biol Biochem 45: 139-146.

Zhou N, Zhao S, Tian CY, 2017. Effect of halotolerant rhizobacteria isolated from halophytes on the growth of sugar beet (*Beta vulgaris* L.) under salt stress. FEMS Microbiol Lett 364 (11). https://doi.org/10.1093/femsle/fnx091.

Zhu JK, 2002. Salt and drought stress signal transduction in plants. Annu Rev Plant Biol 53 (1): 247-273.

4 盐生植物根际土壤中的生长促生菌对提高作物耐盐性研究进展

Ashok Panda, Asish Kumar Parida

[摘要]

农业用地的盐渍化水平不断上升严重制约了农业生产，随着日益增长的世界人口，对粮食安全也构成了十分严重的威胁。据联合国环境规划署报告，近年来全球受到盐渍化威胁的土地和农田面积分别增加了约 20% 和 50%。由于土壤不断遭受盐渍化的危害，导致不能继续用作农业用地的土地面积每年以 1%~2% 的速度增加。发生农业用地土壤盐渍化的区域主要分布在干旱地区。由于盐胁迫对植物的生长、光合作用、蛋白质合成、脂质代谢等许多代谢过程具有十分明显的抑制作用，因此土壤盐渍化正成为导致全球农业产量大幅下降的主要原因。培育耐盐作物品种是满足日益增长的粮食需求和创造可持续农业发展的先决条件。盐生植物根际是植物生长促生菌的储存库，这些微生物能够在高盐胁迫条件下提高农作物对盐胁迫的适应性，并促进农作物的生长发育，即植物生长促生菌能够在逆境条件下对植物的生长起重要的促进作用。植物生长促生菌可以通过直接影响和间接影响两种方式发挥对植物生长的促进作用。直接影响机制包括植物生长激素的生物合成、固氮作用的强化和较高水平的磷酸盐增溶作用；间接影响机制则包括抑制对植物生长起抑制作用的植物病原菌。各种研究表明，从不同盐生植物根际土壤中分离获得的耐盐植物生长促生菌作为生物菌剂，具有在受到盐渍化危害的农田中促进耐盐农作物生长的潜力。植物生长促生菌通过各种机制调控和促进农作物的生长发育，比如通过植物生长激素的调节、基因表达、蛋白质合成和降解以及各种次级代谢产物的合成等。植物生长促生菌能够与吲哚-3-乙酸协同调节 1-氨基环丙烷-1-羧酸脱氨酶的合成，1-氨基环丙烷-1-

---

A. Panda, A. K. Parida (✉), 印度古吉拉特邦巴夫纳格尔中央盐和海洋化学品研究所（CSI-SMCRI）科学和工业研究委员会生物技术和植物学部, 印度古吉拉特邦巴夫纳格尔科学和创新研究院, 科学和工业研究委员会, 中央盐和海洋化学品研究所 (CSIR-CSMCRI), E-mail: A. K. Parida: asishparida@csmcri.org, asishparida@csmcri.res.in

羧酸脱氨酶在农作物遭受到逆境胁迫的应激信号传导过程中发挥作用，并诱导多种应激反应途径发挥作用。有很多研究表明，植物生长促生菌在农作物上接种的实施在促进农业发展和提高全球粮食安全方面是可取的。本章重点讨论了植物生长促生菌的耐盐机制以及植物生长促生菌在提高各种甜土作物耐盐性方面发挥的作用。

[关键词]

1-氨基环丙烷-1-羧酸脱氨酶；盐生植物；吲哚-3-乙酸；植物生长促生菌；植物激素；盐渍化；耐盐作物；铁载体

4.1 前言

目前，盐渍化影响到全世界超过 10 亿 hm^2 的土地，被认为是许多沿海国家严重的环境威胁（Egamberdieva 和 Lugtenberg，2014）。造成土壤盐渍化增加的主要原因是农业生产管理不善、微咸水灌溉和降水量少。淡水供应量的逐渐减少导致灌溉用水使用盐水（Egamberdieva 和 Lugtenberg，2014）。

天然盐渍化是通过自然途径向土壤或地表水体中添加盐分造成的。咸水入侵沿海土地和风积盐是盐分自然积累的主要来源（Manchanda 和 Garg，2008）。此外，盐分从海洋向农业用地的转移是通过风或海水灌溉实现的。另外，盐分还可以通过矿物岩石的风化作用添加到土壤中（Paul 和 Lade，2014）。

土壤盐分增加的人为因素包括盐水过度灌溉，这使得大部分农田因为含盐量的增加而不适合农业耕作（Egamberdieva 和 Lugtenberg，2014）。其他原因还包括过量补充无机肥料和土壤结构的各种变化。所有这些因素将导致土地退化，对自然生态系统造成了巨大负担（Paul 和 Lade，2014）。高浓度盐分主要是通过 Na^+ 的过量积累和土壤水化性质的恶化影响土壤结构的退化。除此之外，土壤中高浓度的 Na^+ 会增加土壤的 pH 值和土壤侵蚀。更重要的是，Na^+ 对其他黏土矿物的分散破坏起着决定性的作用。土壤的分散过程是通过 Na^+ 的侵入来置换土壤内层的 Ca^{2+} 和 Mg^{2+}（Paul 和 Lade，2014）。

由于降水量少和气温高等极端气候变化，良田转化为盐渍化土地的速度变得更快。这导致更多干旱地区的出现，进而有利于积累更多的盐分（Shanker 和 Venkateswaru，2011）。如前所述，盐渍化是指各种盐离子的积累，包括 Mg^{2+}、K^+、Cl^-、Ca^{2+}、CO_3^{2-}、SO_4^{2-}、Na^+ 和 HCO_3^- 等。因此，受盐渍化影响的土地面积和农业废弃用地面积日益增加（Paul 和 Lade，2014）。盐胁迫能够产生活性氧物质，这些物质能够损伤 DNA、RNA 和蛋白质（Habib 等，2016）。活性氧物质还能够导致叶绿素含量的降低以及根分生组织活力的降低，从而抑制根的生长。另外，各种抗氧化酶（如超氧化物歧化酶、过氧化氢酶和抗坏血酸过氧化物酶等）都具有清除活性氧的能力，使活性氧物质维持在耐受限度以内（Jaleelet 等，2009）。超氧化物歧化酶是一种金属酶，通过中和超氧化物自由基保护细胞和各种组织器官免受氧化应激造成的损伤。抗坏血酸过氧化物酶在分解超氧化物歧化酶产生过氧化氢的过程中起着重要作用。超氧化物歧化酶作用产生的过氧化氢被过氧化氢酶水解为水和氧气（Habib 等，2016）。

随着土壤盐渍化程度的提高和农业生产率的下降，培育耐盐作物已成为一项令人期待的科学研究。组织良好的土壤管理实践，包括有效的成本管理和易于付诸实施的方法的制定，是成功复垦盐渍土必须面对的挑战（Munns 和 Tester，2008）。此外，与传统农业相比，必须制定可持续的管理做法以满足未来农业的需要。因此，素有生态友好之称的、在农业中发挥有益作用的，各种微生物已显示出巨大的效益（Rodriguez 和 Redman，2008）。此外，许多研究已经证明，根际微生物和遭受胁迫的植物之间的联系有助于植物适应其所在的微环境（Paul 和 Lade，2014）。根际微生物对不同盐浓度水平下生长的多种农作物有促生长作用，这使开发耐盐作物品种成为可能。

在植物根际土壤中植入对植物生长有益的微生物，能够减轻适应盐渍土环境的作物的胁迫危害。这种方法的成功实施是可能的，因为这些微生物能够与植物相互作物，并且具有广泛的耐受性。在不利的环境条件下，通常根际微生物对植物生长具有有益作用时接种所占的百分比约为 2%～5%。植物生长促生菌能够使土壤生态系统营养丰富，并使作物的产量具有可持续性（Hayat 等，2010）。微生物促进植物生长和缓解应激反应的各种方式包括：①土壤养分的有效流通；②保护植物免受病原菌侵染；③通过对有毒重金属物质的封存来改善土壤结构和质量。微生物还能够降解各种异型生物质（Braud 等，2009）。

在盐渍土壤中使用微生物菌剂来改善植物生长状况，有助于改善盐胁迫下植物的生长，同时提高植物的抗病性（Lugtenberg 等，2013）。与植物根系密切相关的细菌种类对于缓解盐胁迫危害和开拓耐盐技术发展新机遇具有极其重要的意义（Dodd 和 Pérez-Alfocea，2012）。早期的研究报道了微生物在植物应对各种极端环境（如干旱、土壤和地表水盐渍化以及重金属胁迫）反应中的作用。这些方法有助于可持续农业实践的发展。用植物生长促生菌菌株预接种辣椒、番茄、豆类等不同的作物品种，有利于提高植株的生物量和产量，这反映了这些作物对盐胁迫耐受性的增强。

4.2 盐胁迫对植物生理的影响

许多作物的最佳生长状态和农业产量受到盐胁迫的影响（Paul 和 Lade，2014）。各种作物（如谷物、豆类和其他园艺作物）受灌溉水或土壤溶液中普遍存在的盐分的影响很大。有报道称，在盐胁迫下小麦（Egamberdieva，2009）、蚕豆（Rabie 等，2005）、水稻（Xu 等，2011）、玉米和大豆（Essa，2002）的种子萌发率降低。水分吸收减少和必需营养素的不均衡可能是盐胁迫下生长受到抑制的原因（Dolatabadian 等，2011）。植物中某些必需营养素的运输和分配受到根际区域 Na^+ 和 Cl^- 丰度的影响（Heidari 和 Jamshid，2010）。植物主要的生理途径（如蛋白质合成、光合作用和脂质代谢）都会随着盐浓度的升高而改变。普遍存在的盐分导致渗透不平衡和气孔关闭，引起碳同化效率下降（Parida 和 Das，2005）。渗透胁迫对植物的总体影响包括细胞生长的显著减少、叶面积的降低、叶绿素的降解，以及最重要的衰老过程（Paul 和 Lade，2014）。盐胁迫的影响是多方面的，包括激素和离子失衡、蛋白质代谢的改变以及调控核酸代谢的酶活性的下降。这些植株形态和生理的异常归因于盐分离子的渗透不平衡和生理毒性。除了 Na^+，Cl^- 也会扰乱植物的正常生理过程（比如光合作用）。这两种离子在根际的过度积累，导致根中其他矿物离子（如 K^+、NO_3^- 和 $H_2PO_4^-$）的竞争性吸收，同时也引起了养分运输和组织特异性定位的竞争。植物摄入的盐分超过其耐受性（储存能力）时，会在细胞间隙中储存多余的离子，导致植物组织的脱水和死亡。盐分还能够影响根细胞的渗透平衡，使根际周围的土壤溶液渗透势升高，进而影响植物对水分和养分的吸收。盐胁迫的各种主要影响往往会产生各种次要后果，比如渗透不平衡、氧化应激反应和活性氧物质的积累，导致膜或其他生物大分子的崩解。酶和非酶抗氧化系统给予植物保护作用，免受应激反应诱导的氧化损伤。此外，这些抗氧化物质的表达水平可作为植物响应氧化应激程度的指标。

4.3 盐分对根际微生物多样性的影响

根际微生物菌落是土壤组成中不可缺少的动态元素，它作为养分循环和有机质转化的媒介，为植物提供养分以获得最佳生长。大多数微生物由于根际良好的生态环境而定殖于植物根际而不是植物根周围的土壤。植物根际土壤微生物对植物的生长有着直接和间接的影响。当微生物群落调节植物的养分供应时，由于环境因素引起的根际微生物群落变化会影响土壤对植物的养分供应。根际微生物数量的增加有许多作用，特别是在生态系统的正常运行、污染物降解和污染环境的净化方面（Wenzel，2009）。根际微生物群落结构受土壤 NaCl 含量的影响，影响根系分泌物（Nelson 和 Mele，2007）。水稻根际土壤盐分显著降低了假单胞菌等微生物的密度。许多假单胞菌物种生活在盐渍土中，包括施氏假单胞菌（*Pseudomonas stutzeri*）、荧光假单胞菌（*Pseudomonas fluorescens*）、铜绿假单胞菌（*Pseudomonas aeruginosa*）、恶臭假单胞菌（*Pseudomonas putida*）和门多萨假单胞菌（*Pseudomonas mendocina*）（Egamberdieva 和 Lugtenberg，2014）。耐盐固氮菌（*Swaminathania salitolerans*）从与红树林共生的水稻中获得（Loganathan 和 Nair，2004）。除了假单胞菌外，还有很多耐盐的根际细菌，比如黏质沙雷氏菌（*Serratia marcescens*）、铜绿假单胞菌（*Pseudomonas aeruginosa*）和木糖氧化产碱菌（*Achromobacter xylosoxidans*），存在于水稻根际。各种研究表明，在盐渍土中生长的小麦根际是潜在的植物生长促生菌储存库（Egamberdieva 和 Islam，2008）。从铜绿假单胞菌（*Pseudomonas aeruginosa*）在盐渍土中持续存在可以看出，该物种广泛分布于盐渍土生长植物的根际土壤中。微生物能够和植物争夺可用的养分。有益细菌被植物根际的根系分泌物招募到植物中。为了在盐胁迫环境中生存，微生物必须消耗额外的能量排出 Na^+ 来维持其质膜两侧的渗透平衡（Jiang 等，2007）。在盐胁迫条件下，植物利用大部分原本供应微生物的根际分泌物，从而限制细菌的生长。盐胁迫影响细菌表面蛋白质、脂多糖和胞外多糖的组成。此外，盐胁迫会影响调节细菌表面信号传导的膜葡聚糖的排列，从而影响细菌的迁移（Ibekwe 等，2010）。细菌多样性变化是测定盐分对土壤质量影响的生物标志之一。此外，早期土壤盐分检测有助于实施补救措施，以提高作物的存活率和产量（Ibekwe 等，2010）。耐盐根际细菌在高盐条件下需要采取一系列适应性的措施才能维持生存和生长。这些适应机制包括：①改变细胞膜和细胞壁的结构以减少对盐的吸收；②通过调节 K^+/Na^+ 离子转运体和 Na^+/H^+ 反转运体改变细胞内离子浓度来有效调节渗透性；③相容性溶质海藻糖、蔗糖和甘油等的内源生物合成；④能够抵抗高盐浓度的酶的生物合成；⑤产生由胞外多糖组成的生物膜（Sandhya 等，2010b；Qin 等，2016）。除了这些机制外，耐盐根际微生物还有一些其他特性，比如较高的鸟嘌呤-胞嘧啶含量、高水平的低疏水蛋白、形成较少的螺旋结构，这些都可以增强其耐盐特性（Szymańska 等，2016）。

4.4 根际植物生长促生菌的耐盐机制

微生物属于细菌（Bacteria）和古生菌（Archaea）的范畴，它们具有响应外部土壤溶液渗透压变化而保持细胞完整性的独特特性。研究发现，除了盐杆菌科（Halobacteriaceae）的成员外，其他物种也能适应高浓度的 NaCl（Oren，2002）。有研究表明，除了古细菌外，高盐环境中的原核生物多样性还包括不同种类的细菌。在所有纬度地区分离出一系列耐盐细菌进一步证实了这一假设。耐盐细菌可以在浸没在盐溶液中的植物部分生长，例如，在浸泡在7.2%的NaCl溶液中的植物中发现了木糖葡萄球菌（*Staphylococcus xylosus*）（Abou-Elela等，2010）。在中国东北部的一个盐渍化区域发现了一些细菌属，包括喜盐涅斯特连科氏菌（*Nesterenkonia halobia*）、窄食单胞菌属（*Stenotrophomonas*）和利托杆菌（*Litoribacter*）（Shi 等，2012）；在中国西北部沙漠中发现了中慢生根瘤菌（*Mesorhizobium alhagi*）（Zhou 等，2012a）。通过特别的细胞壁组成结构避免高浓度盐分下细胞质的水分流失以抵御高盐浓度，这是已知的古细菌和蓝藻的种群特征。古生菌用以抵抗温度和机械损伤的高抗性类脂膜使其具有耐盐性。与真核生物相比，古生菌的类脂膜使它们在恶劣的环境条件下能够更好地生存。一些盐生细菌的细胞壁含有1个由糖蛋白构成的外层结构，在这个结构中糖蛋白呈二维晶格排列。这些膜大约由40%~50%的糖蛋白组成，并且在NaCl存在的情况下能够保持稳定。类似的，蓝藻膜系统包含两个膜结构，即内膜和外膜。外膜包围周质空间，内膜包围细胞质内容物。另一个类似的膜系统存在于叶绿体中的类囊体膜，能够容纳光合复合物和电子传递链（Ruppel 等，2013）。据报道，隐球菌具有厚的细胞壁，其细胞壁结构保持完整反而需要一定浓度的盐。由聚谷氨酰胺（L-谷氨酰胺）糖复合物组成的嗜盐球菌（*Halococcus*）的细胞壁的完整性因是否存盐分而有很大差异。此外，一些根际细菌还会分泌盐胁迫缓解剂胞外多糖（Upadhyay 等，2011）。这些多糖降低盐胁迫危害的作用在念珠藻属（*Nostoc* sp.）中得到了证实。与对照相比，培养基中的盐分并没有改变组成细菌细胞壁的糖的比例（Yoshimura 等，2012）。细胞内的各种离子通道维持胞内离子浓度并调节渗透势。Na^+/H^+反转运体形成的质子电化学梯度使细胞内的Na^+保持在较低水平。K^+的基本作用包括维持膜的完整性、调节pH值，以及最重要的功能调节基因表达（Hagemann，2011）。Na^+和K^+具有相似的化学和物理性质。Na^+与K^+具有相同大小的水化层，因此在高盐浓度下Na^+与K^+竞争性吸收导致植物中K^+的缺乏。相比之下，细胞基质中K^+维持在最佳水平则表明植物能够从土壤溶液中正常摄取K^+（Hagemann，2011）。

4.5 从不同盐生植物根际分离植物生长促生菌

与土壤和盐生根际有关的微生物能够对植物的生长发育产生有利或不利的影响。接种植物生长促生菌可以直接或间接地改变植物的生长。对于土壤微生物来说，植物根际是一个非常有利的生态位，因为它们能够获得丰富的营养。有益细菌在植物根际能够发挥最大的活性。这是因为根分泌出不同形式的有机化合物作为引诱剂和营养物质招募了一系列细菌（Szymanska 等，2016）。各种土壤微生物能够对植物产生有益的影响，而对土传害虫产生不利的影响，促进作物的可持续生产。不利的生存条件能够使微生物更具耐受性，使它们具有在各种胁迫条件（比如盐胁迫条件）下发挥作用的潜力。尽管如此，各种细菌的多样性和生长还是普遍受到根际盐分的影响（Egamberdiyeva 和 Islam，2008）。根际土壤中的植物-微生物相互作用因物种而异，也不同于非根际土壤中的植物-微生物相互作用。由于具有丰富的营养物质，根际区域是土壤微生物生存的最适宜的生态位（Ahmad 等，2008）。对根际细菌的遗传学和形态学的研究进一步阐明了宿主根系与微生物之间的通信机制（Tripathi 等，2002）。在盐渍化严重的土地上，微生物群落受到很大的影响。然而，耐盐细菌在盐浓度高达 30%的土壤和缺乏盐分的土地上都能大量生长（Khan 等，2016）。因此，它们可以在各种盐生植物低水势的根际土壤中生长（Ruppel 等，2013）。更令人惊讶的是，从盐生植物根际分离出的各种植物生长促生菌在高盐度下仍保持着促进植物生长的特性。从中国东部沿海地区分离的耐盐植物生长促生菌克锡勒氏菌 YCWA18（*Kushneria species YCWA18*）即使在含 20%的 NaCl 溶液中也能够表现出最佳的生长（Zhu 等，2011）。其他已分离的植物生长促生菌也具有耐盐特性，具体表现是他们能在高达 25%的 NaCl 浓度下生长。这些细菌包括节杆菌属（*Arthrobacter*）、盐单胞菌属（*Halomonas*）、门多萨假单胞菌（*Pseudomonas mendocina*）、短小芽孢杆菌（*Bacillus pumilus*）和拉齐斯庞氏硝化菌（*Nitrinicola lacisaponensis*）。除了能够保持在高盐浓度下生长之外，这些微生物还具有其他促进生长的特征，比如有效磷的增溶吲哚-3-乙酸、铁载体和 1-氨基环丙烷-1-羧酸脱氨酶的产生（Tiwari 等，2011）。植物生长促进特性是指在磷缺乏的环境中向寄主提供磷的能力。植物生长促生菌的这种类似植物激素的特性能够刺激植物生长，它们能够为宿主提供足够的铁，并降低 1-氨基环丙烷-1-羧酸脱氨酶的水平（1-氨基环丙烷-1-羧酸是合成乙烯的前体）（Etesami 和 Bettie，2018）。从之前的报道中我们了解到，豆科植物的根瘤菌比豆科植物根际的其他细菌对盐更敏感。从豆科植物根际分离出的细菌种类主要是芽孢杆菌（*Bacillus*，属于革兰氏阳性菌）。在根际植物生长促生菌中，假单胞菌（*Pseudomonas* spp.）是水稻根际中最常见的微生物（Rangarajan 等，2002）。这些假单胞菌微生物具体包括荧光假单胞菌（*Pseudomonas fluorescens*）、类产碱假单胞菌（*Pseudomonas pseudoalcaligenes*）和缺陷假单胞菌（*Pseudomonas diminuta*）。荧光假单胞菌（*Pseudomonas fluorescens*）和类产碱假单胞菌（*Pseudomonas pseudoalcaligenes*）已经在盐渍土和非盐渍土环境中被发现。然而，从野生耐盐水稻中分离得到的微生物物种在盐胁迫条件下具有固氮和增磷的能力。尽管如此，在盐渍土环境中生长的植物生长促生菌缺乏一个明确的分类系统（Loganathan 和 Nair，2004）。除了假单胞菌（*Pseudomonas* spp.）和黄杆菌（*Flavobacterium* spp.）对节杆菌（*Arthrobacter*）和芽孢杆菌（*Bacillus* spp.）的广泛优势外，芽孢杆菌在各种自然生境中也广泛存在。然而，固氮菌（*Azotobacter* spp.）主要是从盐渍土地区分离出来的。这表明芽孢杆菌具有多样的生境范围和高效的适应性。除野生稻品种外，还从臂状盐角草（*Salicornia brachiata*）中分离到多种耐盐细菌。这些细菌包括血杆菌（*Haererehalobacter*）、刘志恒氏菌（*Zhihengliuella*）、放射型根瘤菌（*Rhizobium radiobacter*）和苏拉氏短杆菌（*Brachybacterium saurashtrense*）。据报道，这些微生物通过增强磷的溶解能力、产生吲哚-3-乙酸和合成1-氨基环丙烷-1-羧酸脱氨酶等方式促进臂状盐角草（*Salicornia brachiata*）的生长（Jha 等，2012）。除从印度分离的细菌外，从中国西北地区分离得到的刘志恒氏菌（*Zhihengliuella*）也能够耐受浓度高达25%的NaCl溶液浓度（Zhang 等，2007）。有报道指出，水稻根际定殖了很多种内生固氮菌，比如假单胞菌（*Pseudomonas* spp.）（Jha 等，2009）。此外，从盐角草的内根层中也分离到了类产碱假单胞菌（*Pseudomonas pseudoalcaligenes*）（Ozawa 等，2007）。在从盐生植物碱菀（*Aster tripolium*）中分离出的根际细菌中，36%属于厚壁菌门（Firmicutes），9%属于变形菌门（Proteobacteria），还有55%属于放线菌门（Actinobacteria）（Szymanska 等，2016）。碱菀（*Aster tripolium*）根中分离出的细菌种类仅限于革兰氏阳性菌。革兰氏阳性菌在盐渍化土壤中生存的情况也有报道（Damodaran 等，2013）。植物根际土壤中的细菌主要分布于3个属中：链霉菌属（*Streptomyces*）、芽孢杆菌属（*Bacillus*）和假单胞菌属（*Pseudomonas*）（Bouizgarne，2013）。根际土壤中也发现了草螺菌（*Herbaspirillum*）、固氮螺菌（*Azospirillum*）、类产碱假单胞菌（*Pseudomonas pseudoalcaligenes*）、葡糖醋杆菌（*Gluconacetobacter*）、沙雷氏菌（*Serratia*）等许多其他微生物物种（Szymanska 等，2016）。特征脂质生物标记物（Signature lipid biomarker，SLB）分析对表征根际微生物结构和功能的多样性具有重要意义（Mrozik 等，2014）。从紫穗碱蓬（*Suaeda fruticosa*）中分离到的植物生长促生菌菌株，在300 mmol/L NaCl浓度下仍表现出良好的生长，这表现出了潜在的植物促生特性，并在高盐浓度条件下对其性状进行了筛选（Goswami 等，2014）。从药用植物九味一枝蒿（*Ajuga bracteosa*）中分离根际微生物的研究也有报道（Kumar 等，2012）。由于几丁质酶、离子螯合铁载体和微生物拮抗剂［如氰化氢（HCN）］的合成会经常改变微生物的植物促生特性，因此植物的生长也会受到影响（Zhou 等，2012b）。已经从各种盐生植物包括白花滨藜（*Atriplex leucoclada*）、弗吉尼亚海蓬子（*Salicornia virginica*）和盐角梭梭（*Haloxylon salicornicum*）的根际土壤中分离到芽孢杆菌（*Bacillus*）和滋养节杆菌（*Arthrobacter pascens*）等植物生长促生菌

(Ullah 和 Asghari，2015)。据报道，各种盐生植物的根际土壤中定殖了多种耐盐的植物生长促生菌。这些盐生植物包括比吉洛氏海蓬子（*Salicornia bigelovii*）（Rueda-Puente 等，2010）、盐节木（*Halocnemum strobilaceum*）（Al-Mailem 等，2010）、臂状盐角草（*Salicornia brachiata*）（Jha 等，2012）、海马齿（*Sesuvium portulacastrum*）（Bian 等，2011；Anburaj 等，2012）以及金合欢属（*Acacia* spp.）植物（Boukhatem 等，2012）。除了发现这些植物生长促生菌在盐生植物根际定殖方面的多样性外，它们还在不同的生境（从盐渍化土地到碱土地，再到沙漠）中得到了分离。

4.6 应激条件下植物生长促生菌的促生长机制

植物生长促生菌促进植物生长的机制主要包括生长激素的生物合成和促进养分的吸收，以及从土壤到植物的转运。除此之外，植物生长促生菌诱导的耐盐性极大地影响了植物在恶劣盐渍化条件下的生存（Glick，2010；Sandhya 等，2010a；Ma 等，2011）。然而，属于特定属的自由生存的各种土壤细菌可能对植物的生长有不同程度的影响。为了减轻胁迫对植物的影响，根际细菌可以以不同的方式使用。植物生长促生菌对植物生长具有直接和间接的多维效应。间接效应包括通过分泌抗病原物质，防止植物病原体对植物产生负面影响（Nadeem 等，2013）。植物生长促生菌的促生长特性是通过合成各种次生代谢产物来实现的，这些次生代谢产物对植物的生长促进效应或者增强养分的吸收具有重要的潜力。此外，它们通过增强固氮作用和提高植物对磷利用的有效性来促进寄主植物的生长。

盐生植物的根际包含耐盐和诱导植物生长的微生物定殖的多样性生态。植物生长促生菌通过多种机制促进植物生长，保护植物免受盐渍化和干旱等各种环境胁迫。植物生长促生菌与植物相互作用改善盐胁迫影响的细胞水平机制如图 4.1 所示。

这些机制包括 1-氨基环丙烷-1-羧酸脱氨酶的合成、生物固氮作用的强化、植物激素的产生和胞外多糖的合成。植物生长促生菌在常规生境和极端生境中促进植物生长和促进植物适应极端环境的能力已有很多报道（Hayat 等，2010）。根际细菌能够直接或间接地促进植物生长。其直接机制包括养分固定增溶和植物激素合成等能力的增强。其中的一个间接机制是通过生物防控措施控制植物病原菌在寄主植物中的生长，从而最大限度地减少它们的有害影响（Mapelli 等，2013）。在不同的环境胁迫因子中，盐胁迫对植物生长和作物产量影响最大（Mayak 等，2004）。

盐生植物是世界干旱地区的优势种。盐生植物有相应的方式来应对受高盐度胁迫影响的农业土壤（Ravindran 等，2007）。近年来，大量的研究成果揭示宿

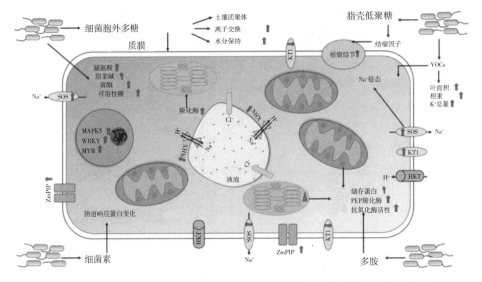

图 4.1 通过植物生长促生菌和植物的相互作用来减轻盐胁迫的细胞机制

注：离子转运体活性的增加维持了渗透平衡，使渗透损伤达到最小。细菌分泌物如多糖，增加了阳离子交换和保水性；多胺，增加储存蛋白的水平和抗氧化活性；细菌素，负责改变应激相关蛋白；脂壳低聚糖，通过增加结瘤因子活性增加结瘤。向上箭头表示活性增加；向下箭头表示活性降低。

主-微生物相互作用的机制以及根际微生物在提高植物对各种非生物胁迫抗性方面的有益作用。这些研究成果包括：①植物-微生物相互作用机制；②盐胁迫条件下植物生长的机制；③可在盐渍化土壤用作生物肥料的潜在菌株（Mapelli 等，2013）。图 4.2 展示了盐胁迫对甜土植物的影响以及植物生长促生菌介导的耐盐机制。这种由植物生长促生菌提供的耐受性称为系统性耐受性（Yang 等，2009）。诱导植物生长的植物生长促生菌可归纳为 3 个类别：生物肥料、植物刺激剂和生物杀虫剂。生物肥料提高了氮和磷对宿主植物的有效性，植物刺激剂诱导植物激素的生物合成，生物杀虫剂降低了病原体侵害宿主植物的风险。多种研究表明，植物生长促生菌在缓解不同作物品种受到的环境胁迫方面发挥了有益的作用（Ali 等，2009；Kohler 等，2009；Sandhya 等，2010a）。接种植物生长促生菌能够通过增加植物激素（如吲哚-3-乙酸）的产生和增加 1-氨基环丙烷-1-羧酸脱氨酶的生物合成促进根系的大量生成，促进植物生长（Dimkpa 等，2009）。

4.6.1 通过增强 1-氨基环丙烷-1-羧酸脱氨酶活性促进植物生长

提高 1-氨基环丙烷-1-羧酸脱氨酶生物合成的量能够降低乙烯的产生，从而增强对不同环境胁迫的耐受性（Lucy 等，2004；Bal 等，2013）。许多报道从增强 1-氨基环丙烷-1-羧酸脱氨酶活性和提高寄主植物耐盐性的角度阐述了植

**图 4.2　耐盐植物生长促生菌的耐盐机理及接种植物生长促生菌
对改善盐胁迫造成不利影响的有益作用**

物生长促生菌对植物生长的有益影响（表 4.1）。乙烯是一种植物应激激素，在遇到各种环境胁迫时作为应激反应释放。盐胁迫能够促使植物增加 1-氨基环丙烷-1-羧酸的生物合成，从而产生过量的乙烯，导致植物组织的各种生理变化（Tank 和 Saraf，2010）。在各种非生物胁迫条件下，植物生长促生菌通过促进 1-氨基环丙烷-1-羧酸脱氨酶的生物合成来促进植物生长。有很多研究报道表明，植物在各种非生物胁迫和病原体侵染情况下，植物生长促生菌通过增强 1-氨基环丙烷-1-羧酸脱氨酶活性来降低乙烯水平，从而促进植物生长。许多革兰氏阴性菌（Wang 等，2000；Babalola 等，2003）、革兰氏阳性菌（Ghosh 等，2003；Belimov 等，2007）和根瘤菌（Ma 等，2003；Uchiumi 等，2004）等都被证实具有 1-氨基环丙烷-1-羧酸脱氨酶活性。藤黄亚甲基杆菌（*Methylobacterium fujisawaense*）、伯克霍尔德氏菌（*Burkholderia*）、青枯雷尔氏菌（*Ralstonia solanacearum*）、肠杆菌属（*Enterobacter*）、土壤杆菌（*Agrobacterium*）以及根瘤菌（*Rhizobium*）等多种微生物的 PGP 能力均有报道（Hontzeas 等，2004；Pandey 等，2005；Blaha 等，2006）。1-氨基环丙烷-1-羧酸脱氨酶通过将乙烯前体（1-氨基环丙烷-1-羧酸）转化为 α-酮丁酸酯和氨减少乙烯的生物合成，从而促进植物生长（Saleem 等，2007）。1-氨基环丙烷-1-羧酸氧化酶对 1-氨基环丙烷-1-羧酸具有更高的亲和能力，通过与 1-氨基环丙烷-1-羧酸脱氨酶的竞争性结合，能够进一步减少乙烯的生物合成，从而导致植物内源性乙烯的减少（Saleem 等，2007）。如果这个假设成立，那么植物中1-氨基环丙烷-1-羧酸脱氨酶的含量应该比1-氨基环丙烷-1-羧酸氧化酶的含量至少高100倍。而植物生长激素吲哚-3-乙酸生物合成的前体是色氨酸，吲哚-3-乙酸在植物生长促生菌中合成后分泌到环境中被植物吸收。由此，植物从外界环境中吸收的吲哚-3-乙酸也添加到现有的内源性吲哚-3-乙酸库中，诱导植物生长。此外，吲哚-3-乙酸还能够通过调节1-氨基环丙烷-1-羧酸合成酶的活性来促进S-腺苷蛋氨酸（SAM）转化为1-氨基环丙烷-1-羧酸（Saleem 等，2007）。根瘤菌1-氨基环丙烷-1-羧酸脱氨酶活性能促进植物根系产生和分泌1-氨基环丙烷-1-羧酸，这为微生物提供了一种独特的氮素来源来促进植物根系生长。环境胁迫下不同微生物1-氨基环丙烷-1-羧酸脱氨酶活性的变化对植物修复具有积极作用。根际微生物的1-氨基环丙烷-1-羧酸脱氨酶活性能够通过根系分泌物和有害元素的生物转化来诱导植物生长和根降解。除此之外，重金属的植物修复是通过各种盐生植物根际土壤中的根际细菌实现的（Singh 和 Jha，2016；Singh 等，2015）。产生1-氨基环丙烷-1-羧酸脱氨酶的根际细菌通过实施减轻各种环境胁迫的不利影响、延缓衰老、保护植物免受病原菌侵染和诱导豆科植物结瘤等不同策略，来促进植物生长（Ali 等，2012）。目前已经分离到具有1-氨基环丙烷-1-羧酸脱氨酶活性、适应盐胁迫环境的植物生长促生菌，并用来缓解盐胁迫对各种作物生长带来的负面影响。例如，与对照相比，接种荧光假单胞菌N3（*Pseudomonas fluorescens* N3）和恶臭假单胞菌（*Pseudomonas putida*）（具有1-氨基环丙烷-1-羧酸脱氨酶活性的植物生长促生菌）后，玉米的茎和根的生长分别提高了2.3倍和3.3倍（Khan 等，2016）。同样的，在豆科作物中接种具有1-氨基环丙烷-1-羧酸脱氨酶活性的根际细菌能够促进结瘤的研究也有报道（Shaharoona，2006）。此外，已在小麦中证实植物生长促生菌菌株巴西固氮螺菌FP2（*Azospirillum brasilense* FP2）能够降低1-氨基环丙烷-1-羧酸氧化酶的活性（Calielneto 等，2014）。此外，从海榄雌（*Avicennia marina*）中分离到的植物生长促生菌菌株马里卡假单胞菌（*Pseudomonas maricaloris*）具有很高的1-氨基环丙烷-1-羧酸脱氨酶活性（El-Tarabily 和 Youssef，2010）。接种这些菌株后，在盐胁迫下，幼苗的1-氨基环丙烷-1-羧酸含量下降并表现出较高的生长速率（El Tarabily 和 Youssef，2010）。接种从盐生植物中分离出的刘志恒氏菌（*Zhihengliuella alba*）、碘短杆菌（*Brevibacterium iodinum*）和地衣芽孢杆菌（*Bacillus licheniformis*）后，辣椒植株中乙烯的含量分别降低了44%、53%和57%。

表 4.1 植物生长促生菌通过调整 1-氨基环丙烷-1-羧酸脱氨酶活性来提高多种甜土作物品种的耐盐能力

植物品种	植物生长促生菌	促生效果	参考文献
芸苔（*Brassica campestris*）	藤黄亚甲基杆菌（*Methylobacterium fujisawaense*）	促进根伸长	Madhaiyan 等（2006）
	表皮短杆菌 RS15（*Brevibacterium epidermidis* RS15）	促进根长和干物质积累	Siddikee 等（2010）
	环状芽孢杆菌（*Bacillus circulans*）、坚强芽孢杆菌（*Bacillus firmus*）、球孢芽孢杆菌（*Bacillus globisporus*）	增加根和茎的长度；增加新鲜和干燥的生物量	Ghosh 等（2003）
香石竹（*Dianthus caryophyllus*）	巴西固氮螺菌（*Azospirillum brasilense*）	接种的插条产生最长的根	Li 等（2005）
豌豆（*Pisum sativum*）	豌豆根瘤菌（*Rhizobium leguminosarum*）	增强植物结瘤能力	Ma 等（2003）
玉米（*Zea mays*）	阪崎肠杆菌（*Enterobacter sakazakii*）	增强农艺性状	Babalola 等（2003）
	假单胞菌（*Pseudomonas*）	促进根伸长	Shaharoona 等（2006）
	丁香假单胞菌（*Pseudomonas syringae*）、产气肠杆菌（*Enterobacter aerogenes*）、荧光假单胞菌（*Pseudomonas fluorescens*）	高相对水量；高叶绿素；低脯氨酸含量	Nadeem 等（2007）
水稻（*Oryza sativa*）	贪噬菌（*Variovorax*）和红球菌（*Rhodococcus*）	增加根和茎长；增加根鲜生物量	Bal 等（2013）
	肠杆菌属（*Enterobacter*）	生长参数和各种抗氧化酶活性的增加	Sarkar 等（2017）
番茄（*Lycopersicon esculentum*）	假单胞菌（*Pseudomonas brassicacearum*）	根系生物量增加	Belimov 等（2007）
	短小芽孢杆菌（*Bacillus pumilus*）、枯草芽孢杆菌（*Bacillus subtilis*）	在盐碱条件下提高活力指数	Damodaran 等（2013）
	无色杆菌（*Achromobacter*）	在盐碱条件下提高光合作用的速率（与未接种的植物相比）	Mayak 等（2004）
莴苣（*Lactuca sativa*）	固氮螺菌（*Azospirillum*）	盐分胁迫下的种子萌发和营养生长	Barassi 等（2006）
落花生（*Arachis hypogaea*）	荧光假单胞菌（*Pseudomonas fluorescens*）	在盐碱条件下生长更大（与未接种的植物相比）	Saravanakumar 和 Samiyappan（2007）
小麦（*Triticum aestivum*）	假单胞菌（*Pseudomonas*）和沙雷氏菌（*Serratia*）	改进的地上部和根部生长和生长促进参数	Zahir 等（2009）

(续表)

植物品种	植物生长促生菌	促生效果	参考文献
燕麦（Avena sativa）	克雷伯氏菌（Klebsiella）	增加生长参数；降低各种胁迫指标，如丙二醛和脯氨酸的水平	Sapre 等（2018）
	不动杆菌（Acinetobacter）	生长参数和抗氧化酶活性增加	Xun 等（2015）

4.6.2 提高内源激素含量促进生长

植物激素在调节植物对各种环境胁迫的应激反应、维持正常的植物生理和提高植物对盐度和其他环境胁迫的耐受性方面至关重要（Ryu 和 Cho，2015）。植物对盐胁迫的反应是在分子、生化和生理方面引起各种变化。这些变化的程度随着盐胁迫的频率和强度、环境条件、土壤的物理和化学性质以及植物的年龄而变化（Kumari 等，2015）。外源激素的应用有助于改善盐胁迫产生的不利影响（Zahir 等，2010）。植物的激素状态受到土壤微生物的调节，这些微生物释放出的外源植物激素能够提高植物的耐盐性（Ilangumaran 和 Smith，2017）。外源植物激素可改变内源植物激素水平，为应对盐胁迫提供了一种方法（Ilangumaran 和 Smith，2017）。调节植物激素平衡是植物生长促生菌促进植物生长的策略之一（Tsukanova 等，2017）。在植物激素中，大多数耐盐的植物生长促生菌诱导吲哚-3-乙酸的生物合成，从而促进受到盐胁迫幼苗的生长（Tiwari 等，2011）。植物生长促生菌还能通过改变根-芽的激素信号传导来提高作物耐盐性（Yang 等，2009）。耐盐根瘤菌诱导植物产生较高含量的吲哚-3-乙酸，这是导致在沿海土壤中植物根系生长增加的原因（Siddeke 等，2011）。从耐盐牧豆树属植物（*Prosopis strombulifera*）的根中分离出了植物生长促生菌（Sgroy 等，2009；Tiwari 等，2011；Piccoli 等，2011；Bian 等，2011）。盐胁迫诱导产生的脱落酸（ABA）能够通过解淀粉芽孢杆菌 SQR9（*Bacillus amyloliquefaciens* SQR9）抵消，它通过提高叶绿素和总可溶性糖含量、增加各种抗氧化酶的活性，提高玉米对盐胁迫的耐受性（Chen 等，2016）。微生物细胞分裂素也可以提高植物的耐盐性，但是报道很少。植物生长促生菌能够缓解因盐胁迫诱导的小麦休眠，使小麦的生长提高了52%（Egamberdieva，2009）。Arkhipova 等（2007）报道了在水分胁迫条件下，枯草芽孢杆菌（*Bacillus subtilis*）产生的由细胞分裂素能够改善莴苣幼苗的生长。过量乙烯的合成是由产生1-氨基环丙烷-1-羧酸脱氨酶的植物生长促生菌调节的（Glick 等，2007）。在盐胁迫条件下，接种恶臭假单胞菌 UW4（*Pseudomonas putida* UW4）的番茄的生物量有所提高（Yan 等，2014）。吲哚-

3-乙酸在植物上的积累能够诱导 1-氨基环丙烷-1-羧酸合成酶基因的活性，导致 1-氨基环丙烷-1-羧酸大量积累。高浓度的 1-氨基环丙烷-1-羧酸导致乙烯的大量产生。具有 1-氨基环丙烷-1-羧酸脱氨酶活性的植物生长促生菌通过分解过量的 1-氨基环丙烷-1-羧酸来减少其积累，进而减少不同环境胁迫下乙烯的生成。一些研究已经证明了细菌脱落酸在影响植物脱落酸状态中的作用。在小麦中接种碱湖迪茨氏菌 STR1 (*Dietzia natronolimnaea* STR1) 能够上调不同的脱落酸响应基因，从而调节脱落酸信号级联调控（Bharti 等，2002）。同样地，脱落酸产生菌洋葱伯克霍尔德菌 SE4 (*Burkholderia cepacia* SE4) 和原小单孢菌 SE188 (*Promicromonospora* sp. SE188) 在盐胁迫条件下能够促进黄瓜地上部生物量的积累（Kang 等，2014）。在盐胁迫条件下，接种了恶臭假单胞菌 Rs-198 (*Pseudomonas putida* Rs-198) 的棉花幼苗脱落酸含量较低，生物量积累增加（Yao 等，2010）。接种了原玻璃蝇节杆菌 SA3 (*Arthrobacter protophormiae* SA3) 的小麦表现出更强的耐盐性，植株内吲哚-3-乙酸水平升高，但脱落酸和 1-氨基环丙烷-1-羧酸含量减少。

4.6.3 通过产生铁载体的植物生长促生菌提高植物对微量营养元素利用的有效性

铁载体是铁的螯合物，对铁有很高的亲和力。铁载体通过与铁形成铁-铁载体复合体能够使铁离子更容易被植物吸收。在盐渍土和碱性土中，可供植物吸收的铁是有限的。植物所需的铁和其他微量营养素在这种土壤中的低有效性导致植物缺铁（Abbas 等，2015）。植物生长促生菌的铁载体生产能力能够增加植物对微量营养元素的吸收能力（Navarro-Torre 等，2017；Zhou 等，2017）。细菌的铁载体被植物用作铁的来源。此外，细菌铁载体还可以有效地防止有害微生物的扩散及其对植物生长的有害影响（Shailendra Singh，2015）。已有研究报道了植物生长促生菌合成铁载体的能力（Kuffner 等，2008）。随着生长素的产生，在含有金属污染土壤中，含有铁载体的土壤和根际细菌能够促进植物的生长。有报道指出，假单胞菌属菌株 BE3dil (*Pseudomonas* sp. BE3dil)、兰黑紫色杆菌 (*Janthinobacterium lividum*)、黏质沙雷氏菌 (*Serratia marcescens*) 和冷黄杆菌 (*Flavobacterium fridimaris*) 等细菌具有产生铁载体的能力（Kuffner 等，2008）。之前有研究建议可以使用能够产生铁载体的植物生长促生菌促进植物生长（Glick，2003）。

4.7 植物生长促生菌在耐盐作物品种开发中的应用

已经有一些研究报道指出，使用不同的耐盐植物生长促生菌能够改善盐碱条件下作物的生长（Mayak 等，2004；Shukla 等，2012；Qin 等，2016）（表 4.2）。当前，各种环境胁迫因子（如盐胁迫）已经导致发展中国家农业生产能力的下降（Shukla 等，2012）。盐胁迫能够改变植物的许多基本生理过程，影响它们的产量。各种因素对植物的生长有不同的负面影响，在高盐度下植物的生长减缓，最重要的是养分的吸收和分配不平衡，进一步增加了大量必需元素（如 K^+、Ca^{2+} 和 NO_3^-）与普遍存在的 Na^+ 和 Cl^- 之间的竞争。为了解决这一矛盾，在各种农作物中施用能够促进植物产生耐盐能力的植物生长促生菌是与传统方法一起使用的革命性技术（Jha 等，2012）。正如前面所讨论的，耐盐植物生长促生菌改善植物生长的机制有如下几种：①抗氧化防御系统清除活性氧的有效调控（Jha 和 Subramanian，2014；Qin 等，2016）；②提高大气中氮的固定，从而提高对植物的氮营养供应；更好的 P 增溶，并增加铁载体的产量；③通过选择性吸收离子以维持较高的 K^+/Na^+ 比，从而将 Na^+ 和 Cl^- 的毒性限制在植物的耐受范围以内（Sukla 等，2012；Etesami，2018）；④胞外聚合物质（Extracellular polymeric substances，EPS）的分泌，形成一个称为"根鞘"的物理屏障，它能够减少根对 Na^+ 的吸收，并向上转移到芽。耐盐植物生长促生菌通过产生胞外聚合物质改善土壤结构。通过促进土壤的聚集，进而增加了土壤的保水能力和植物获取养分的能力。细菌的一个功能特性是它们含有羟基、巯基、羧基和磷酰基等官能团。这些官能团能够与 Na^+ 结合以减少 Na^+ 的吸收和转运（Nunkaew 等，2014）。各种耐盐细菌，比如变异盐单胞菌（*Halomonas variabilis*）、肠杆菌（*Enterobacter*）、类芽孢杆菌（*Paenibacillus*）、莱比托游动球菌（*Planococcus rifietoensis*）和嗜水气单胞菌（*Aeromonas hydrophila*）等，都能产生胞外聚合物质并且诱导生物膜合成（Qurashi 和 Sabri，2012；Khan 等，2016）。各种研究表明，植物生长促生菌在减轻花生、水稻、番茄和红辣椒等多种作物的盐害胁迫方面发挥了重要作用（Shukla 等，2012；Bal 等，2013；Upadhyay 和 Singh，2015）。在高盐条件下生长的花生中，接种了有益植物生长促生菌的植株，其叶绿素、脯氨酸和丙二醛的水平以及其他生理参数都有所升高（Shukla 等，2012）。此外，还有研究报道，接种耐盐植物生长促生菌[如不动杆菌（*Acinetobacter*）]的燕麦植株的耐盐性也得到提高（Xun 等，2015）。在大麦（Chang 等，2014）和燕麦（Sapre 等，2018）的耐盐性研究中，接种不同植物生长促生菌的植株比未接种的对照植株表现出更强的耐盐性。在水稻上接种耐盐肠杆菌植物生长促生菌，能够减轻盐胁迫对植株带来的负面影响，促进植株的生长，提高水稻的耐盐性（Sarkar 等，2017）。有报道指出，接种了肠杆菌（*Enterobacter*）植物生长促生菌的水稻，其发芽率、幼苗生长率、根茎长度和生物量以及叶绿素的含量都得到增加。接种肠杆菌 P23 菌株（*Enterobacter* strain P23）的植株，生长能力增强，可能与菌株的植物促生特性有关。在盐胁迫条件下，随着接种植物生长促生菌的植物旺盛生长，植株内的各种抗氧化酶的活性也都在降低，表明在接种植物生长促生菌后植株体内的活性氧物质产生减少（Sarkar 等，2017）。在接种根瘤菌（*Rhizobium* sp.）和沙雷氏菌（*Serratia* sp.）的生菜中（Lee，2005），接种节杆菌（*Arthrobacter* sp.）和枯草芽孢杆菌（*Bacillus subtilis*）的小麦中（Upadhyay 等，2012），以及在盐胁迫条件下接种了短小芽孢杆菌（*Bacillus pumilus*）和类产碱假单胞菌（*Pseudomonas pseudoalcaligenes*）的水稻中（Jha 和 Subramanian，2014）都得到了类似的研究结果。除了限制活性氧的产生外，植物生长促生菌还增加了吲哚-3-乙酸的生物合成，进而诱导植物生长（最重要的是促进根系生长）。Kim 等（2014）证明，通过接种肠杆菌 EJ01（*Enterobacter* sp. EJ01）能够增强番茄的耐盐性。植物耐盐能力的增强很有可能是因为在受到盐胁迫的时候优先激活了各种应激反应因子的转录（Kim 等，2014）。

表 4.2　植物生长促生菌与各种作物的互作和寄主植物的反应

作物	植物生长促生菌	分离植物生长促生菌的盐生植物/地点	植物对植物生长促生菌的反应	参考文献
番茄（*Solanum lycopersicum*）	皮氏无色杆菌 ARV8（*Achromobacter piechaudii* ARV8）	枸杞（*Lycium shawii*）	通过降低乙烯产量提高番茄鲜重、干重和水分利用效率	Mayak 等（2004）
	固氮菌属（*Azotobacter*）、沙福芽孢杆菌（*Bacillus safensis*）、枯草芽孢杆菌（*Bacillus subtilis*）、根瘤菌（*Rhizobium*）	盐生禾草（Halophyte grasses）	吲哚-3-乙酸、铁载体和磷酸盐增溶能力增加	Damodaran 等（2013）
玉米（*Zea mays*）	短小芽孢杆菌（*Bacillus pumilus*）、变黄节杆菌（*Arthrobacter aurescens*）、滋养节杆菌（*Arthrobacter pascens*）	盐角梭梭（*Haloxylon salicornicum*）、胡枝子（*Lespedeza bicolor*）、弗吉尼亚海蓬子（*Salicornia virginica*）	生长参数增加；叶绿素和糖含量增加；抗氧化酶如过氧化氢酶和超氧化物歧化酶活性增加	Ullah 和 Asghari（2015）
茄子	枯草芽孢杆菌（*Bacillus subtilis*）	—	通过微生物产生的生长素改变植物诱导机制提高植物生长素合成	Bochow 等（2001）
绿豆（*Vigna radiata*）	丁香假单胞菌（*Pseudomonas syringae*）、荧光假单胞菌（*Pseudomonas fluorescens*）	盐害绿豆（*Vigna radiata*）	增加生长参数，如茎和根鲜重；提高水分利用效率	Ahmad 等（2012）

(续表)

作物	植物生长促生菌	分离植物生长促生菌的盐生植物/地点	植物对植物生长促生菌的反应	参考文献
小麦（Triticum aestivum）	罕见芽孢杆菌（Bacillus insolitus）、嗜水气单胞菌（Aeromonas hydrophila）	盐害小麦	改进了接种样品的生长参数（与未接种样品相比）；植株体内较高的 Ca^{2+} 水平有助于降低 Na^+ 吸收	Ashraf 等（2004）
	枯草芽孢杆菌（Bacillus subtilis）、节杆菌（Arthrobacter）	恒河平原东部盐渍区	接种的样本显示抗氧化酶活性较低，如抗坏血酸过氧化物酶、过氧化氢酶和谷胱甘肽还原酶，表明活性氧的产生较低	Upadhyay 等（2012）
萝卜（Raphanus sativus）	克卢氏葡萄球菌（Staphylococcus kloosii）、玫瑰考克氏菌（Kocuria Erythromyxa）	在高盐土壤上自然生长的植物	盐胁迫条件下细菌处理对植物生长有积极影响，叶绿素含量增加	Hayat 等（2010）
辣椒（Capsicum annuum）	碘短杆菌（Brevibacterium iodinum）、地衣芽孢杆菌（Bacillus licheniformis）	印度黄海西岸土壤	调控乙烯合成，促进植物生长和耐盐性	Siddikee 等（2011）
燕麦	不动杆菌（Acinetobacter）、克雷伯氏菌（Klebsiella）	盐害小麦	促进生长；减少电解质渗漏；降低丙二醛水平	Sapre 等（2018）
水稻（Oryza sativa）	肠杆菌（Enterobacter）	盐害水稻	增加磷酸盐溶解、吲哚-3-乙酸生成和铁载体生成	Kim 等（2014）

4.8 结论和展望

植物生长促生菌从遭受盐渍胁迫的植物根际中分离出来，能够减轻盐胁迫的有害影响并且增加萌发率、茎和根的长度、干物质产量以及农业生产产量。植物生长促生菌还作为生物肥料和生物控制剂在有机农场中使用。土壤微生物多样性的研究是满足全球粮食需求的关键，由于盐胁迫普遍性存在，并且严重性不断增加，造成了土壤质量加剧恶化，导致农业生产产量的降低。植物进化出了有效的策略，增强面对非生物胁迫的抵抗能力，比如高盐胁迫和干旱。从盐生植物根际分离出的植物生长促生菌能够以经济有效的方式减轻盐胁迫造成的负面影响，这是培育耐盐作物品种急需的有效方式。因此，给农作物接种耐盐的植物生长促生菌是盐渍土区域农业可持续发展的一项有效策略。对植物生长促生菌开展研究将是开辟耐盐农作物发展的新途径。目前的研究正在加深我们对微生物-盐生植物相互作用的基本认识，并将这些知识应用在耐盐作物品种培育方面。揭示植物生长促生菌耐盐的分子机制将使我们能够培养出在盐胁迫条件下具有促进植物生长能力的细菌。此外，植物生长促生菌在盐胁迫条件下刺激相关作物生长的效率必须经过长时间的评估。除了对耐盐作物品种的生产有用外，植物生长促生菌还可用于盐渍农田的植物脱盐、植物修复和各种病原菌的生物防治。由于开发的生物菌剂可以用来改善盐渍条件下作物的生长，因此必须制定快速且高效的能够鉴定可使用的植物生长促生菌菌株的策略。随着农业用地越来越多地遭受到各种环境胁迫影响，开发耐盐作物品种已成为当务之急，利用耐盐植物生长促生菌为实现这一目标奠定了基础。

[致谢]

A. K. P. 受到印度政府科学和技术部（DST）、科学和工程研究委员会（自主号：SB/SO/PS-14/2014）的资助。本章手稿已分配给科学和工业研究委员会-中央海洋化学品研究所（CSIR-CSMCRI），注册号为 PRIS 074/2018。

参考文献

Abbas G, Saqib M, Akhtar J, Anwar M, 2015. Interactive effects of salinity and iron deficiency on different rice genotypes. J Plant Nutr Soil Sci 178：306-311. https://doi.org/10.1002/jpln.201400358.

Abou-Elela SI, Kamel MM, Fawzy ME, 2010. Biological treatment of saline wastewater using a salt-tolerant microorganism. Desalination 250：1-5. https://doi.org/10.1016/j.desal.2009.03.022.

Ahmad F, Ahmad I, Khan MS, 2008. Screening of free-living rhizospheric bacteria for their multiple plant growth promoting activities. Microbiol Res 163：173-181. https://doi.org/10.1016/j.micres.2006.04.001.

Ahmad M, Zahir ZA, Asghar HN, Arshad M, 2012. The combined application of rhizobial strains and plant growth promoting rhizobacteria improves growth and productivity of mung bean (*Vigna radiata* L.) under salt-stressed conditions. Ann Microbiol 62：1321-1330. https://doi.org/10.1007/s13213-011-0380-9.

Ali SZ, Sandhya V, Grover M et al., 2009. *Pseudomonas* sp. strain AKM-P6 enhances tolerance of sorghum seedlings to elevated temperatures. Biol Fertil Soils 46：45-55. https://doi.org/10.1007/s00374-009-0404-9.

Ali S, Charles TC, Glick BR, 2012. Delay offlower senescence by bacterial endophytes expressing 1-aminocyclopropane-1-carboxylate deaminase. J Appl Microbiol 113: 1139-1144. https://doi. org/10. 1111/j. 1365-2672. 2012. 05409. x.

Al-Mailem DM, Sorkhoh NA, Marafie M et al., 2010. Oil phytoremediation potential of hypersaline coasts of the Arabian Gulf using rhizosphere technology. Bioresour Technol 101: 5786-5792. https://doi. org/10. 1016/j. biortech. 2010. 02. 082.

Anburaj R, Nabeel MA, Sivakumar T, Kathiresan K, 2012. The role of rhizobacteria in salinity effects on biochemical constituents of the halophyte *Sesuvium portulacastrum*. Russ J Plant Physiol 59: 115-119. https://doi. org/10. 1134/S1021443712010025.

Arkhipova TN, Prinsen E, Veselov SU et al., 2007. Cytokinin producing bacteria enhance plant growth in drying soil. Plant Soil 292: 305-315. https://doi. org/10. 1007/s11104-007-9233-5.

Ashraf M, Hasnain S, Berge O, Mahmood T, 2004. Inoculating wheat seedlings with exopolysaccharide-producing bacteria restricts sodium uptake and stimulates plant growth under salt stress. Biol Fertil Soils 40: 157-162. https://doi. org/10. 1007/s00374-004-0766-y.

Babalola OO, Osir EO, Sanni AI et al., 2003. Amplification of 1-amino-cyclopropane-1-carboxylic (ACC) deaminase from plant growth promoting rhizobacteria in *Striga*-infested soil. Afr J Biotechnol 2: 157-160. https://doi. org/10. 4314/ajb. v2i6. 14791.

Barassi CA, Ayrault G, Creus CM et al., 2006. Seed inoculation with *Azospirillum mitigates* NaCl effects on lettuce. Sci Hortic (Amsterdam) 109: 8-14. https://doi. org/10. 1016/j. scienta. 2006. 02. 025.

Bal HB, Nayak L, Das S, Adhya TK, 2013. Isolation of ACC deaminase producing PGPR from rice rhizosphere and evaluating their plant growth promoting activity under salt stress. Plant Soil 366: 93-105. https://doi. org/10. 1007/s11104-012-1402-5.

Belimov AA, Dodd IC, Safronova VI et al., 2007. *Pseudomonas brassicacearum* strain Am3 containing 1-aminocyclopropane-1-carboxylate deaminase can show both pathogenic and growth-promoting properties in its interaction with tomato. J Exp Bot 58: 1485-1495. https:// doi. org/10. 1093/jxb/erm010.

Bharti N, Pandey SS, Barnawal D et al., 2016. Plant growth promoting rhizobacteria *Dietzia natronolimnaea* modulates the expression of stress responsive genes providing protection of wheat from salinity stress. Sci Rep 6: 1-16. https://doi. org/10. 1038/srep34768.

Bian G, Zhang Y, Qin S et al., 2011. Isolation and biodiversity of heavy metal tolerant endophytic bacteria from halotolerant plant species located in coastal shoal of Nantong. Wei Sheng Wu Xue Bao 51: 1538-1547.

Blaha D, Prigent-Combaret C, Mirza MS, Moënne-Loccoz Y, 2006. Phylogeny of the 1-aminocyclopropane-1-carboxylic acid deaminase-encoding gene acdS in phytobeneficial and pathogenic proteobacteria and relation with strain biogeography. FEMS Microbiol Ecol 56: 455-470. https://doi. org/10. 1111/j. 1574-6941. 2006. 00082. x.

Bochow H, El-Sayed SF, Junge H et al., 2001. Use of *Bacillus subtilis* as biocontrol agent. IV. Saltstress tolerance induction by *Bacillus subtilis* FZB24 seed treatment in tropical vegetable field crops, and its mode of action. J Plant Dis Prot 108: 21-30. https://doi. org/10. 2307/43215378.

Bouizgarne B, 2013. Bacteria for plant growth promotion and disease management. In: Maheshwari DK (ed) Bacteria in agrobiology: disease management. Heidelberg: Springer, 15-47.

Boukhatem ZF, Domergue O, Bekki A et al., 2012. Symbiotic characterization and diversity of rhizobia associated with native and introduced acacias in arid and semi-arid regions in Algeria. FEMS Microbiol Ecol 80: 534 – 547. https://doi.org/10.1111/j.1574-6941.2012.01315.x.

Braud A, Jézéquel K, Bazot S, Lebeau T, 2009. Enhanced phytoextraction of an agricultural Crand Pb-contaminated soil by bioaugmentation with siderophore- producing bacteria. Chemosphere 74: 280–286. https://doi.org/10.1016/j.chemosphere.2008.09.013.

Camilios-Neto D, Bonato P, Wassem R et al., 2014. Dual RNA-seq transcriptional analysis of wheat roots colonized by *Azospirillum brasilense* reveals up-regulation of nutrient acquisition and cell cycle genes. BMC Genomics 15: 1–13. https://doi.org/10.1186/1471-2164-15-378.

Chang P, Gerhardt KE, Huang XD et al., 2014. Plant growth-promoting bacteria facilitate the growth of barley and oats in salt-impacted soil: implications for phytoremediation of saline soils. Int J Phytoremediation 16: 1133 – 1147. https://doi.org/10.1080/15226514.2013.821447.

Chen L, Liu Y, Wu G et al., 2016. Induced maize salt tolerance by rhizosphere inoculation of *Bacillus amyloliquefaciens* SQR9. Physiol Plant. https://doi.org/10.1111/ppl.12441.

Damodaran T, Sah V, Rai RB et al., 2013. Isolation of salt tolerant endophytic and rhizospheric bacteria by natural selection and screening for promising plant growth-promoting rhizobacteria (PGPR) and growth vigour in tomato under sodic environment. Afr J Microbiol Res 7: 5082–5089. https://doi.org/10.5897/AJMR2013.6003.

Dimkpa C, Weinand T, Asch F, 2009. Plant-rhizobacteria interactions alleviate abiotic stress conditions. Plant Cell Environ 32: 1682–1694. https://doi.org/10.1111/j.1365-3040.2009.02028.x.

Dodd IC, Pérez-Alfocea F, 2012. Microbial amelioration of crop salinity stress. J Exp Bot 63: 3415–3428. https://doi.org/10.1093/jxb/ers033.

Dolatabadian A, Modarres Sanavy SAM, Ghanati F, 2011. Effect of salinity on growth, xylem structure and anatomical characteristics of soybean. Not Sci Biol 3: 41–45.

Egamberdieva D, 2009. Alleviation of salt stress by plant growth regulators and IAA producing bacteria in wheat. Acta Physiol Plant 31: 861–864. https://doi.org/10.1007/s11738-009-0297-0.

Egamberdieva D, Lugtenberg B, 2014. Use of plant growth-promoting rhizobacteria to alleviate salinity stress in plants. In: Miransari M (ed) Use of microbes for the alleviation of soil stresses, vol 1. Springer, New York, pp. 73–96.

Egamberdiyeva D, Islam KR, 2008. Salt-tolerant rhizobacteria: plant growth promoting traits and physiological characterization within ecologically stressed environments. In: Ahmad I, Pichtel J, Hayat S (eds)(eds) Plant-bacteria interactions: strategies and techniques to promote plant growth. Wiley, Weinheim, pp. 257–281.

El-Tarabily KA, Youssef T, 2010. Enhancement of morphological, anatomical and physiological characteristics of seedlings of the mangrove *Avicennia marina* inoculated with a native phosphate-solubilizing isolate of *Oceanobacillus picturae* under greenhouse conditions. Plant Soil 332: 147–162. https://doi.org/10.1007/s11104-010-0280-y.

Essa TA, 2002. Effect of salinity stress on growth and nutrient composition of three soybean (*Glycine max* L. Merrill) cultivars. J Agron Crop Sci 188: 86–93. https://doi.org/10.

1046/j. 1439-037X. 2002. 00537. x.

Etesami H, 2018. Can interaction between silicon and plant growth promoting rhizobacteria benefit in alleviating abiotic and biotic stresses in crop plants? Agric Ecosyst Environ 253: 98-112. https: //doi. org/10. 1016/j. agee. 2017. 11. 007.

Etesami H, Beattie GA, 2018. Mining halophytes for plant growth – promoting halotolerant bacteria to enhance the salinity tolerance of non-halophytic crops. Front Microbiol 9: 148. https: //doi. org/ 10. 3389/fmicb. 2018. 00148.

Ghosh S, Penterman JN, Little RD et al., 2003. Three newly isolated plant growth-promoting bacilli facilitate the seedling growth of canola, *Brassica campestris*. Plant Physiol Biochem 41: 277-281. https: //doi. org/10. 1016/S0981-9428 (03) 00019-6.

Glick BR, 2003. Phytoremediation: synergistic use of plants and bacteria to clean up the environment. Biotechnol Adv 21: 383-393. https: //doi. org/10. 1016/S0734-9750 (03) 00055-7.

Glick BR, 2010. Using soil bacteria to facilitate phytoremediation. Biotechnol Adv 28: 367-374. https: //doi. org/10. 1016/j. biotechadv. 2010. 02. 001.

Glick BR, Cheng ZY, Czarny J, Duan J, 2007. Promotion of plant growth by ACC deaminase-producing soil bacteria. Eur J Plant Pathol 119: 329-339. https: //doi. org/10. 1007/s10658-007-9162-4.

Goswami D, Dhandhukia P, Patel P, Thakker JN, 2014. Screening of PGPR from saline desert of Kutch: growth promotion in *Arachis hypogaea* by *Bacillus licheniformis* A2. Microbiol Res 169: 66-75. https: //doi. org/10. 1016/j. micres. 2013. 07. 004.

Habib SH, Kausar H, Saud HM, 2016. Plant growth-promoting rhizobacteria enhance salinity stress tolerance in okra through ROS-scavenging enzymes. Biomed Res Int 2016: 6284547. https: //doi. org/10. 1155/2016/6284547.

Hagemann M, 2011. Molecular biology of cyanobacterial salt acclimation. FEMS Microbiol Rev 35: 87-123. https: //doi. org/10. 1111/j. 1574-6976. 2010. 00234. x.

Hayat R, Ali S, Amara U et al., 2010. Soil beneficial bacteria and their role in plant growth promotion: a review. Ann Microbiol 60: 579-598. https: //doi. org/10. 1007/s13213-010-0117-1.

Heidari M, Jamshid P, 2010. Interaction between salinity and potassium on grain yield, carbohydrate content and nutrient uptake in pearl millet. J Agric Biol Sci 5: 39-46.

Hontzeas N, Zoidakis J, Glick BR, Abu-Omar MM, 2004. Expression and characterization of 1-aminocyclopropane-1-carboxylate deaminase from the rhizobacterium *Pseudomonas putida* UW4: a key enzyme in bacterial plant growth promotion. Biochim Biophys Acta—Proteins Proteomics 1703: 11-19. https: //doi. org/10. 1016/j. bbapap. 2004. 09. 015.

Ibekwe AM, Papiernik SK, Yang CH, 2010. Influence of soil fumigation by methyl bromide and methyl iodide on rhizosphere and phyllosphere microbial community structure. J Environ Sci Heal—Part B Pestic Food Contam Agric Wastes 45: 427-436. https: //doi. org/10. 1080/ 03601231003800131.

Ilangumaran G, Smith DL, 2017. Plant growth promoting rhizobacteria in amelioration of salinity stress: a systems biology perspective. Front Plant Sci 8: 1-14. https: //doi. org/10. 3389/fpls. 2017. 01768.

Jaleel CA, Riadh K, Gopi R et al., 2009. Antioxidant defense responses: physiological plasticity in higher plants under abiotic constraints. Acta Physiol Plant 31: 427 - 436. https://doi.org/10.1007/s11738-009-0275-6.

Jha Y, Subramanian RB, 2014. PGPR regulate caspase-like activity, programmed cell death, and antioxidant enzyme activity in paddy under salinity. Physiol Mol Biol Plants 20: 201-207. https://doi.org/10.1007/s12298-014-0224-8.

Jha B, Thakur MC, Gontia I et al., 2009. Isolation, partial identification and application of diazotrophic rhizobacteria from traditional Indian rice cultivars. Eur J Soil Biol 45: 62-72. https://doi.org/10.1016/j.ejsobi.2008.06.007.

Jha B, Gontia I, Hartmann A, 2012. The roots of the halophyte *Salicornia brachiata* are a source of new halotolerant diazotrophic bacteria with plant growth-promoting potential. Plant Soil 356: 265-277. https://doi.org/10.1007/s11104-011-0877-9.

Jiang H, Dong H, Yu B et al., 2007. Microbial response to salinity change in Lake Chaka, a hypersaline lake on Tibetan plateau. Environ Microbiol 9: 2603-2621. https://doi.org/10.1111/j.1462-2920.2007.01377.x.

Kang SM, Khan AL, Waqas M et al., 2014. Plant growth-promoting rhizobacteria reduce adverse effects of salinity and osmotic stress by regulating phytohormones and antioxidants in *Cucumis sativus*. J Plant Interact 9: 673 - 682. https://doi.org/10.1080/17429145.2014.894587.

Khan M, Böer B, Öztürk M et al., 2016. Sabkha ecosystems. Springer, Cham.

Kim K, Jang YJ, Lee SM et al., 2014. Alleviation of salt stress by *Enterobacter* sp. EJ01 in tomato and *Arabidopsis* is accompanied by up-regulation of conserved salinity responsive factors in plants. Mol Cells 37: 109-117. https://doi.org/10.14348/molcells.2014.2239.

Kohler J, Hernández JA, Caravaca F, Roldán A, 2009. Induction of antioxidant enzymes is involved in the greater effectiveness of a PGPR versus AM fungi with respect to increasing the tolerance of lettuce to severe salt stress. Environ Exp Bot 65: 245-252. https://doi.org/10.1016/j.envexpbot.2008.09.008.

Kuffner M, Puschenreiter M, Wieshammer G et al., 2008. Rhizosphere bacteria affect growth and metal uptake of heavy metal accumulating willows. Plant Soil 304: 35-44. https://doi.org/10.1007/s11104-007-9517-9.

Kumar G, Kanaujia N, Bafana A, 2012. Functional and phylogenetic diversity of root-associated bacteria of *Ajuga bracteosa* in Kangra Valley. Microbiol Res 167: 220-225. https://doi.org/10.1016/j.micres.2011.09.001.

Kumari A, Das P, Parida AK, Agarwal PK, 2015. Proteomics, metabolomics, and ionomics perspectives of salinity tolerance in halophytes. Front Plant Sci 6: 1-20. https://doi.org/10.3389/fpls.2015.00537.

Lee KD, Han HS, Lee KD, 2005. Plant growth promoting rhizobacteria effect on antioxidant status, photosynthesis, mineral uptake and growth of lettuce under soil salinity. Res J Agric Biol Sci 1: 210-215.

Li Q, Saleh-Lakha S, Glick BR, 2005. The effect of native and ACC deaminase-containing *Azospirillum brasilense* Cd1843 on the rooting of carnation cuttings. Can J Microbiol 51: 511-514. https://doi.org/10.1139/w05-027.

Loganathan P, Nair S, 2004. *Swaminathania salitolerans* gen. nov., sp. nov., a salt-tolerant, nitrogen-fixing and phosphate-solubilizing bacterium from wild rice (*Porteresia coarctata*

Tateoka). Int J Syst Evol Microbiol 54: 1185-1190. https://doi.org/10.1099/ijs.0.02817-0.

Lucy M, Reed E, Glick BR, 2004. Applications of free living plant growth-promoting rhizobacteria. Antonie Van Leeuwenhoek 86: 1-25. https://doi.org/10.1023/B: ANTO.0000024903.10757.6e.

Lugtenberg BJJ, Malfanova N, Kamilova F, Berg G, 2013. Plant growth promotion by microbes. In: de Bruijn FJ (ed) (ed) Molecular microbial ecology of the rhizosphere, 1st edn. Wiley Blackwell, Singapore, pp. 561-574.

Ma W, Sebestianova SB, Sebestian J et al., 2003. Prevalence of 1-aminocyclopropane-1-carboxylate deaminase in *Rhizobium* spp. Antonie van Leeuwenhoek, Int J Gen Mol Microbiol 83: 285-291. https://doi.org/10.1023/A: 1023360919140.

Ma Y, Prasad MNV, Rajkumar M, Freitas H, 2011. Plant growth promoting rhizobacteria and endophytes accelerate phytoremediation of metalliferous soils. Biotechnol Adv 29: 248-258. https://doi.org/10.1016/j.biotechadv.2010.12.001.

Madhaiyan M, Poonguzhali S, Ryu J, Sa T, 2006. Regulation of ethylene levels in canola (*Brassica campestris*) by 1-aminocyclopropane-1-carboxylate deaminase-containing *Methylobacterium fujisawaense*. Planta 224: 268-278. https://doi.org/10.1007/s00425-005-0211-y.

Manchanda G, Garg N, 2008. Salinity and its effects on the functional biology of legumes. Acta Physiol Plant 30: 595-618.

Mapelli F, Marasco R, Rolli E et al., 2013. Potential for plant growth promotion of rhizobacteria associated with *Salicornia* growing in Tunisian hypersaline soils. Biomed Res Int 2013: 248078. https://doi.org/10.1155/2013/248078.

Mayak S, Tirosh T, Glick BR, 2004. Plant growth-promoting bacteria confer resistance in tomato plants to salt stress. Plant Physiol Biochem 42: 565-572. https://doi.org/10.1016/j.plaphy.2004.05.009.

Mrozik A, Nowak A, Piotrowska-Seget Z, 2014. Microbial diversity in waters, sediments and microbial mats evaluated using fatty acid-based methods. Int J Environ Sci Technol 11: 1487-1496. https://doi.org/10.1007/s13762-013-0449-z.

Munns R, Tester M, 2008. Mechanisms of salinity tolerance. Annu Rev Plant Biol 59: 651-681. https://doi.org/10.1146/annurev.arplant.59.032607.092911.

Nadeem SM, Zahir ZA, Naveed M, Arshad M, 2007. Preliminary investigations on inducing salt tolerance in maize through inoculation with rhizobacteria containing ACC deaminase activity. Can J Microbiol 53: 1141-1149. https://doi.org/10.1139/W07-081.

Nadeem SM, Zahir ZA, Naveed M, Nawaz S, 2013. Mitigation of salinity-induced negative impact on the growth and yield of wheat by plant growth-promoting rhizobacteria in naturally saline conditions. Ann Microbiol 63: 225-232. https://doi.org/10.1007/s13213-012-0465-0.

Navarro-Torre S, Barcia-Piedras JM, Mateos-Naranjo E et al., 2017. Assessing the role of endophytic bacteria in the halophyte *Arthrocnemum macrostachyum* salt tolerance. Plant Biol 19: 249-256. https://doi.org/10.1111/plb.12521.

Nelson DR, Mele PM, 2007. Subtle changes in rhizosphere microbial community structure in response to increased boron and sodium chloride concentrations. Soil Biol Biochem 39: 340-351.

https://doi.org/10.1016/j.soilbio.2006.08.004.

Nunkaew T, Kantachote D, Nitoda T et al., 2014. Characterization of exopolymeric substances from selected *Rhodopseudomonas palustris* strains and their ability to adsorb sodium ions. Carbohydr Polym 115: 334-341. https://doi.org/10.1016/j.carbpol.2014.08.099.

Oren A, 2002. Molecular ecology of extremely halophilic archaea and bacteria. FEMS Microbiol Ecol 39: 1-7. https://doi.org/10.1071/FP12355.

Ozawa T, Wu J, Fujii S, 2007. Effect of inoculation with a strain of *Pseudomonas pseudoalcaligenes* isolated from the endorhizosphere of *Salicornia europea* on salt tolerance of the glasswort. Soil Sci Plant Nutr 53: 12-16. https://doi.org/10.1111/j.1747-0765.2007.00098.x.

Pandey P, Kang SC, Maheshwari DK, 2005. Isolation of endophytic plant growth promoting *Burkholderia* sp. MSSP from root nodules of *Mimosa pudica*. Curr Sci 89: 177-180.

Parida AK, Das AB, 2005. Salt tolerance and salinity effects on plants: a review. Ecotoxicol Environ Saf 60: 324-349. https://doi.org/10.1016/j.ecoenv.2004.06.010.

Paul D, Lade H, 2014. Plant-growth-promoting rhizobacteria to improve crop growth in saline soils: a review. Agron Sustain Dev 34: 737-752. https://doi.org/10.1007/s13593-014-0233-6.

Piccoli P, Travaglia C, Cohen A et al., 2011. An endophytic bacterium isolated from roots of the halophyte *Prosopis strombulifera* produces ABA, IAA, gibberellins A1 and A3 and jasmonic acid in chemically-defined culture medium. Plant Growth Regul 64: 207-210. https://doi.org/10.1007/s10725-010-9536-z.

Qin Y, Druzhinina IS, Pan X, Yuan Z, 2016. Microbially mediated plant salt tolerance and microbiome-based solutions for saline agriculture. Biotechnol Adv 34: 1245-1259. https://doi.org/10.1016/j.biotechadv.2016.08.005.

Qurashi AW, Sabri AN, 2012. Bacterial exopolysaccharide and biofilm formation stimulate chickpea growth and soil aggregation under salt stress. Braz J Microbiol 43: 1183-1191. https://doi.org/10.1590/S1517-83822012000300046.

Rabie GH, Aboul-Nasr MB, Al-Humiany A, 2005. Increased salinity tolerance of cowpea plants by dual inoculation of an *Arbuscular mycorrhizal* fungus *Glomus clarum* and a nitrogen-fixer *Azospirillum brasilense*. Mycobiology 33: 51-60. https://doi.org/10.4489/myco.2005.33.1.051.

Rangarajan S, Saleena LM, Nair S, 2002. Diversity of *Pseudomonas* spp. isolated from rice rhizosphere populations grown along a salinity gradient. Microb Ecol 43: 280-289. https://doi.org/10.1007/s00248-002-2004-1.

Ravindran KC, Venkatesan K, Balakrishnan V et al., 2007. Restoration of saline land by halophytes for Indian soils. Soil Biol Biochem 39: 2661-2664. https://doi.org/10.1016/j.soilbio.2007.02.005.

Rodriguez R, Redman R, 2008. More than 400 million years of evolution and some plants still can't make it on their own: plant stress tolerance via fungal symbiosis. J Exp Bot 59: 1109-1114. https://doi.org/10.1093/jxb/erm342.

Rueda-Puente EO, Castellanos-Cervantes T, Díaz De León-Álvarez JL, Preciado P, Almaguer Vargas G, 2010. Bacterial community of rhizosphere associated to the annual halophyte *Salicornia bigelovii* (Torr.). Terra Latinoam 28: 345-353.

Ruppel S, Franken P, Witzel K, 2013. Properties of the halophyte microbiome and their implications for plant salt tolerance. Funct Plant Biol 40: 940 - 951. https://doi.org/10.1071/FP12355.

Ryu H, Cho YG, 2015. Plant hormones in salt stress tolerance. J Plant Biol 58: 147-155. https://doi.org/10.1007/s12374-015-0103-z.

Saleem M, Arshad M, Hussain S, Bhatti AS, 2007. Perspective of plant growth promoting rhizobacteria (PGPR) containing ACC deaminase in stress agriculture. J Ind Microbiol Biotechnol 34: 635-648. https://doi.org/10.1007/s10295-007-0240-6.

Sandhya V, Ali SZ, Grover M et al., 2010a. Effect of plant growth promoting *Pseudomonas* spp. on compatible solutes, antioxidant status and plant growth of maize under drought stress. Plant Growth Regul 62: 21-30. https://doi.org/10.1007/s10725-010-9479-4.

Sandhya V, Ali SZ, Venkateswarlu B et al., 2010b. Effect of osmotic stress on plant growth promoting *Pseudomonas* spp. Arch Microbiol 192: 867 - 876. https://doi.org/10.1007/s00203-010-0613-5.

Sapre S, Gontia-Mishra I, Tiwari S, 2018. *Klebsiella* sp. confers enhanced tolerance to salinity and plant growth promotion in oat seedlings (*Avena sativa*). Microbiol Res 206: 25-32. https://doi.org/10.1016/j.micres.2017.09.009.

Saravanakumar D, Samiyappan R, 2007. ACC deaminase from *Pseudomonas fluorescens* mediated saline resistance in groundnut (*Arachis hypogaea*) plants. J Appl Microbiol 102: 1283-1292. https://doi.org/10.1111/j.1365-2672.2006.03179.x.

Sarkar A, Ghosh PK, Pramanik K et al., 2017. A halotolerant *Enterobacter* sp. displaying ACC deaminase activity promotes rice seedling growth under salt stress. Res Microbiol. https://doi.org/10.1016/j.resmic.2017.08.005.

Sgroy V, Cassán F, Masciarelli O et al., 2009. Isolation and characterization of endophytic plant growth – promoting (PGPB) or stress homeostasis – regulating (PSHB) bacteria associated to the halophyte *Prosopis strombulifera*. Appl Microbiol Biotechnol 85: 371-381. https://doi.org/10.1007/s00253-009-2116-3.

Shaharoona B, Arshad M, Zahir ZA, Khalid A, 2006. Performance of *Pseudomonas* spp. containing ACC-deaminase for improving growth and yield of maize (*Zea mays* L.) in the presence of nitrogenous fertilizer. Soil Biol Biochem 38: 2971-2975. https://doi.org/10.1016/j.soilbio.2006.03.024.

Shailendra Singh GG, 2015. Plant growth promoting rhizobacteria (PGPR): current and future prospects for development of sustainable agriculture. J Microb Biochem Technol 07: 96-102. https://doi.org/10.4172/1948-5948.1000188.

Shanker AK, Venkateswarlu B, 2011. Abiotic stress in plants—mechanisms and adaptations, 1st edn. InTechOpen, Rijeka.

Shi W, Takano T, Liu S, 2012. Isolation and characterization of novel bacterial taxa from extreme alkali-saline soil. World J Microbiol Biotechnol 28: 2147-2157. https://doi.org/10.1007/s11274-012-1020-7.

Shukla PS, Agarwal PK, Jha B, 2012. Improved salinity tolerance of *Arachis hypogaea* (L.) by the interaction of halotolerant plant-growth-promoting rhizobacteria. J Plant Growth Regul 31: 195-206. https://doi.org/10.1007/s00344-011-9231-y.

Siddikee MA, Glick BR, Chauhan PS et al., 2011. Enhancement of growth and salt tolerance of red pepper seedlings (*Capsicum annuum* L.) by regulating stress ethylene synthesis with halotolerant bacteria containing 1-aminocyclopropane-1-carboxylic acid deaminase activity. Plant Physiol Biochem 49: 427-434. https://doi.org/10.1016/j.plaphy.2011.01.015.

Siddikee MA, Chauhan PS, Anandham R et al., 2010. Isolation, characterization, and use for plant growth promotion under salt stress, of ACC deaminase-producing halotolerant bacteria derived from coastal soil. J Microbiol Biotechnol 20: 1577-1584. https://doi.org/10.4014/jmb.1007.07011.

Singh RP, Jha PN, 2016. A halotolerant bacterium *Bacillus licheniformis* HSW-16 augments induced systemic tolerance to salt stress in wheat plant (*Triticum aestivum*). Front Plant Sci 7: 1-18. https://doi.org/10.3389/fpls.2016.01890.

Singh RP, Shelke GM, Kumar A, Jha PN, 2015. Biochemistry and genetics of ACC deaminase: a weapon to "stress ethylene" produced in plants. Front Microbiol 6: 1-14. https://doi.org/10.3389/fmicb.2015.00937.

Szymańska S, Płociniczak T, Piotrowska-Seget Z et al., 2016. Metabolic potential and community structure of endophytic and rhizosphere bacteria associated with the roots of the halophyte *Aster tripolium* L. Microbiol Res 182: 68-79. https://doi.org/10.1016/j.micres.2015.09.007.

Tank N, Saraf M, 2010. Salinity-resistant plant growth promoting rhizobacteria ameliorates sodium chloride stress on tomato plants. J Plant Interact 5: 51-58. https://doi.org/10.1080/17429140903125848.

Tiwari S, Singh P, Tiwari R et al., 2011. Salt-tolerant rhizobacteria-mediated induced tolerance in wheat (*Triticum aestivum*) and chemical diversity in rhizosphere enhance plant growth. Biol Fertil Soils 47: 907-916. https://doi.org/10.1007/s00374-011-0598-5.

Tripathi AK, Verma SC, Ron EZ, 2002. Molecular characterization of a salt-tolerant bacterial community in the rice rhizosphere. Res Microbiol 153: 579-584. https://doi.org/10.1016/S0923-2508(02)01371-2.

Tsukanova KA, Chebotar V, Meyer JJM, Bibikova TN, 2017. Effect of plant growth-promoting rhizobacteria on plant hormone homeostasis. S Afr J Bot 113: 91-102. https://doi.org/10.1016/j.sajb.2017.07.007.

Uchiumi T, Ohwada T, Itakura M et al., 2004. Expression islands clustered on the symbiosis island of the *Mesorhizobium loti* genome. J Bacteriol 186: 2439-2448. https://doi.org/10.1128/JB.186.8.2439-2448.2004.

Ullah S, Asghari B, 2015. Isolation of plant-growth-promoting rhizobacteria from rhizospheric soil of halophytes and its impact on maize (*Zea mays* L.) under induced soil salinity. Can J Microbiol 61: 1-34. https://doi.org/10.1139/cjm-2014-0668.

Upadhyay SK, Singh DP, 2015. Effect of salt-tolerant plant growth-promoting rhizobacteria on wheat plants and soil health in a saline environment. Plant Biol 17: 288-293. https://doi.org/10.1111/plb.12173.

Upadhyay SK, Singh JS, Singh DP, 2011. Exopolysaccharide-producing plant growth-promoting rhizobacteria under salinity condition. Pedosphere 21: 214-222. https://doi.org/10.1016/S1002-0160(11)60120-3.

Upadhyay SK, Singh JS, Saxena AK, Singh DP, 2012. Impact of PGPR inoculation on growth and antioxidant status of wheat under saline conditions. Plant Biol 14: 605-611. https://doi.org/10.1111/j.1438-8677.2011.00533.x.

Wang C, Knill E, Glick BR, Défago G, 2000. Effect of transferring 1-aminocyclopropane-1-carboxylic acid (ACC) deaminase genes into *Pseudomonas fluorescens* strain CHA0 and its *gac*A derivative CHA96 on their growth-promoting and disease-suppressive capacities. Can J Microbiol 46: 898-907. https: //doi. org/10. 1139/w00-071.

Wenzel WW, 2009. Rhizosphere processes and management in plant-assisted bioremediation (phytoremediation) of soils. Plant Soil 321: 385-408. https: //doi. org/10. 1007/s11104-008-9686-1.

Xu GY, Rocha PSCF, Wang ML et al., 2011. A novel rice calmodulin-like gene, *OsMSR*2, enhances drought and salt tolerance and increases ABA sensitivity in *Arabidopsis*. Planta 234: 47-59. https: //doi. org/10. 1007/s00425-011-1386-z.

Xun F, Xie B, Liu S, Guo C, 2015. Effect of plant growth-promoting bacteria (PGPR) and *Arbuscular mycorrhizal* fungi (AMF) inoculation on oats in saline-alkali soil contaminated by petroleum to enhance phytoremediation. Environ Sci Pollut Res 22: 598-608. https: //doi. org/10. 1007/s11356-014-3396-4.

Yan J, Smith MD, Glick BR, Liang Y, 2014. Effects of ACC deaminase containing rhizobacteria on plant growth and expression of Toc GTPases in tomato (*Solanum lycopersicum*) under salt stress. Botany 92: 775-781. https: //doi. org/10. 1139/cjb-2014-0038.

Yang J, Kloepper JW, Ryu CM, 2009. Rhizosphere bacteria help plants tolerate abiotic stress. Trends Plant Sci 14: 1-4. https: //doi. org/10. 1016/j. tplants. 2008. 10. 004.

Yao L, Wu Z, Zheng Y et al., 2010. Growth promotion and protection against salt stress by *Pseudomonas putida* Rs-198 on cotton. Eur J Soil Biol 46: 49-54. https: //doi. org/10. 1016/j. ejsobi. 2009. 11. 002.

Yoshimura H, Kotake T, Aohara T et al., 2012. The role of extracellular polysaccharides produced by the terrestrial cyanobacterium *Nostoc* sp. strain HK-01 in NaCl tolerance. J Appl Phycol 24: 237-243. https: //doi. org/10. 1007/s10811-011-9672-5.

Zahir ZA, Ghani U, Naveed M et al., 2009. Comparative effectiveness of *Pseudomonas* and *Serratia* sp. containing ACC-deaminase for improving growth and yield of wheat (*Triticum aestivum* L.) under salt-stressed conditions. Arch Microbiol 191: 415-424. https: //doi. org/10. 1007/s00203-009-0466-y.

Zahir ZA, Shah MK, Naveed M, Akhter MJ, 2010. Substrate-dependent auxin production by *Rhizobium phaseoli* improves the growth and yield of *Vigna radiata* L. under salt stress conditions. J Microbiol Biotechnol 20: 1288-1294. https: //doi. org/10. 4014/jmb. 1002. 02010.

Zhang YQ, Schumann P, Yu LY et al., 2007. *Zhihengliuella halotolerans* gen. nov., sp. nov., a novel member of the family Micrococcaceae. Int J Syst Evol Microbiol 57: 1018-1023. https: // doi. org/10. 1099/ijs. 0. 64528-0.

Zhou M, Chen W, Chen H, Wei G, 2012a. Draft genome sequence of *Mesorhizobium alhagi* CCNWXJ12-2 T, a novel salt-resistant species isolated from the desert of northwestern China. J Bacteriol 194: 1261-1262. https: //doi. org/10. 1128/JB. 06635-11.

Zhou T, Chen D, Li C et al., 2012b. Isolation and characterization of *Pseudomonas brassicacearum* J12 as an antagonist against *Ralstonia solanacearum* and identification of its antimicrobial components. Microbiol Res 167: 388-394. https: //doi. org/10. 1016/j. micres. 2012. 01. 003.

Zhou N, Zhao S, Tian CY, 2017. Effect of halotolerant rhizobacteria isolated from halophytes on the growth of sugar beet (*Beta vulgaris* L.) under salt stress. FEMS Microbiol Lett 364: 1-8. https://doi.org/10.1093/femsle/fnx091.

Zhu F, Qu L, Hong X, Sun X, 2011. Isolation and characterization of a phosphate-solubilizing halophilic bacterium *Kushneria* sp. YCWA18 from Daqiao saltern on the coast of yellow sea of China. Evid-Based Complement Altern Med 2011: 615032. https://doi.org/10.1155/2011/615032.

5 耐盐植物生长促生真菌和细菌提高盐胁迫下作物养分利用效率的替代策略

Hassan Etesami, Hossein Ali Alikhani

[摘要]

目前,面临世界人口不断增长的巨大压力,由于水和可耕作土壤资源的限制导致世界上可耕作土地的面积不断减少以及人们对农产品需求的不断增加,在恶劣的自然环境条件下进行农作物生产和管理以实现农作物生长和产量的最大潜力,从而生产足够的食物已成为世界上农业生产的当务之急。盐胁迫是制约干旱和半干旱地区农产品生产和减少全球耕地面积的主要非生物逆境胁迫之一。在盐渍化土壤中生长的农作物,由于过量吸收Na^+、Cl^-等离子,会降低对K、Ca、Mg、N和P等大量营养素和Fe、Zn、Cu、Mn和B等微量营养素的吸收,从而导致植物对营养元素吸收的不平衡。为了满足在盐胁迫条件下生长的作物对营养元素的需求,大量和微量营养元素通常以化肥的形式施用到土壤中,供农作物吸收利用。但这些化学肥料的生产是一个高能耗的过程,并且从长期来看化肥的过度施用会对环境富营养化、土壤理化质量以及环境碳循环有很大的负面影响。这些潜在的环境问题促使人们不断寻求环境友好且可持续发展的、能够为农作物提供养分的方式和方法。目前在土壤-微生物方面的热点研究就提供了一个比较理想的方法,研究发现,利用耐盐植物生长促生菌是缓解遭受盐胁迫农作物营养元素缺乏的一种效果良好且生态友好的手段。这些微生物通过固氮、增磷和溶钾等机制将植物所需的关键营养元素(N、P、K和微量营养素等)变得易于植物吸收,从而提高农作物对大量和中微量营养元素的生物利用度。因此,耐盐植物生长促生菌的使用也被视为一种替代性、创新性和生态环境友好的方法,其在农业上的应用可以减少昂贵且对环境不友好的化学肥料的使用。耐盐植物生长促生菌代表了一种能够提高植物对养分的生物利用程度,并且可以促进盐渍土农业发展的很有前景的方法。本章讨论了盐分胁迫下植物生长促生菌和丛枝菌根真菌

H. Etesami (✉), H. A. Alikhani, 伊朗德黑兰大学农业与自然资源学院土壤科学系,E-mail: H. Etesami: hassanetesami@ut.ac.ir

(*Arbuscular mycorrhizal*，AM）影响农作物对土壤 N、P、K 和微量元素等养分的生物利用有效性的机制。

[关键词]

丛枝菌根真菌；植物生长促生菌；盐胁迫；农作物；植物-微生物相互作用；盐渍土农业

5.1 前言

目前世界上的人口正在以每年 8 000 万左右的速度增长，因此粮食短缺危机是 21 世纪在农业生产上要面临的一个重大问题。粮食安全是世界经济可持续发展的重要指标之一，因此在农业生产中必须寻找恰当的方式来满足整个社会对粮食的需求。鉴于土壤盐渍化是造成目前全球农业用地面积不断下降的主要原因，为了满足全球粮食安全需求，除了提高土地单位面积产量（最大产量）的方法之外，很难再找到一种能够大幅度增产全球粮食的有效方式（Etesami 和 BeaTee，2018）。一方面全球对粮食的需求不断增加，另一方面世界上的可用耕地面积由于盐渍化威胁不断减少，而近几十年粮食单产在世界各个国家和地区基本上已经达到了当地的最高水平，后续提升幅度有限。因此，发展盐土农业，利用盐渍化土地生产主要的农作物以满足全球不断增长的粮食需求，成为解决粮食危机的一个可选项。但是土壤盐分是限制农作物生长和农产品生产的重要因素之一，在世界范围内受盐渍化影响的耕地总面积在不断上升。据估计，目前世界上约有 30%~50% 的土地处在盐渍化胁迫之下。在伊朗，约有 50% 的农业可耕用地面临着土壤盐渍化的威胁。在盐渍化条件下，土壤溶液中 Na^+ 的浓度通常要高于一般中微量营养元素，这就会导致在盐渍化土壤上生长的农作物发生营养失衡。一般来说，土壤或灌溉水中过量的盐分会导致植物面临盐胁迫威胁。土壤盐胁迫是由于土壤溶液中积累了过量的阳离子和阴离子而产生的，与发生土壤盐渍化的机制无关。土壤溶液中的这些盐分主要包括 Na^+ 和 Cl^- 等离子，还包括一些碳酸氢根离子（HCO_3^-）、硫酸根离子（SO_4^{2-}）、钙离子（Ca^{2+}）、镁离子（Mg^{2+}）、硼酸根离子（BO_3^{3-}）等，也有少量的硝酸根离子（NO_3^-）（Sha Valli Khan 等，2014）。土壤溶液中的这些阳离子和阴离子在土壤中会阻碍农作物对其他必需营养元素的吸收，例如 N、P、K、Fe、Cu 和 Zn 等（Giri 等，2007；Munns 和 Tester，2008；Tester 和 Davenport，2003）。因此在缺少水分和土壤盐分过度积累的农业种植区域，盐胁迫导致的农作物减产可以达到 10%~60%（FAO，2005）。因此，传统的农业生产中为了保证遭受盐胁迫危害的农作物对必需营养元素的需求量，往往需要定期大量施用化学肥料。在这些耕作土壤发生盐渍化的农业地区，当地农民也正是这样做的。通过在土地上投入更多的化肥、种子和水来补偿盐胁迫对农作物正常生长造成的破坏，而这些投入成本有时会占到农民种植农作物收入的65%，甚至更高。同时，除了增加生产成本以外，化学肥料的过量使用还会导致许多生态问题和环境问题，包括土壤理化性质下降、水体富营养化、土壤养分失衡、土壤板结、农产品品质下降等，也会危害人类和其他生物的健康（Adesemoye 和 Kloepper，2009）。虽然化学肥料的投入会在一定程度上补偿农作物因盐胁迫而造成的产量损失，但在盐渍化土壤中使用化学肥料对农业的可持续发展并不是最好的解决方案。因为过量的化学肥料的投入反而会进一步增加土壤中盐分离子的积累，使土壤的盐渍化程度变得更高。大多数农作物几乎都是甜土植物，对盐分十分敏感，并不能在高盐浓度下正常生存（Siddiqui 等，2009）。一些研究结果表明，在盐渍化土壤中施加农作物有益微生物能够明显地缓解农作物的盐胁迫危害（Etesami 和 Beattie，2018）。因此，近年来，许多研究者已在考虑使用植物生长促进微生物（plant growthpromoting microorganisms，PGPM）作为生物肥料或生物防治剂应用到盐渍化土地的农业生产中。可喜的是，这些微生物对不同农作物的促生长效果已在温室和田间等不同的试验环境中得到了初步验证。植物生长促生菌的使用相对于传统农业中大量化学肥料的施用是一种替代的、创新的，且生态环境友好的方法，这种方法可以减少昂贵且污染环境的化学肥料的使用（Rashid 等，2016）。植物生长促生菌能够通过多种机制来促进农作物的生长，并提高产量（Etesami 和 Maheshwari，2018）。

植物生长促生菌是一类生活在植物根际的细菌，能够通过直接和（或）间接的方式改善植物的生长。在间接作用机制中，植物生长促生菌可以通过特定的机制间接消除植物病原体对植物的有害影响，从而促进植物的正常生长。直接作用机制是植物生长促生菌促进植物生长的主要方式。这些细菌能够通过固氮、增磷和溶钾等方式提高土壤溶液中农作物所必需的营养元素；并且还能合成和分泌对植物生长具有刺激性的代谢产物，例如植物激素类物质吲哚-3-乙酸、细胞分裂素和赤霉素等，从而促进植物的生长。植物生长促生菌能够通过合成、分泌多种有机酸和无机酸来溶解附着在土壤颗粒上的难溶性 P 和 K 元素，通过合成铁-铁载体等方式增加土壤溶液中 Fe 等元素的保有量，以及通过合成 1-氨基环丙烷-1-羧酸脱氨酶的方式来减少盐胁迫下植物体内的乙烯产量等方式促进植物良好的生长（Etesami，2018；Etesami 和 Maheshwari，2018；Gamalero 和 Glick，2011，Glick，2014）。目前有许多细菌属中的细菌已被报道具有促进植物生长的能力，这些细菌属具体包括：假单胞菌属（*Pseudomonas*）、固氮螺菌属（*Azospirillum*）、固氮菌属（*Azotobacter*）、克雷伯氏菌属（*Klebsiella*）、肠杆菌属（*Enter-*

obacter)、产碱杆菌属（*Alcaligenes*）、节杆菌属（*Arthrobacter*）、伯克霍尔德氏菌属（*Burkholderia*）、芽孢杆菌属（*Bacillus*）、沙雷氏菌属（*Serratia*）、欧文氏菌属（*Erwinia*）、黄杆菌属（*Flavobacterium*）和根瘤菌属（*Rhizobium*）等（Egamberdiyeva，2005；Glick，2014）。

另一类与植物生长和发育非常密切的微生物是真菌。真菌和细菌一样，可以在植物的根际、根面以及根内组织（内生菌）中定殖。丛枝菌根真菌（*Arbuscular mycorrhizal*，AM）是农业可耕地土壤中含量最为丰富的真菌之一。丛枝菌根真菌的植物宿主非常广泛，它能够在83%的双子叶植物和79%的单子叶植物的根际系统中与植物形成共生关系（Peterson等，2004）。丛枝菌根真菌在促进植物对养分的吸收利用方面的能力只有在营养缺乏的生态系统中才能够体现（Cardoso和Kuyper，2006）。这些丛枝菌根真菌对植物的促生长作用也可以通过多种机制实现（Owen等，2015；Rawat和Tecwari，2011；Waller等，2005）。

盐胁迫除了对植物的生长产生负面影响外，还对包括细菌和丛枝菌根真菌在内的微生物产生负面影响（Greaves，1922），这些负面影响包括降低微生物定殖能力、孢子萌发、真菌菌丝生长以及菌根共生体的形成等（Giri等，2007；McMillen等，1998；Porcel等，2012；Sheng，2008）。此外，还有报道称，植物生长促生菌促进植物生长的能力，比如溶解磷酸盐的能力（Sharma等，2013），或者产生1-氨基环丙烷-1-羧酸脱氨酶的能力（Upadhyay等，2009）等，也都受到盐胁迫等环境条件的制约（Sánchez Porro等，2009）。因此，只有本身就能抵抗这些逆境胁迫环境胁迫条件的微生物才能更好地通过各种机制发挥促进植物生长的作用（Etesami和Beattie，2018；Etesami和Maheshwari，2018）。众所周知，从盐胁迫环境或盐生植物中分离得到的微生物在盐胁迫条件下能够很好地发挥其促进植物生长（Plant growth-promoting，PGP）的特性（Etesami和Beattie，2018）。而这一类微生物则通常具有能够在盐胁迫环境中维持生存的策略（Vreeland，1987）。

嗜盐菌（*Halophiles*）是一种喜盐微生物，这一类微生物能够在相对较高的盐浓度下良好生长。嗜盐菌几乎存在于所有主要的微生物类群中（包括古生菌、细菌和真菌等）中，它们的进化分类主要取决于这类微生物的耐盐水平和正常生长所需要的盐浓度（Ma等，2010）。耐盐细菌和嗜盐细菌在生理层面上是两种不同的微生物。耐盐微生物能够耐受一定程度的盐胁迫，通常具有在盐胁迫环境中维持生存的机制（Vreeland，1987）。耐盐细菌的耐盐机制主要包括细胞内Na^+的外向排泄，胞外多糖的产生，细胞内相容性化合物的积累（包括脯氨酸、海藻糖、甘氨酸、甜菜碱、蔗糖和甘油等），蛋白质和酶等生物大分子对高浓度盐环境的适应以及积累 K^+ 以提高微生物耐受高浓度的盐胁迫等（Ruppel 等，2013）。目前，有很多研究报道指出，这一类耐盐植物生长促生菌已经从盐生植物（Etesami 和 Beattie，2018）和受到盐胁迫危害的土壤（Hingole 和 Pathak，2016；Orhan 和 Gulluce，2015）中得到了分离和鉴定。这些植物生长促生菌可以提高对盐分敏感农作对盐胁迫的抗性，并改善这些农作物在盐胁迫条件下的生长发育和最终产量（Etesami 和 Beattie，2018）。因此，具有固氮和磷酸盐增溶能力的耐盐植物生长促生菌有可能作为生物肥料进行施用，以提高生长在盐渍土环境下的农作物的生长和产量（Delgado Garcia 等，2013）。此外，这些植物生长促生菌还可以产生生物表面活性剂和胞外多糖等，提高农作物应对逆境环境胁迫（如盐胁迫和干旱）的能力（Margesin 和 Schinner，2001）。

丛枝菌根真菌（*Arbuscular mycorrhizal*）也广泛存在于各种自然生态系统和受到盐胁迫危害的可耕作土壤中（Sengupta 和 Chaudhuri，2002），丛枝菌根真菌被认为是具有发挥缓解植物盐胁迫危害能力的适宜候选菌（Garg 和 Chandel，2011）。丛枝菌根真菌属真菌如球囊霉属（*Glomus* sp.）和丛枝菌根真菌（*Glomus mosseae*）也存在于自然形成的盐渍化环境中（Evelin 等，2009）。有报道指出，盐胁迫并不能阻止丛枝菌根真菌的孢子形成和其在农作物根际的定殖（Aliasgharzadeh 等，2001）。有研究表明，在盐胁迫条件下接种丛枝菌根真菌的植物比未接种丛枝菌根真菌的植物生长得更好（Feng 等，2002；Giri 等，2007；Hajiboland 等，2010；Porcel 等，2012；Sannazzaro 等，2007；Zuccarini 和 Okurowska，2008）。这就说明这一类真菌的确能够改善并促进农作物在盐胁迫条件下的生长活力，从而减轻和缓解农作物遭受到的盐胁迫危害（Evelin 等，2009；Garg 和 Chandel，2011；Giri 等，2007；Kaya 等，2009；Kumar 等，2010；Porras Soriano 等，2009）。一般来说，丛枝菌根真菌也是通过多种机制（包括营养、生化以及生理效应的结合）来促进和提高植物的抗盐能力（Evelin 等，2009）。具体作用机制包括以下 7 个方面。

（1）促进植物细胞积累渗透调节物质（例如脯氨酸、甘氨酸和甜菜碱等）（Garg 和 Manchanda，2009）。这一类丛枝菌根真菌还可以在菌丝体中合成海藻糖，海藻糖对丛枝菌根真菌而言是一种主要的储能碳水化合物，也可以用作农作物生产过程中的非生物胁迫保护剂（Garg 和 Chandel，2011；Ocón 等，2007）。

（2）改善宿主农作物对土壤溶液中矿物质微量元素和常量营养素的吸收（例如 N、P、Zn 和 Fe 等）（Asghari，2012；Miransari 和 Smith，2008）和维持农作物细胞中的离子平衡（Giri 等，2007；Wu 等，2010）。

（3）提高农作物光合速率和水分利用效率（water-use efficiency，WUE）

(Colla 等，2008；Wu 等，2010）。

（4）改善宿主农作物在盐胁迫环境下的生理过程。例如，通过提高宿主农作物根系的水分传导效率和水分利用效率以促使细胞内获得更高的 K^+/Na^+ 比率，促使某些植物组织中形成 Na^+ 储运的分区结构等，并有利地调节细胞碳水化合物的合成和消耗平衡以及种类组成等，通过这些生理层面的微调提高宿主农作物对盐胁迫的耐受能力（Al-Karaki，2006；Giri 等，2003；Ruiz-Lozano，2003）。

（5）保持生物膜的完整性。这有利于在细胞液泡内进行 Na^+ 等区室化和选择性吸收离子（Rinaldelli 和 Mancuso，1996）。

（6）通过促进宿主农作物的生长来稀释作物体内相应的 Na^+ 浓度，以降低盐胁迫对农作物生长的影响（Al-Karaki，2006）。

（7）通过增加或选择性吸收养分，对盐胁迫条件下生长的农作物矿质营养需求产生积极影响（Porcel 等，2012；Al-karaki 和 Clark，1998）。真菌生物广泛的菌丝系统可吸收 N、P、K、Zn、Cu、Fe 等营养物质。丛枝菌根真菌吸收的营养不仅用于自身生长发育，有一部分还会被输送给宿主农作物以提高对盐胁迫的抵御能力（Gosling 等，2006）。

目前，已经有很多研究者报道了近年来对农业生产具有重要作用的微生物在非生物胁迫逆境中促进农作物生长（Grover 等，2011；Paul，2013；Paul 和 Lade，2014；Venkateswarlu 等，2008；Yang 等，2009）。植物生长促生菌和丛枝菌根真菌都可以改善非生物逆境胁迫环境下植物的生长性能，因此可以直接或间接地提高农作物生产的产量（Banik 等，2006；Barassi 等，2006；Chakraborty 等，2015；Choudhary，2012；Choudhar 等，2015；Dahmardeh，2009；Damodaran 等，2014；Davies 等，2011；Del Amor Francisco 和 Cuadra-Crespo，2012；Dimkpa 等，2009；Dolkar 等，2018；Egamberdieva 等，2008；Etesami 和 Alikhani，2016b；Fu 等，2010；Gray 和 Smith，2005；Hamilton 等，2016；Kaymak 等，2009；Khan 等，2012；Mayak 等，2004b；Milošević 等，2012；Nadeem 等，2007；Paul，2012；Paul Nair，2008；Ramadoss 等，2013；Rojas-Tapias 等，2012；Shrivastava 和 Kumar，2015；Soleimani 等，2011；Tiwari 等，2011；Upadhyay 等，2009；Yang 等，2009；Yao 等，2010；Yildirim 和 Taylor，2005）。本章节的主要目的是介绍和讨论有关盐胁迫条件下植物生长促生菌和丛枝菌根真菌影响宿主农作物对土壤养分（主要是 N、P、K 和 Fe 等）生物利用的机理。

5.2 盐胁迫

无论是自然过程还是人为干预过程造成的土壤盐渍化，都是各种环境胁迫因子中对农业生产造成较大压力的因素。因为土壤的盐渍化会使可耕地变为不可耕土地,特别是在世界干旱和半干旱地区,严重影响世界上农业发展的可持续性以及农作物的生产(Flowers,2004;Munns,2005)。有统计表明,在世界范围内因土壤盐渍化每年减少的可用于农业生产的土地面积占整个耕地面积的1%~2%。目前,地球上大约7%的土地和20%左右的农用耕地都遭受到盐胁迫的不利影响。土壤盐渍化影响着发达国家和发展中国家的大片土地。农业生产用地的集约化发展和各种不利自然条件的叠加,加速了世界范围内许多地区农用土壤的盐渍化进程。如果土壤饱和溶液的电导率大于4 dS/m(America,2001),即大致相当于约40 mmol/L的NaCl溶液,那么这种土壤就被认为是盐渍化土壤(Munns和Tester,2008)。世界上各个国家和地区的可耕作农用土地的盐渍化程度由于灌溉用水不当、过度施肥和自然条件下的沙漠化过程变得越来越严重(Munns和Tester,2008)。联合国粮农组织报告称,目前世界上超过10亿hm^2的土地正在遭受盐胁迫的影响(Ahmad,2013;FAO,2008)。同时,由于全球气候变化异常,遭受盐胁迫影响的土地面积呈现出逐渐增加的趋势(Shrivastav和Kumar,2015)。据估计,到2050年全世界50%以上的可耕农业用地都将面临盐渍化的威胁(Jamil等,2011)。

  盐胁迫在各个方面限制农作物的生长发育和最终的粮食产量,对农作物的发芽率、植株的生活力和作物的粮食产量都会产生非常不利的影响,比如对谷物中的水稻、小麦和玉米,牧草中的三叶草以及园艺作物中的马铃薯和番茄等(Munns和Tester,2008)。上述这些农作物相对容易受到盐胁迫影响。土壤盐渍化过程中的盐分来源既可能溶解于灌溉水中,也可能存在于土壤颗粒中,随土壤溶液浸出。而从植物方面来讲,植物对盐胁迫的响应非常复杂,既有与干旱胁迫效应相似的地方,也有盐胁迫产生的离子毒害等特殊效应。为了更好地了解植物遭受盐胁迫的性质,有必要首先对盐胁迫进行定义。根据Shannon和Grive(1998)的说法,盐胁迫是指土壤溶液中可溶性盐和矿物质元素的浓度过高,导致植物根区(根际)盐分积累增加,渗透势增加,从而使植物变得难以从根周附近吸收水分的现象。而土壤盐渍化是指土壤表层的可溶性盐积累增加,从而使土壤表层失去作为植物生长介质的潜力的过程。在干旱和半干旱地区,土壤中盐分的过度积累主要是由于降雨的缺乏和地表蒸发量过高形成的。除此之外,土壤盐渍化也可能是由于其他原因造成的,包括过度使用化学肥料、灌溉用水水质不佳、森林过度砍伐导致地表裸露增加等。土壤盐渍化也会导致土壤溶液发生渗透效应(Munns,2005;Munns和Tester,2008),因此导致植物吸收水分发生困难而产生缺水效用,进而在细胞中产生过量的活性氧物质,比如过氧化物阴离子、羟基自由基、过氧化氢和单线态氧等。这些活性氧物质无论是哪一种都会扰乱植物正常的生理代谢，会导致植物细胞质膜和内膜系统发生损伤，从而失去正常的生理功能（Parida 和 Das，2005）。另外，盐胁迫还能够限制植物正常的生长发育，导致植物在生理、形态以及生化特性等方面发生非常显著的变化。一般来说，盐胁迫会降低许多农作物的产量，这是因为盐胁迫会抑制植物的光合效率、蛋白质合成过程、脂质代谢过程、叶绿素合成以及气孔导度和蒸腾速率发生变化，植物的正常生长过程遭到干扰，因此生物量降低，最终降低了粮食产量。盐胁迫条件下，农作物为了抵御细胞内过量活性氧物质积累对细胞正常功能造成的危害，会在细胞内大量合成超氧化物歧化酶（SOD）、过氧化物酶（POD）、抗坏血酸过氧化物酶（APX）、过氧化氢酶（CAT）等用以清除活性氧物质。同时，作为应激反应，遭受到盐胁迫危害的植物也会适当增加细胞内渗透调节物质的含量，包括脯氨酸、甜菜碱、总游离氨基酸以及可溶性糖等。种子的萌发期是植物最重要的生命期之一，对植物的生长发育起着非常重要的作用，而盐胁迫对种子萌发的抑制作用是农作物在盐渍化土地上进行生产的又一大障碍。不仅如此，盐胁迫还会导致农作物细胞内植物激素（特别是吲哚-3-乙酸和乙烯）合成和分布的不平衡，从而影响植物的生长发育（Kay 等，2009）。土壤溶液中高浓度的可溶性盐和一些盐分离子（比如 Na^+ 和 Cl^-）会导致植物根系对矿物质离子的吸收发生紊乱，从而导致植物营养失衡和（或）离子毒性（Munns，2005；Munn 和 Tester，2008）。盐胁迫通过阻碍植物根系对必需营养元素的吸收和转移来影响植物的正常生长（Shrivastava 和 Kumar，2015），并且还会降低土壤中营养元素（包括 N、P、K、Ca、Mg、Fe、Cu、Zn 等）的可用性（Barea 等，2005；Moradi 等，2011）。有研究指出，在植物培养基中添加 NaCl 培养植物，随着培养基中 NaCl 浓度的增加，植物的根和芽中几乎所有的大量营养元素和微量营养元素的含量都明显降低（Paul 和 Lade，2014）。

一般来说，盐胁迫对植物造成的最重要的危害是营养失衡和（或）离子毒性。植物营养吸收的不平衡又会反过来抑制植物正常的生长发育，造成植物生长异常和衰退。因此，在盐胁迫条件下，增强对植物营养的管理是克服盐胁迫危害最实用和最简单的方法。植物的营养状况主要影响其对不利环境条件，特别是非生物逆境胁迫因子的适应能力。目前，一些研究表明，暴露在逆境环境因子胁迫下的植物需要更多的养分来适应和抵御胁迫造成的危害，以减少这些胁迫因素对植物生长的负面影响。除了对植物细胞产生离子毒性之外，土壤溶液中的养分失衡还会影响植物吸收水分的能力，从而影响植物的正常生长和生存（Etesami，2018；Etesami 和 Maheshwari，2018）。目前，已经有研究指出，耐盐微生物能够在高盐和高渗条件下良好生长，并且能够满足盐胁迫下植物对营养元素的正常需求（图 5.1 和表 5.1）（Etesami 和 Beattie，2018）。

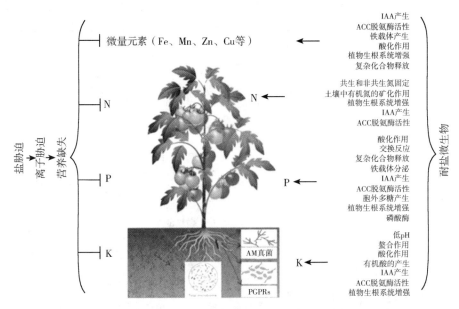

图 5.1 植物生长促生菌和菌根真菌在盐胁迫下促进养分获取的作用概述

表 5.1 植物生长促生菌和丛枝菌根真菌在提高盐胁迫植物养分有效性方面的潜在应用

营养元素	试验植物	微生物	微生物对盐胁迫植物的改良作用	参考文献
N	木豆（Cajanus cajan L.）	丛枝菌根真菌（Glomus mosseae）	与丛枝菌根真菌的共生关系使盐胁迫下根瘤的植株干质量和固氮能力显著提高。接种丛枝菌根真菌的植物在盐碱和非盐碱条件下的海藻糖酶活性较低	Garg 和 Chandel（2011）
P、K、Zn 和 Cu	阿拉伯金合欢（Acacia nilotica）	束状球囊霉菌（Glomus fasciculatum）	与非菌根植物相比，菌根植物在所有盐度水平下都保持了更大的根和茎生物量。接种丛枝菌根真菌的植物比未接种丛枝菌根真菌的植物具有更高的 P、Zn 和 Cu 浓度。菌根植物在不同盐度下积累了更高的 K 浓度。接种菌根的植株芽组织中 Na 含量较低。菌根真菌缓解了盐渍土对植物生长的不利影响，这可能主要与改善磷营养有关。在盐胁迫条件下，菌根植物根冠组织中 K^+/Na^+ 比值的提高可能有助于保护 K 介导的酶过程的破坏	Giri 等（2007）

(续表)

营养元素	试验植物	微生物	微生物对盐胁迫植物的改良作用	参考文献
P、K 和 Zn	大豆（*Glycine max* L.）	丛枝菌根菌（*Glomus etunicatum*）	接种丛枝菌根菌（*Glomus etunicatum*）的大豆植株鲜重、干重、根脯氨酸，N、K、Zn 含量均显著高于未接种丛枝菌根真菌的植株	Sharifi 等（2007）
N、P 和 Mg	印度田菁（*Sesbania aegyptiaca*）和大花田菁（*Sesbania grandiflora*）	大果球囊霉菌（*Glomus macrocarpum*）	菌根苗比盐渍土中的非菌根苗具有更高的根冠干重。接种菌根的幼苗叶片叶绿素含量高于未接种菌根的幼苗。菌根植物的根瘤数明显高于非菌根植物。菌根苗组织中 P、N、Mg 含量显著高于非菌根苗，而 Na 含量显著低于非菌根苗	Giri 和 Mukerji（2004）
N	木豆［*Cajanus cajan*（L.）Mill sp.］	丛枝菌根真菌（*Glomus mosseae*）	丛枝菌根真菌显著改善盐胁迫下的结瘤、血红蛋白含量和固氮酶活性。菌根胁迫下超氧化物歧化酶、过氧化氢酶、过氧化物酶、谷胱甘肽还原酶活性显著升高	Garg 和 Manchanda（2008）
P	苦味叶下珠（*Phyllanthus amarus*）	不动杆菌（*Acinetobacter*）和芽孢杆菌（*Bacillus* sp.）	与未接种的对照相比，单独或联合使用两种细菌都能提高活力指数、发芽率、植物生物量、磷含量、植物酚含量、自由基清除和抗氧化活性	Joe 等（2016）
N、K 和 P	小麦（*Triticum aestivum* L.）	短小芽孢杆菌（*Bacillus pumilus*）、海水芽孢杆菌（*Bacillus aquimaris*）、亚砷酸芽孢杆菌（*Bacillus arsenicus*）、节杆菌（*Arthrobacter* sp.）、蜡状芽孢杆菌（*Bacillus cereus*）、门多萨假单胞菌（*Pseudomonas mendocina*）和枯草芽孢杆菌（*Bacillus subtilis*）	接种 PGPR 后，小麦根系干重、地上部生物量、脯氨酸、可溶性总糖积累量，叶片 N、K、P 含量均显著高于未接种植物生长促生菌的植株。结果表明，播后 60d 和 90d，小麦叶片中 Na 含量下降幅度最大，达 23% 左右，比对照增产 17.8% 左右	Upadhyay 和 Singh（2015）

(续表)

营养元素	试验植物	微生物	微生物对盐胁迫植物的改良作用	参考文献
N 和 P	比吉洛氏海蓬子 (*Salicornia bigelovii*)	盐生固氮螺菌 (*Azospirillum halopraeferens*)、巴西固氮螺菌 (*Azospirillum brasilense*)、河口弧菌 (*Vibrio aestuarianus*)、解蛋白弧菌 (*Vibrio proteolyticus*)、地衣芽孢杆菌 (*Bacillus licheniformis*) 和叶杆菌 (*Phyllobacterium* sp.)	生长季末，种子中的氮和蛋白质含量显著增加。细菌处理后植物叶片中磷含量显著增加	Bashan 等 (2000)
P	海榄雌（咸水矮让木）(*Avicennia marina*)	墓画大洋芽孢杆菌 (*Oceanobacillus picturae*)	与未接种的植株相比，墓画大洋芽孢杆菌显著增加有效磷，降低 pH，显著提高根和芽的养分吸收参数，增加茎周、木质部导管数目、平均木质部直径和水力加权木质部导管直径	El-Tarabily 和 Youssef (2010)
N、P、K 和 Ca	落花生 (*Arachis hypogaea* L.)	苏拉氏短杆菌 JG-06 [*Brachybacterium saurashtrense* (JG-06)]、乳酪短杆菌 JG-08 (*Brevibacterium casei* (JG-08)) 和血杆菌 JG-11 (*Haererohalobacter* (JG-11)	接种植株的株长、根长、茎干重、根干重和总生物量均显著高于未接种植株。接种植物生长促生菌的植株长势好，水分充足，而未接种植物生长促生菌的植株叶片在 100 mmol/L NaCl 下有失水现象。接种植株的 K^+/Na^+ 比值和 Ca、P、N 含量也较高	Shukla 等 (2012)
N、P 和 K	玉米 (*Zeya mays* L.)	假单胞菌 (*Pseudomonas*) 和肠杆菌 (*Enterobacter* spp.)	在胁迫条件下，与对照相比，接种植株对 N、P 和 K 的吸收更多，K^+/Na^+ 比值更高	Nadeem 等 (2009)
N	花生	克雷伯氏菌 (*Klebsiella*)、假单胞菌 (*Pseudomonas*)、土壤杆菌 (*Agrobacterium*) 和苍白杆菌属 (*Ochrobactrum* sp.)	与未接种的对照相比，总氮含量显著增加（高达 76%）。盐胁迫下，接种花生幼苗比未接种花生幼苗保持离子内稳态，积累较少的活性氧，生长增强	Sharma 等 (2016)

(续表)

营养元素	试验植物	微生物	微生物对盐胁迫植物的改良作用	参考文献
K 和 Ca	小麦	地衣芽孢杆菌 HSW-16（Bacillus licheniformis HSW-16）	地衣芽孢杆菌 HSW-16（Bacillus licheniformis HSW-16）的接种保护了小麦植株免受 NaCl 的生长抑制，提高了植株的根长、茎长、鲜重和干重（6%~38%）。对植物样品进行离子分析表明，在不同浓度的 NaCl 溶液中，细菌接种降低了植物体内 Na^+ 含量（51%）的积累，提高了植物体内 K^+（68%）和 Ca^{2+} 含量（32%）	Singh 和 Jha（2016）
N、K、P、Ca、Mg、S、Mn、Cu 和 Fe	草莓	枯草芽孢杆菌 EY2（Bacillus subtilis EY2）、萎缩芽孢杆菌 EY6（Bacillus atrophaeus EY6）、球形芽孢杆菌（Bacillus spharicus）、克卢氏葡萄球菌 EY37（Staphylococcus kloosii EY37）和玫瑰考克氏菌 EY43（Kocuria Erythromyxa EY43）	接种植物生长促生菌能显著提高草莓植株的生长、叶绿素含量、营养元素含量和产量。接种植株中 N、K、P、Ca、Mg、S、Mn、Cu 和 Fe 的含量最高	Karlidag 等（2013）
K、Ca、Mg 和 P	豌豆（Pisum sativum L. cv. Alderman）	争论贪噬菌 5C-2（Variovorax paradoxus 5C-2）	争论贪噬菌 5C-2（Variovorax paradoxus 5C-2）增加了 K、Ca、Mg 和 P 的吸收和根冠 K^+ 流，但降低了 Na^+ 流，增加了 Na^+ 在根中的沉积。因此，接种争论贪噬菌 5C-2（Variovorax paradoxus 5C-2）后，地上部 K^+/Na^+ 比值增加	Wang 等（2016）
N、P 和 K	小麦	恶臭假单胞菌（Pseudomonas putida）、阴沟肠杆菌属（Enterobacter cloacae）、无花果沙雷氏菌（Serratia ficaria）和荧光假单胞菌（Pseudomonas fluorescens）	接种植物生长促生菌能显著提高小麦的生长和产量。接种植株也改善了小麦植株的营养状况。接种植株的 Na 含量低，N、P、K 含量升高	Nadeem 等（2013）

(续表)

营养元素	试验植物	微生物	微生物对盐胁迫植物的改良作用	参考文献
N、P、K、Ca 和 Fe	棉花	居植物柔武氏菌 (Raoultella planticola)	棉花植株 N、P、K、Ca 和 Fe 的积累量在居植物柔武氏菌 (Raoultella planticola) 处理下显著增加，而 Na^+ 的吸收量则下降	Wu 等（2012）
K、Ca 和 Mg	玉米 (Zea mays L)	固氮菌 C5 (Azotobacter sp. C5)	接种植株叶绿素、Mg、K、Ca、总多酚、脯氨酸含量增加	Rojas Tapias 等（2012）

5.3 盐胁迫下植物生长促生菌和菌根真菌介导的氮素利用有效性的提高

氮素（N）是植物生长过程中最重要的营养元素。尽管大气中氮的含量占绝大多数，约达到78%，但土壤中的氮素主要以有机态存在，植物无法直接吸收利用。土壤中氮素的缺乏会导致植物生长异常和产量的显著下降。如上所述，土壤盐渍化对植物生长的不利影响之一是改变植物对营养元素的吸收（随着土壤溶液中盐分的增加，植物对营养元素的吸收减少）（Cheng-Song 等，2010；Rawal 和 Kuligod，2014）。有研究表明，随着土壤盐渍化程度的升高，植物对氮素的吸收和（或）积累能力逐渐降低（Feigin，1985）。而能够在一定程度上减轻盐胁迫对植物生长的有害影响，提高农作物产量的一个有效途径就是使用化肥。根据 Rawal 和 Kuligod（2014）的研究，在盐胁迫条件下，大量施用氮肥可以明显提高农作物的产量。一般来说，在大多数盐渍化土壤和干旱、半干旱地区的土壤中，可供植物吸收和利用的氮素的确是短缺的，而这种短缺并不能仅通过使用化肥来弥补。另外，通过生物固氮作用（Biological nitrogen fixation，BNF）在土壤中固定氮素供农作物利用是一种比施用化学肥料更为有效和具有可持续发展潜力的方式（Shamseldin 和 Werner，2005）。大气中的氮气通过生物固氮作用转化为能够被植物利用的形式，这对生长在盐渍化土壤上的农作物很有利。植物生长促生菌的生物固氮作用是利用一种称固氮酶的复合酶系统，将大气中的氮气通过一系列生物和化学反应，转化成可供植物吸收利用的铵根离子（NH_4^+）（Kim 和 Rees，1994）。生物固氮作用这一过程对生态系统和盐胁迫下农作物的重要性在于节省化肥的使用，并且还能够提供满足植物生长所需要的氮素营养，从而提高植物对盐胁迫的耐受性，促进农作物的生长，最终提高粮食产量。此外，与使用化肥不同的是，通过生物固氮作用向盐渍化土壤中输入氮素，即便过量输入也不会造成环境问题。同时，有研究指出，随着豆科植物种植的增加，土壤中氮素的含量会更加丰富，豆科植物种植的氮素剩余效应将有助于后续种植农作物的生长（Shamseldin 和 Werner，2005）。因此，在规划农业的可持续发展系统时，使用这种具有生物固氮作用的微生物和（或）植物，尤其是面对盐渍化土壤，是非常必要的。

众所周知，大多数豆科植物属于甜土植物范围，对盐胁迫十分敏感（Dulormne 等，2010；Garg 和 Chandel，2011；Jebara 等，2010；Khadri 等，2006；López 和 Lluch，2008）。同时，这些豆科的健壮生长和高产需要依赖于大量的氮肥供给（Chalk 等，2010；Jebara 等，2010）。盐胁迫除了对植物的生长产生负面影响外，还能够对植物对氮素的吸收过程产生不利影响，如抑制植物根瘤的生长（Abdelmoumen 和 El Idrissi，2009），抑制根瘤和植物根系的共生活动（Dulormne 等，2010；Jebara 等，2010），影响固氮菌固氮酶的活性（Jebara 等，2010），阻碍根瘤组织呼吸等（Dulormne 等，2010；López 和 Lluch，2008）。在盐胁迫条件下，豆科农作物植株中的豆类血红蛋白、乙炔还原酶活性以及植株的氮含量都会受到抑制（Garg 和 Chandel，2011），这些豆科农作物包括大豆（van Hoorn 等，2001）、豇豆属曼戈（*Vigna mungo* L. Hepper）（Mensah 和 Ihenyen，2009）、菜豆（*Phaseolus vulgaris*）、蚕豆（*Vicia faba*）等（Rabie 和 Almadini，2005；Egamberdieva 等，2013；Paul 和 Lade，2014）。

盐胁迫除了影响农作物对氮素养分的吸收和利用外，还能够影响根瘤菌与豆科农作物共生体系的建立以及固氮酶等的活性，生物固氮作用过程中的固氮酶复合体对盐胁迫非常敏感，功能易受盐胁迫的影响（Jebara 等，2010）。众所周知，逆境环境胁迫，尤其是盐胁迫对根瘤菌在农作物根际定殖的数量、生物固氮作用的能力以及产生一些对农作物有益处的代谢物的能力等方面都有很大的负面影响（Duzan 等；Sánchez Porro 等，2009；Upadhyay 等，2009；Yoon 等，2001）。耕作土壤中含有各种类型的根瘤菌，其生物固氮能力以及应对不同环境胁迫的能力都不尽相同，因此对宿主农作物生长的促进效果也大不一样（Rehman 和 Nautiyal，2002）。为了获得最佳的促进农作物生长的根瘤菌作为生物肥料，必须对各个耕作地区和不同农作物中的根瘤菌菌株进行分离和鉴定。同时，还需要通过试验等手段摸索证实这些分离出来的、对农作物生长有促进作用的根瘤菌发挥正常作用所需要的不同的气候条件。因此，筛选和鉴定能在逆境胁迫条件下正常生长、促进农作物生长和提高最终产量的耐盐根瘤菌，可能是解决土壤盐渍化威胁农作物安全生产的有效手段。

近年来，菌根真菌与根瘤菌联合接种农作物是能够提高盐胁迫下农作物生长和微生物固氮能力的有效策略（Chalk 等，2006；Franzini 等，2010；Garg 和 Chandel，2011）。丛枝菌根真菌在进化过程中演化出了多种不同的机制来抵御盐胁迫对农作物根瘤生长和固氮酶活性等的有害影响，从而减缓了盐胁迫引起的根瘤衰老（Garg 和 Chandel，2011；Garg 和 Manchanda，2008、2009；Goss 和 De Varennes，2002；Patreze 和 Cordeiro，2004）。这些机制包括：(i) 通过促进宿主农作物的生长和增加根系生物量来提高宿主农作物对盐胁迫的耐受能力（Garg 和 Chandel，2011）；(ii) 增强宿主农作物对氮素和磷素营养以及其他低迁移率营养素，例如 Fe、Cu 和 Zn 等的吸收（Kaya 等，2009a；Miransari 和 Smith，2008）；(iii) 减少遭受盐胁迫农作物对 Na^+ 的吸收（Al-Karaki，2006；Chakraborty 等，2008；Giri 等，2007）；(iv) 通过抑制丛枝菌根真菌中的海藻糖酶活性，并在细胞中增加总可溶性糖的含量，来刺激在盐胁迫条件下丛枝菌根真菌细胞中海藻糖的积累。海藻糖是一种能够起到渗透调节保护作用的二糖碳水化合物，能够通过渗透调节作用来增强农作物耐受盐胁迫的程度（Garg 和 Chandel，2011）。

遭受盐胁迫的农作物增强对盐胁迫耐受能力的一个重要机制就是在体内积累一些小分子有机化合物，这些物质与胞内各种生理活动具有很强的相容性，即浓度大小不影响细胞的正常生理活动，这类化合物统称为渗透调节保护剂（Contreras-Cornejo 等，2009；Cortina 和 Culiáñez-Macià，2005）。这些渗透调节保护剂在多种非生物逆境胁迫环境中，尤其是盐胁迫中对增强农作物的抗逆性发挥着重要的作用（Contreras-Cornejo 等，2009；Elbein 等，2003；Fernandez 等，2010；Garg 和 Chandel，2011）。海藻糖（α-D-吡喃葡糖基-1,1-α-D-吡喃葡糖苷）即是在与豆科植物共生的固氮细菌（如根瘤菌）中分离鉴定到的一种渗透调节保护物质（Müller 等，2001）。海藻糖等具有相容性的渗透调节保护物质在维持根瘤菌的固氮作用、调节固氮酶复合体的活性，以及对提高整个农作物对干旱胁迫和盐胁迫的耐受性等方面起着非常重要的作用（Farías-Rodríguez 等，1998；Garg 和 Chandel，2011；López 等，2008；Zacarías 等，2004）。目前，已经有研究证明，丛枝菌根真菌在盐胁迫条件下能够促进农作物植株氮素含量的增加。例如，Giri 和 Mukerji（2004）在一项研究中表明，在盐胁迫条件下接种丛枝菌根真菌的植物比未接种的植物能够吸收更多的氮素，并且根系会产生更多的根瘤。盐胁迫条件下，植株氮素营养吸收的增加有助于农作物减少对 Na^+ 的吸收。农作物植株既可以保持稳定的叶绿素含量，又可以因为减少对 Na^+ 的吸收从而减少盐胁迫对光合作用的抑制，从而有利于植物的正常生长（Giri 和 Mukerji，2004）。另外，丛枝菌根真菌结瘤和固氮能力的提高可能是因为接种丛枝菌根真菌提高了农作物对磷素的吸收和利用，同时在根系土壤环境中的一些农作物生长所必需的微量营养素也变得易于被吸收,既能够促进植物的生长,又对丛枝菌根真菌和农作物互作的固氮系统产生了积极的影响(Founue 等,2002)。

非生物逆境胁迫等环境条件会影响根瘤菌菌株胞外多糖的产生。在逆境条件下根瘤菌会选择产生更多的胞外多糖以抵御不利的环境条件对生长造成的负面影响(Ashraf 等,2004;Delavechia 等,2003)。据报道,耐盐和耐旱的根瘤菌菌株能够产生更多的胞外多糖,通过胞外多糖含量的增加耐受不良的生长环境以维持自身的正常生长(Ashraf 等,2004;Delavechia 等,2003)。另外,根瘤菌在土壤中的存活情况还取决于这些菌株与宿主农作物结合并建立共生关系的能力。这其中,胞外多糖对宿主农作物识别根瘤具有特别重要的意义(Werner,1992)。根瘤菌胞外多糖在与宿主植物的相互识别和保护细菌细胞免受环境胁迫危害的过程中起着非常重要的作用(Werner,1992)。因此,能够产生胞外多糖的耐盐根瘤菌不仅能耐受更高浓度的盐胁迫,还能够比一般的耐盐根瘤菌更加成功地与宿主农作物建立共生关系。据报道,在盐渍化和干旱半干旱地区进行的一些试验中,从遭受盐渍化威胁的土壤中分离出的耐盐和耐干旱根瘤菌菌株与农作物的共生效率比其他耐逆性一般的根瘤菌高很多(Shamseldin 和 Werner,2004,2005)。

促进植物生长的基本营养元素是由植物的根系从土壤中吸收而来(Mills 等,1996),因此良好的根系生长是促进植物生长和发育的先决条件。植物根系的形态学改变在应对逆境胁迫所产生的应激反应机制中起着非常重要的作用(Carmen 和 Roberto,2011)。植物生长促生菌能够通过产生植物激素(如吲哚-3-乙酸和1-氨基环丙烷-1-羧酸脱氨酶)来诱导植物根系的生长。除了能够产生吲哚-3-乙酸之外,植物生长促生菌产生的赤霉素和其他类型的植物生长调节剂也都能够促进植物根长、根表面积和根尖数量的增加,从而增强植物对营养元素的吸收,进而改善逆境胁迫条件下植物的生长状态(Egamberdieva 和 Kucharova,2009;Etesami 和 Maheshwari,2018)。植物生长促生菌介导的宿主农作物根系增殖的现象已经得到了很多研究的证明,而且农作物根系的增殖也能够在抵御盐渍化胁迫中起作用(Diby 等,2005)。据报道,接种不同种类的植物生长促生菌也会导致农作物产生更多的侧根和根毛,从而增强农作物本身对非生物逆境胁迫的耐受性(Etesami 和 Maheshwari,2018;Paul 和 Lade,2014)。促进农作物根系的生长会产生更大的根表面积,从而促进植物对水分和各种营养元素的吸收(Diby 等,2005;Paul 和 Sarma,2006),也有利于缓解植物遭受各种逆境胁迫反应(Chakraborty 等,2006;Hamdia 等,2004;Long 等,2008;Paul 和 Sarma,2006)。吲哚-3-乙酸产生菌能够参与豆科植物根瘤菌根瘤形成的许多过程,包括参与根瘤细胞的形态建成,通过吲哚-3-乙酸的积累而参与根瘤的起始和分化,对形成的根瘤数目的多少,维管束的形成,以及细胞的分裂和分化等都有非常重要的作用（Etesami 等，2015；Etesami 和 Beattie，2017）。其中，形成的根瘤数目多少、维管束的形成以及根瘤细胞的分裂和分化过程对农作物根瘤的形成十分重要（Glick，2012；Theunis，2005）。此外，由于吲哚-3-乙酸产生菌能够增加农作物的根系，以及促进根瘤的形成和生长，因此可以为各种植物生长促生菌提供更多的活性位点，并且促进这一类有益微生物的定殖。

有研究表明，植物生长促生菌产生吲哚-3-乙酸的水平受到所处生长环境盐浓度的调节。低浓度盐胁迫（如在 100 mmol/L NaCl 胁迫水平）下，植物生长促生菌能够产生更多的吲哚-3-乙酸（Albacete 等，2008），而在较高浓度的盐胁迫水平（如 300 mmol/L NaCl）下，其产生的吲哚-3-乙酸则会受到抑制（Dunlap 和 Binzel，1996）。植物生长促生菌产生的吲哚-3-乙酸水平的高低会影响农作物根系细胞的生长和伸长。众所周知，从盐胁迫环境中分离到的细菌能够在盐分存在的情况下良好地生长，并产生大量的吲哚-3-乙酸（Etesami 和 Beattie，2018）。Sadeghi 等（2012）在一项研究中证明，链霉菌（*Streptomyces*）分离物能够促进小麦的生长，进一步的试验证实，分离物中含有大量的吲哚-3-乙酸，说明链霉菌在盐胁迫下具有产生吲哚-3-乙酸的能力。另外，吲哚-3-乙酸能够参与植物侧根的发育，因此有研究指出，通过提供吲哚-3-乙酸来调节植物侧根的生长能够有效地提高植物耐受逆境胁迫的能力（Bian 等，2011；Marasco 等，2012；Naz 等，2009；Piccoli 等，2011；Sgroy 等，2009；Siddake 等，2010；Tiwari 等，2011）。节杆菌（*Arthrobacter* sp.）、拉齐斯庞氏硝化菌（*Nitrinicola lacisaponensis*）、短杆菌（*Brachybacterium* sp.）、苏拉氏短杆菌（*Brachybacterium saurashtrense*）、乳酪短杆菌（*Brevibacterium casei*）、耐盐芽孢杆菌（*Bacillus halotolerans*）、血杆菌（*Haererohalobacter* sp.）、克雷伯氏菌（*Klebsiella* sp.）、假单胞菌（*Pseudomonas* sp.）、施氏假单胞菌（*Pseudomonas stutzeri*）、类产碱假单胞菌（*Pseudomonas pseudoalcaligenes*）、恶臭假单胞菌（*Pseudomonas putida*）、门多萨假单胞菌（*Pseudomonas mendocina*）、根癌土壤杆菌（*Agrobacterium tumefaciens*）、人苍白杆菌（*Ochrobactrum anthropi*）、芽孢杆菌（*Bacillus* sp.）、蜡状芽孢杆菌（*Bacillus cereus*）、短小芽孢杆菌（*Bacillus pumilus*）、简单芽孢杆菌（*Bacillus simplex*）、蕈状芽孢杆菌（*Bacillus mycoides*）、海水芽孢杆菌（*Bacillus aquimaris*）、地衣芽孢杆菌（*Bacillus licheniformis*）、蕈状芽孢杆菌（*Bacillus mycoides*）、枯草芽孢杆菌（*Bacillus subtilis*）、海水芽孢八叠球菌（*Sporosarcina aquimarina*）、黏质沙雷氏菌（*Serratia marcescens*）、普城沙雷菌（*Serratia plymuthica*）、深海螺旋菌（*Thalassospira permensis*）、不动杆菌（*Acinetobacter*）、微杆菌属（*Microbacterium* sp.）、氧化微杆菌（*Microbacterium oxydans*）、链霉菌属（*Streptomyces* sp.）、亲和素链霉菌（*Streptomyces avidinii*）、欧洲疮痂链

霉菌（*Streptomyces europaeiscabiei*）、脱叶链霉菌（*Streptomyces exfoliatus*）、赭褐链霉菌（*Streptomyces umbrinus*）、灰平链霉菌（*Streptomyces griseoplanus*）、红球菌（*Rhodococcus*）、红城红球菌（*Rhodococcus erythropolis*）、藤黄微球菌（*Micrococcus luteus*）、嗜盐土地芽孢杆菌（*Terribacillus halophilus*）、诺卡氏菌（*Nocardia* sp.）、盐单胞菌属（*Halomonas* sp.）、伸长盐单胞菌（*Halomonas elongata*）、广盐盐单胞菌（*Halomonas eurihalina*）、西奈盐单胞菌（*Halomonas sinaiensis*）、喜海水盐单胞菌（*Halomonas halmophila*）、嗜盐单胞菌（*Halomonas ilicicola*）、印度盐单胞菌（*Halomonas indalina*）、变异盐单胞菌（*Halomonas variabilis*）、新疆盐单胞菌（*Halomonas xinjiangensis*）、泰香盐单胞菌（*Halomonas taeheungii*）、嗜盐芽孢杆菌（*Halobacillus trueperi*）、喜盐涅斯特连科氏菌（*Nesterenkonia halobia*）、墓画大洋芽孢杆菌（*Oceanobacillus picturae*）、克锡勒氏菌（*Kushneria*）、橄榄枝芽孢杆菌（*Virgibacillus olivae*）、纺锤形赖氨酸芽孢杆菌（*Lysinibacillus fusiformis*）、木糖氧化产碱菌（*Achromobacter xylosoxidans*）、中慢生根瘤菌（*Mesorhizobium* sp.）、刘志恒氏菌（*Zhihengliuella* sp.）、溶藻弧菌（*Vibrio alginolyticus*）和阪崎克罗诺杆菌（*Cronobacter sakazakii*）等都是从盐生植物中分离出的，能够耐受盐胁迫，且可以产生吲哚-3-乙酸的一些植物生长促生菌（Gontia 等, 2011; Mapelli 等, 2013; Sgroy 等, 2009; Sharma 等, 2016; Shukla 等, 2012; Tiwari 等, 2011）。在这些菌种的报道中均有提到，耐盐植物生长促生菌具有产生吲哚-3-乙酸能力的根瘤菌，有助于提高植物对盐胁迫的耐受性，增加植物根系的生长，并提高对土壤中营养元素的吸收，从而增加植物体内氮等元素的含量。

  植物在遭受盐胁迫等逆境胁迫之后的应激反应之一就是细胞内乙烯水平的升高（乙烯胁迫）。乙烯胁迫能够降低植物的生长速率，并引起衰老反应，最终会导致农作物产量的降低，严重时甚至引起农作物的死亡，导致绝产（Etesami 和 Maheshwari, 2018; Glick, 2005）。Siddake 等（2011）还指出，因盐胁迫等逆境胁迫农作物产生的乙烯能够抑制根系的生长，从而减少了农作物对水分和营养元素等的吸收，最终农作物的生长受到抑制。此外，先前的研究人员还发现，添加外源乙烯也能够严重抑制豆科植物根系上根瘤的形成，并影响其功能的发挥（Peters 和 Crist Estes, 1989）。乙烯能够抑制农作物根系的生长和伸长，因此对能够形成根瘤菌的大多数豆科农作物而言，细胞内尤其是根系细胞内乙烯含量的提高会抑制大多数豆科植物形成根瘤（Etesami 等, 2015; Sugawara 等, 2006）。此外，乙烯还是一种诱导植物系统性抗性的信号分子，乙烯的产生能够降低各种植物生长促生菌在农作物的内生定殖（Iniguez 等, 2005）。乙烯的过量产生也会抑制农作物根系在伸长、侧根生长和根毛形成等（Belimov 等, 2009; Mayak 等, 2004; Saleem 等, 2007），其结果就是导致农作物根系上的结瘤数减少。一些土壤细菌能够产生1-氨基环丙烷-1-羧酸脱氨酶，这种酶能够将乙烯合成的底物1-氨基环丙烷-1-羧酸转化为α-酮丁酸盐和铵。1-氨基环丙烷-1-羧酸脱氨酶产生菌就可以通过降低植物中1-氨基环丙烷-1-羧酸的含量，减少植物中因逆境胁迫而过量产生的乙烯（Glick，2014）。很多微生物都能够产生1-氨基环丙烷-1-羧酸脱氨酶，其特性和功能在许多土壤微生物（如细菌、真菌和内生菌）中都得到了广泛的研究。同时，这一特性也是植物生长促生菌中最为常见的一个基本功能（Etesami 和 Maheshwari，2018；Glick，2005）。这些植物生长促生菌包括无色杆菌属（Achromobacter）、嗜酸菌属（Acidovorax）、产碱杆菌属（Alcaligenes）、肠杆菌属（Enterobacter）、克雷伯氏菌属（Klebsiella）、亚甲基杆菌（Methylobacterium）、假单胞菌属（Pseudomonas）、根瘤菌属（Rhizobium）和贪噬菌（Variovorax）等（Esquivel Cote 等，2010）。Ahmed 等（2004）发现，1-氨基环丙烷-1-羧酸脱氨酶产生菌能够降低农作物中乙烯的水平，从而提高了小麦的产量、根重、根长以及氮素吸收能力，并且还提高了秸秆的产量。

目前已经证明，盐胁迫会导致植物生长促生菌失去产生1-氨基环丙烷-1-羧酸脱氨酶的能力（Upadhyay 等，2009）。然而，也有研究结果表明，有大量能够产生1-氨基环丙烷-1-羧酸脱氨酶的耐盐植物生长促生菌均能够在盐胁迫环境中很好地生存，其产生1-氨基环丙烷-1-羧酸脱氨酶降解1-氨基环丙烷-1-羧酸的特性有助于农作物通过降低细胞中乙烯的水平来克服逆境胁迫对正常生长的影响（Mayak 等，2004）。从盐生植物和盐渍化环境中分离出的耐盐细菌也能够产生1-氨基环丙烷-1-羧酸脱氨酶，其促进农作物生长的特性也有很多研究与报道（Jha 等，2012；Siddake 等，2010；Zhou 等，2017）。盐生植物的根系可能是产1-氨基环丙烷-1-羧酸脱氨酶细菌的候选定殖位置，在盐生植物根系定殖的这类细菌能够促进农作物的生长和耐盐能力（Etesami 和 Beattie，2018）。例如，从臂状盐角草（Salicornia brachiata）根系分离出的新型重氮营养型耐盐菌具有1-氨基环丙烷-1-羧酸脱氨酶的活性，分离出的根际菌包括苏拉氏短杆菌（Brachybacterium saurashtrense）、乳酪短杆菌（Brevibacterium casei）、阪崎克罗诺杆菌（Cronobacter sakazakii）、血杆菌（Haererehalobacter）、盐单胞菌属（Halomonas）、中慢生根瘤菌（Mesorhizobium alhagi）、假单胞菌（Pseudomonas）、放射型根瘤菌（Rhizobium radiobacter）、弧菌（Vibrio）和刘志恒氏菌（Zhihengliuella）（Jha 等，2012）。此外，能够产生1-氨基环丙烷-1-羧酸脱氨酶的耐盐植物生长促生菌还包括云南微球菌（Micrococcus yunnanensis）、莱比托游动球菌（Planococcus rifietoensis）、争论贪噬菌（Variovorax paradoxus）、苏拉氏短杆菌（Brachybacterium saurashtrense）、克雷伯氏菌属（Klebsiella sp.）、假单胞菌属（Pseudomonas sp.）、施氏假单胞菌（Pseudomonas stutzeri）、恶臭假单胞菌（*Pseudomonas putida*）、根癌土壤杆菌（*Agrobacterium tumefaciens*）、人苍白杆菌（*Ochrobactrum anthropi*）、黏质沙雷氏菌属（*Serratia marcescens*）、普城沙雷菌属（*Serratia plymuthica*）、深海螺旋菌（*Thalassospira permensis*）、微杆菌属（*Microbacterium sp.*）、芽孢杆菌属（*Bacillus sp.*）、蜡状芽孢杆菌属（*Bacillus cereus*）、地衣芽孢杆菌属（*Bacillus licheniformis*）、短小芽孢杆菌（*Bacillus. Pumilus*）、枯草芽孢杆菌（*Bacillus subtilis*）、蕈状芽孢杆菌（*Bacillus mycoides*）、普城沙雷菌（*Serratia plymuthica*）、氧化微杆菌（*Microbacterium oxydans*）、微杆菌属（*Microbacterium sp.*）、链霉菌属（*Streptomyces sp.*）、红球菌属（*Rhodococcus sp.*）、红城红球菌（*Rhodococcus erythropolis*）、诺卡氏菌（*Nocardia sp.*）、盐单胞菌属（*Halomonas*）、新疆盐单胞菌（*Halomonas xinjiangensis*）、耐盐短杆菌（*Brevibacterium halotolerans*）、木糖氧化产碱菌（*Achromobacter xylosoxidans*）、刘志恒氏菌（*Zhihengliuella*）、中慢生根瘤菌（*Mesorhizobium alhagi*）、短杆菌（*Brachybacterium sp.*）、溶藻弧菌（*Vibrio alginolyticus*）、乳酪短杆菌（*Brevibacterium casei*）、阪崎克罗诺杆菌（*Cronobacter sakazakii*）、类产碱假单胞菌（*Pseudomonas pseudoalcaligenes*）、血杆菌（*Haererehalobacter sp.*）、成团泛菌（*Pantoea agglomerans*）、栖稻泛菌（*Pantoea oryzihabitans*）和盐单胞菌属（*Halomonas*）等（Gontia 等，2011；Jha 等，2012；Mapelli 等，2013；Sgroy 等，2009；Sharma 等，2016；Szymańska 等，2016；Teng 等，2010；Zhou 等，2017）。这些耐盐植物生长促生菌都是从盐生植物中分离得到的。一般而言，耐盐植物生长促生菌产生吲哚-3-乙酸和1-氨基环丙烷-1-羧酸脱氨酶是生长在盐渍土环境中的农作物抵御盐胁迫危害的一种重要的耐盐工具（Etesami 和 Bettie，2018；Etesami 和 Maheshwari，2018）。

5.4 植物生长促生菌和丛枝菌根真菌介导的盐胁迫下磷利用有效性的提高

磷（P）是植物和微生物生长发育的必需大量元素之一。磷的缺乏会限制植物正常的生长和发育。与氮素可以通过固定空气中的氮气（N_2）不同，磷素营养并不能从大气中获得（Katznelson 等，1962）。农作物根、茎、芽的分化发育和生长成熟，豆类作物中氮素的固定，细胞中大分子物质的生物合成，农作物产品质量的改善，生命代谢中能量的生产和转移，光合作用以及农作物的抗病性均与磷素营养有关（Roychoudhury 和 Kaushik，1989）。土壤中磷素的供应和农作物对磷营养元素吸收有效性的缺乏，也会严重限制农作物对氮素的吸收和利用（Pereira 和 Bliss，1989）。磷酸盐化合物与氮素相比不易溶于水，因此磷素营养不易从土壤中浸出。通常情况下，土壤中磷素的含量很低。这是因为土壤中的磷素营养容易与土壤中其他成分发生化学反应，生成一些相对不易溶于水的化合物，因此不能够被植物直接吸收利用。所以磷素营养含量的高低是评价土壤是否肥沃的重要指标之一，对土壤肥力非常重要。然而，值得关注的是，土壤中磷素的总含量要远远大于速效磷的含量，但速效磷在植物生长过程中发挥的作用比总磷更重要。磷素在土壤中的存在方式有很多种，主要存在于矿物土壤中或吸附在土壤矿物表面，有时也作为一种相对可溶的沉积物或以有机态形式吸附在土壤矿物表面。磷素有时还可以作为微生物生物量的一部分或与一些有机物等混合存在。农业应用的磷素是从磷矿中提取的，以磷肥的形式在生产中利用，目前超过80%的磷矿储量都用于肥料的生产（Owen 等，2015）。磷矿是一种不可再生资源，大量的挖掘开采和利用会使磷矿资源越来越少，因此投入农业领域的磷肥也会受到抑制，成本也会越来越高，这对于世界各个地方的农业生产都非常严峻（Van Vuuren 等，2010）。由于磷矿资源的有限性，磷矿的开采和磷肥在农业上的推广利用是不可持续的，这将使磷肥的生产和利用在不久的将来面临困境。此外，在生产实践中，磷的有效回收率通常为 10%~25%。即使用到土壤中的磷肥约有 75% 通过与土壤中的金属离子（如 Fe^{3+}、Al^{3+} 和 Ca^{2+} 等）形成络合物从而转化成不溶于水的形式，难以被农作物吸收利用（Gyaneshwar 等，2002；Stevenson 和 Cole，1999）。

农作物从土壤溶液中吸收的磷素主要是土壤溶液中溶解的正磷酸盐的一价和二价离子（$H_2PO_4^-$ 和 HPO_4^{2-}）。然而这些离子在土壤溶液中的浓度通常都小于 0.1 mg/L，因此基本上都可以忽略不计（Marschner，1995）。所以，为了维持农作物正常的生长和生理需求，应该向土壤溶液中持续不断地补充磷肥，否则农作物将会因为没有足够的磷素供应，而对其正常生长造成影响。土壤溶液中各种形式磷离子的浓度与环境 pH 值密切相关。在酸性环境下，土壤溶液中 $H_2PO_4^-$ 离子含量增多；而在 pH 值接近甚至稍微高于 7 的环境时，$H_2PO_4^-$ 离子的含量则几乎没有，HPO_4^{2-} 离子的含量较多。根据多地、多点的测量和试验结果，土壤溶液的 pH 值为 5.5~7 时，土壤中有效磷的含量较多，基本上能够满足大多数农作物对磷素营养的需求。然而石灰性土壤中的磷很容易转化为不溶性的钙镁等磷酸盐化合物；而在酸性土壤中的磷素则容易与铁和铝发生反应，形成铁和铝的磷酸盐，从而降低农作物对磷素的正常吸收。一般来说，土壤溶液中磷素的不同形态受土壤自然理化性质等因素的影响，这些因素包括 pH 值、可溶性和交换性阳离子（主要是 Ca^{2+}、Mg^{2+} 和 Fe^{2+} 等）含量以及土壤颗粒类型等（Penfold，2001）。

盐胁迫降低土壤中农作物对磷的吸收或积累的方式有很多种，其中最主要的方式之一就是降低农作物对磷素吸收的有效性（Etesami 和 Maheshwari，2018；

Paul 和 Lade，2014）。这是因为磷酸根离子能够与 Ca^{2+} 发生化学反应，形成不易溶于水的磷酸盐沉淀（Navarro 等，2001；Parida 和 Das，2005；Rogers 等，2003）。

此外，盐渍化土壤中农作物对磷素利用有效性的降低可能是由于土壤溶液中离子强度效应降低了磷素的活性所致。另外一个方面，在土壤中磷素极易与钙素形成不易溶于水的 Ca-P 沉淀物质，并通过土壤颗粒的吸附过程限制磷素在土壤溶液中的浓度。在盐渍化土壤中，由于高浓度阳离子（如 Ca^{2+} 和 Mg^{2+}）的存在，磷素经常会在受盐渍化影响的土壤中与这些阳离子形成不溶性的阳离子-磷化合物沉淀（Etesami 和 Maheshwari，2018）。此外，很多研究人员的研究也表明，盐胁迫可以降低植物对磷素营养的吸收（Barea 等，2005；Patel 等，2010）。例如，一项研究表明，盐胁迫降低了观赏植物金合欢（*Acacia gerrardii*）对磷素营养的吸收（Patel 等，2010）。除与土壤溶液中的阳离子形成不溶性阳离子-磷化合物沉淀外，盐胁迫下植物对磷素吸收的减少还很有可能与根系的发育不良有关（Rawal 和 Kuligod，2014）。

在盐渍化土壤或非盐渍化土壤中，满足农作物磷需求的有效途径之一是施用化肥，但由于商业化肥资源有限，成本较高，并且施用化肥还会对自然环境造成严重危害，因此商业化肥料在盐渍化土壤中的应用并不利于可持续发展。同时，大部分磷肥作为农作物养分施入土壤之后还会在盐胁迫环境下固化为不溶性的阳离子-磷化合物沉淀等，在盐渍化土壤中，磷素肥料的利用效率会大打折扣。但是，在过去的几十年中，世界上农业生产中磷酸盐肥料的使用一直在增加。近几十年来，人们发现土壤微生物（细菌、真菌、放线菌，甚至藻类）能够促进土壤中磷酸盐沉淀的一些溶解过程，从而影响土壤溶液中磷的转化和存在状态，溶解态的磷素最终有利于农作物对磷素营养的吸收和利用（Richardson，2001；Sharma 等，2013）。

众所周知，不同的植物生长促生菌可以增加农作物对营养元素吸收的有效性。例如，溶磷菌（Phosphate-solubilizing microorganisms，PSM）能够通过不同的机制为农作物提供可吸收利用的磷素营养（Sharma 等，2013）。有证据表明，溶磷细菌（Phosphate-solubilizing bacteria，PSB）在土壤中能够将不溶性的磷素溶解，并将其转化为可供农作物吸收和利用的状态。同时，土壤中大量的微生物体量对磷素的应用也可以有效地阻止土壤中磷素的固定作用或沉淀物质的产生（Khan 等，2007）。

溶磷菌是土壤微生物种群中重要的组成部分（Kucey 等，1989）。在土壤中，细菌和真菌占溶磷菌的比例最高，因此它们被分为溶磷细菌（Phosphate-solubilizing bacteria，PSB）和溶磷真菌（Phosphate-solubilizing fungi，PSF）（Whitelaw

等，1997）。溶磷细菌能够溶解土壤中不溶性的磷素化合物沉淀，从这些化合物中释放出可供农作物吸收利用的磷素（Sharma 等，2013）。目前，发现的重要的溶磷细菌类型包括凝集肠杆菌（*Enterobacter agglomerans*）、环状芽孢杆菌（*Bacillus circulans*）、枯草芽孢杆菌（*Bacillus subtilis*）、巨大芽孢杆菌（*Bacillus megaterium* var. *phosphaticum*）、恶臭假单胞菌（*Pseudomonas putida*）、荧光假单胞菌（*Pseudomonas fluorescens*）、纹状体假单胞菌（*Pseudomonas striata*）、泛菌属（*Pantoea*）、多黏类芽孢杆菌（*Paenibacillus polymyxa*）、豆科三叶草根瘤菌（*Rhizobium leguminosarum* pv. *trifolii*）、运动黄色杆菌（*Xanthobacter agilis*）和克雷伯氏菌属（*Klebsiella* sp.）等，此外，放射型根瘤菌（*Rhizobium radiobacter*）、曲霉属真菌（*Aspergillus*）和拟青霉属（*Paecilomyces*）是溶磷真菌的重要类型（Sharma 等，2013）。另据报道，与溶磷细菌相比，溶磷真菌在磷素溶解的过程中能够产生更多的酸性物质（Venkateswaru 等，1984），因此其在土壤中发生溶磷作用的作用范围也比溶磷细菌更大一些，说明溶磷真菌对土壤中的磷素增溶作用可能更为重要（Kucey，1983）。然而，考虑到细菌的增殖周期更短一些，其在土壤中具有较高的种群密度，因此与真菌相比，细菌在土壤中的磷酸盐溶解效果方面可能更为有效。

微生物在农作物根际进行的磷素增溶作用是植物生长促生菌（包括溶磷细菌）向农作物提供养分的最重要机制。溶磷细菌在土壤磷素循环的3个主要组成部分中起着重要作用，这3个过程分别为：磷素的溶解-沉淀、磷素的吸收-解吸（土壤颗粒表面与磷素的相互作用）以及磷素的矿化-固定（通过生物转化将有机磷转化为无机磷）（Gyaneshwar 等，2002；Jones 和 Oburger，2011；Owen 等，2015）。溶磷细菌通过对磷素营养的增溶和矿化反应等与土壤相互作用，将磷素固定在周围环境的微生物生物量中，或形成相对有效的有机磷储存形态，这对土壤溶液可溶性磷库和土壤全磷之间的分配起着关键作用。一般来说，溶磷细菌通过产生和释放自身代谢物质（如有机酸等），利用这些有机酸上的羟基和羧基螯合与磷素形成不溶性沉淀的阳离子（主要是钙），将不溶性磷酸盐化合物转化为可溶性的形式（Miller 等，2010；Sagoe 等，1998）。

溶磷细菌可以通过多种机制将土壤中农作物无法吸收获取的磷素转化为可供农作物吸收利用的形态，这些机制主要包括以下几个方面。

（1）通过产生各种有机酸物质释放 H^+ 从土壤中吸收 NH_4^+（NH_4^+ 吸收过程中，H^+ 离子从农作物根中释放），从而对农作物根际区域微环境土壤进行酸化（Illmer 和 Schinner，1992）。

（2）交换反应。在这个过程中，溶磷细菌产生的小分子量有机阴离子（如琥珀酸根、柠檬酸根、葡萄糖酸根、α-酮葡萄糖酸根和草酸根等）在土壤颗粒磷素的结合位点上将磷素分子置换，形成可供农作物吸收利用的可溶性形态（Chen 等，2006；Jones 和 Oburger，2011；Zhang 等，2014）。溶磷细菌产生的有机酸/有机阴离子等（如乳酸、酒石酸、天冬氨酸、葡萄糖酸、草酸和柠檬酸）也能够降低农作物根际土壤微环境的 pH 值，并且释放的有机酸/有机阴离子等能够与不溶性的磷素竞争土壤颗粒上的吸附位点，并与结合磷素形成不溶性化合物的相关金属离子（主要是 Ca^{2+}、Al^{3+} 和 Fe^{3+}），形成可溶性金属络合物，从而将磷素酸根释放出来供农作物吸收和利用（Sharma 等，2013）。

（3）进行呼吸作用时释放二氧化碳。微生物通过呼吸作用释放的二氧化碳可以溶解于土壤孔隙中的水中，形成的碳酸可以通过降低农作物根系周围的 pH 值，有效地溶解固态不溶性磷素（Marschner，1995）。

（4）从土壤微粒中去除和同化磷酸盐，降低土壤溶液中的磷酸根水平，从而刺激磷酸钙（Ca-P）的间接溶解，以重新建立磷素平衡（Halvorson 等，1990）。

（5）通过释放螯合化合物，如铁载体、酶、酚、氨基酸、糖和有机酸阴离子等，增加土壤溶液中磷素的水平。这些化合物可以将与磷酸盐结合高氧化态金属离子还原为较低的氧化状态，从而释放出多余的磷酸根离子，达到增加土壤溶液中磷素含量的目的（Kim 等，1997）。

（6）微生物细胞碱性和酸性磷酸酶或植酸酶（水解酶）的分泌能够将有机态的磷酸盐转化为无机态的形式，从而增加磷酸盐的溶解性，释放更多可以被农作物直接吸收利用的磷素（Etesami 和 Maheshwari，2018；Franco Correa 等，2010；Gyaneshwar 等，2002；Owen 等，2015；Sharma 等，2013）。

据报道，有机态磷矿化形成无机态磷的主要机制是酸性磷酸酶的产生（Khan 等，2009；Sharma 等，2013）。芽孢杆菌属（*Bacillus*）、伯克霍尔德氏菌（*Burkholderia*）、肠杆菌属（*Enterobacter*）、假单胞菌属（*Pseudomonas*）、根瘤菌属（*Rhizobium*）、沙雷氏菌属（*Serratia*）和葡萄球菌属（*Staphylococcus*）等微生物菌种是参与有机态磷矿化的最重要的细菌类别（Shedova 等，2008）。

有研究表明，溶磷细菌还具有其他促进植物生长的特征（Etesami 和 Maheshwari，2018；Sharma 等，2013）。基于科学发现，植物生长促生菌对植物生长的促进机制可能有很多种（Etesami 和 Maheshwari，2018）。目前已有研究表明，在保证农作物有充足的磷素营养的情况下，农作物的生长和产量的增加与施用溶磷细菌无关（Ponguguzhali 等，2008）。事实上目前认为，溶磷菌能够通过多种途径控制农作物病原体的生长，并促进农作物本身的生长发育。这些途径包括产生铁载体、合成和分泌 1-氨基环丙烷-1-羧酸脱氨酶和吲哚-3-乙酸等（Han 和 Lee，2006；Sharma 等，2006）。例如，Gulati 等（2010）报道，属于不动杆菌属（*Acinetobacter*）的一个溶磷菌株 BIHB723 能够产生很多促进植物生长的代谢物质，包括吲哚-3-乙酸、1-氨基环丙烷-1-羧酸脱氨酶、铁载体和氨等。农作物的生长除了依赖溶磷菌在农作物根系微环境介质中溶解磷素外，还依赖于根系的生长。根系对农作物的生长和发育尤其重要，因为根系是农作物吸收水分和营养元素的主要场所（Etesami 和 Alikhani，2016；Fageria 和 Moreira，2011）。农作物对磷素的吸收与根系的根密度成正比，因此只要增加农作物的生根系统，就可以增加农作物从土壤中吸收磷素的能力（Grant 等，2001）。例如，Iqbal Hussain 等（2013）研究指出，农作物对磷素吸收的增加可能是由于特定的微生物促进了农作物根系的生长或促进了根毛的延长。有报道指出，吲哚-3-乙酸含量低至一定范围能够促进农作物主根延长；而当吲哚-3-乙酸的含量增加到一定程度之后，则会促进侧根的形成和数量增加（Xie 等，1996）。同时，保持低浓度的乙烯也可以促进农作物根系的生长。因此，能够产生 1-氨基环丙烷-1-羧酸脱氨酶的溶磷细菌能够通过将乙烯水解成氨和 α-酮丁酸盐这一生理过程来维持农作物体内较低的乙烯浓度，从而促进农作物根系的生长（Pereira 和 Castro，2014）。尤其是在营养缺乏（如磷素缺乏）的状态下，溶磷细菌中的这些促进农作物生长的机制发挥的促进农作物生长的作用则会更加明显（Etesami 和 Maheshwari，2018）。

  目前，科学研究已经基本证实，在包括盐胁迫在内的非生物逆境环境胁迫下，各种溶磷菌对农作物生长的影响和细菌本身的活性受到严重影响（Johri 等，1999）。即各种溶磷菌对磷酸盐的增溶能力在遭受到盐胁迫时降低，各种溶磷菌释放的无机磷素含量随 NaCl 浓度的增加而不断减少（Cherif Silini 等，2013；Sánchez Porro 等，2009；Srinivasan 等，2012）。然而也有很多研究表明，从盐渍环境或盐生植物中分离得到的耐盐微生物在遭受盐胁迫时，仍能够保持良好的溶解磷酸盐的能力，尤其是溶磷菌能够很好地分解受到盐胁迫危害的土壤中的磷酸钙，从而增加了土壤中的可溶性磷素营养的含量（Etesami 和 Beattie，2018）。有研究报道也能从胁迫环境中分离出溶磷菌（Chen，2006；Etesami 和 Beattie，2018；Etesami 和 Maheshwari，2018）。例如，有研究人员从中国东海岸的盐湖沉积物中分离出一株克锡勒氏菌（*Kushneria sinocanin*），这株菌具有很好的磷酸盐增溶作用（Zhu 等，2011）。在另外一项研究中，Srinivasan 等（2012）研究指出，随着盐浓度增加至 800 mmol/L NaCl 后，溶磷菌菌株，如气球菌属菌株 PSB-CRG1-1（*Aerococcus* sp. strain PSBCRG1-1）、铜绿假单胞菌菌株 PSBI3-1（*Pseudomonas aeruginosa* strain PSBI3-1）以及交替单胞菌菌株 PSBCRG（*Alteromonas* sp. PSBCRG）等，溶解磷酸钙从而释放的无机磷含量增加，而菌株纹状体假单胞菌（*Pseudomonas striata*）在受到盐胁迫后其磷素增溶作用反而显著降低。

也有研究指出，从蝎节木属植物（*Arthrocnemum indicum*）、盐节木（*Salicornia strobilacea*）以及盐角草属植物（*Salicornia brachiate*）等盐生植物中能够分离得到大量的溶磷耐盐菌。这些溶磷耐盐细菌包括克雷伯氏菌属（*Klebsiella* sp.）、假单胞菌属（*Pseudomonas* sp.）、施氏假单胞菌（*Pseudomonas stutzeri*）、根癌土壤杆菌（*Agrobacterium tumefaciens*）、人苍白杆菌（*Ochrobactrum anthropi*）、伸长盐单胞菌（*Halomonas elongata*）、广盐盐单胞菌（*Halomonas eurihalina*）、西奈盐单胞菌（*Halomonas sinaiensis*）、喜海水盐单胞菌（*Halomonas halmophila*）、嗜盐单胞菌（*Halomonas ilicicola*）、印度盐单胞菌（*Halomonas indalina*）、变异盐单胞菌（*Halomonas variabilis*）、新疆盐单胞菌（*Halomonas xinjiangensis*）、泰香盐单胞菌（*Halomonas taeheungii*）、刘志恒氏菌（*Zhihengliuella* sp.）、溶藻弧菌属（*Vibrio alginolyticu*）、乳酪短杆菌（*Brevibacterium casei*）、嗜盐土地芽孢杆菌（*Terribacillus halophilus*）、嗜盐芽孢杆菌（*Halobacillus trueperi*）、喜盐涅斯特连科氏菌（*Nesterenkonia halobia*）、墓画大洋芽孢杆菌（*Oceanobacillus picturae*）、橄榄枝芽孢杆菌（*Virgibacillus olivae*）、死海色盐杆菌（*Chromohalobacter marismortui*）、血杆菌（*Haererehalobacter* sp.）、阪崎克罗诺杆菌（*Cronobacter sakazakii*）和需盐色盐杆菌（*Chromohalobacter salexigens*）等（Jha 等，2012；Mapelli 等，2013；Sharma 等，2016）。这些溶磷耐盐细菌也能为农作物的生长提供磷素营养，从而使农作物在盐渍化条件下良好地生长。例如，在盐胁迫下番茄植株接种溶磷菌皮氏无色杆菌（*Achromobacter piechaudii*）后，植株的磷含量明显增加（Mayak 等，2004）。另外一项类似的研究也表明，在田间试验中，接种了溶磷菌海水芽孢杆菌（*Bacillus aquimaris*）的小麦，在盐胁迫条件下其叶片中的磷含量也显著增加（Upadhyay 和 Singh，2015）。通过对海榄雌（*Avicennia marina*）根际进行微生物的筛选，分离和鉴定出了 129 株具有能够溶解岩石磷酸盐能力的细菌（El Tarabily 和 Youssef，2010）。从盐生植物中分离鉴定得到的溶磷细菌，如节杆菌（*Arthrobacter*）、芽孢杆菌（*Bacillus*）、固氮螺菌（*Azospirillum*）、弧菌（*Vibrio*）、叶杆菌（*Phyllobacterium*）和墓画大洋芽孢杆菌（*Oceanobacillus picturae*）等，能够很好地溶解 $Ca_3(PO_4)_2$、$FePO_4$ 和 $AlPO_4$（Banerjee 等，2010；Bashan 等，2000；El Tarabily 和 Youssef，2010；Yasmin 和 Bano，2011）。溶磷菌在盐胁迫下对不溶性磷化物的溶解作用能够增加生长在周围的宿主植物和非宿主植物（包括盐生植物和甜土植物）生长微环境中的磷素含量。此外，有研究报道指出，当盐生植物比吉洛氏海蓬子（*Salicornia bigelovii*）接种溶磷耐盐细菌[弧菌（*Vibrio*）、固氮螺菌属（*Azospirillum*）、芽孢杆菌（*Bacillus*）和叶杆菌（*Phyllobacterium*）]后，叶片中的磷素含量也比未接种的植物高很多（Bashan 等，2000）。

如上所述，除了细菌外，真菌对不溶性磷酸盐化合物的增溶能力目前也有很多文献报道，溶磷真菌的数量约占真菌总种群种类的 0.1%~0.5%（Kucey，1983）。众所周知，溶磷真菌与宿主植物联合生长会增加宿主农作物根系磷素含量（Smith 和 Read，2010）。有研究指出，丛枝菌根真菌根内球囊霉菌（*Glomus intraradices*）和珠状巨孢囊霉菌（*Gigaspora margarita*）能够促进在相对贫瘠土壤中长叶车前（*Plantago lanceolata*）的生长。这些丛枝菌根真菌能够增加农作物对 N、P 和 K 等营养元素的吸收，但不同的真菌类型对各种营养元素的吸收效果也是不同的（Veresoglou 等，2011）。丛枝菌根真菌对农作物磷素吸收和利用的促进作用是丛枝菌根真菌在盐渍化农业中重要的应用价值之一。丛枝菌根真菌能够在农作物根系生长的贫瘠带向外延伸，在对根系贫瘠带以外的磷素溶解中发挥着重要的作用。此外，还有研究人员提出了丛枝菌根真菌溶解磷素的其他辅助机制。丛枝菌根真菌菌丝吸收磷素的最低浓度阈值要低于农作物根系，因此丛枝菌根真菌对磷素营养的敏感度要比农作物高，它们可以更好地吸收和利用土壤环境中的磷素营养。并且丛枝菌根真菌菌丝具有进入土壤颗粒狭小空间微环境的能力，因此这些丛枝菌根真菌可以获得农作物等非菌根植物无法获得的无机和有机态磷素（Zwetsloot 等，2016）。这些真菌通过分泌有机酸和磷酸酶等能够溶解岩石中无机磷的物质，从而增加土壤中的磷素含量。这些真菌菌类包括曲霉属真菌（*Aspergillus*）、拟青霉属真菌（*Paecilomyces*），属于球囊菌门（Glomeromycota），根内嗜根菌（*Rhizophagus intraradices*）和摩西管柄囊霉（*Funneliformis mosseae*）等菌根真菌（Krüger 等，2012）。它们能够提高土壤中磷素的溶解能力，从而促进不同农作物对磷素的吸收（Barrow 和 Osuna，2002；Cozzolino 等，2013 年；Koide 和 Kabir，2000；Owen 等，2015；Richardson 和 Simpson，2011；Smith 等，2011；Kirsten 等，1996；Williams 等，2013）。此外，这些丛枝菌根真菌还可以通过在土壤中形成一个广泛的菌丝网络，增加养分固定的能力。例如通过菌丝网络的构建，可以减缓土壤中 PO_4^{3-} 的扩散速度（Owen 等，2015），从而为农作物吸收磷素营养争取了时间，即增加了农作物吸收磷素营养的能力；此外，通过菌丝网络的构建，还能够促使农作物根系的生长，促使农作物根系由磷素耗尽区向外围扩散生长，以获得更多的磷素营养（Porcel 等，2012）。真菌菌丝在功能上类似于农作物根系的根毛，也可以作为吸附其他微生物群落生长的载体（Andrade 等，1998）。这样的菌丝复合物在有机态磷素矿化和溶解难降解的营养复合物等方面具有更好的潜力。这些营养复合物降解后会通过菌丝在土壤中富集，并最终被农作物吸收和利用（Owen 等，2015；Toljander 等，2007）。

据报道，溶磷真菌也已从盐胁迫环境中得到了分离和鉴定。例如，从遭受盐渍化胁迫影响的土壤中分离出的溶磷真菌，如曲霉属真菌 PSFNRH－2

(*Aspergillus* sp. PSFNRH-2)，无论 NaCl 浓度如何，随着 NaCl 浓度的增加（最高可达 1.0 mmol/L），其磷增溶效果显著增加（Srinivasan 等，2012）。有研究结果表明，与菌根真菌共生的农作物生长的改善主要与菌根真菌介导的宿主农作物磷素营养的增强有关，因此在盐胁迫条件下在农作物根系定殖丛枝菌根真菌是增强农作物耐盐能力的重要策略（Al-Karaki，2000；Giri 等，2003；Hirrel 和 Gerdemann，1980；Ojala 等，1983；Pond 等，1984；Poss 等，1985）。有研究指出，丛枝根菌真菌通过提高植物对磷素吸收的有效性来缓解 Na^+ 和 Cl^- 等盐胁迫对农作物生长造成的负面影响（Pfetffer 和 Bloss，1988；Poss 等，1985）。耐盐溶磷真菌和植物生长促生菌在盐胁迫下溶解难溶性磷化合物促进农作物的生长具体表现包括以下几个方面：①保持农作物细胞液泡膜的完整性，防止 Na^+ 和 Cl^- 等离子干扰细胞正常的生长和代谢途径（Cantrell 和 Linderman，2001）；②促进液泡内的离子区域化分隔和离子选择性摄入（Rinaldelli 和 Mancuso，1996）。

5.5 植物生长促生菌和丛枝根菌真菌介导的盐胁迫下钾利用有效性的提高

钾素（K）是植物继氮素和磷素之后的必需的大量元素之一，是继氮素之后植物中最丰富的营养元素。钾素被认为是土壤肥力和植物生长的关键参数，如果没有足够的钾素，植物的根系就会发育不良，植株生长缓慢，种子成熟后变小，产量较低（Parmar 和 Sindhu，2013）。此外，钾素在农作物的光合作用、细胞分裂和生长、蛋白质合成、作物产量和质量以及农作物抗病虫害能力提高等方面都发挥着非常重要的作用（Saber 和 Zanati，1984）。钾素还能激活农作物细胞内 80 多种不同的酶，而这些酶负责整个植株能量代谢、淀粉合成、硝酸盐还原、光合作用和糖降解等几乎所有重要的植物生理过程。如果农作物没有吸收足够的钾素，细胞的生命活动就会衰弱，水分就会从细胞中流失并最终枯萎。缺钾的农作物的细胞壁也会发育不良，导致细胞形态和结构异常，最终储存在细胞中的蛋白质和淀粉等物质含量降低。同时，细胞壁发育不良也容易成为真菌孢子入侵的目标，从而发生严重的病害危害（Meena 等，2014，2015）。

根据文献报道，土壤中超过 90% 的钾素都是以不溶性的岩石和各种钾矿石的形式存在（Etesami 等，2017；Parmar 和 Sindhu，2013），因此土壤中绝大部分的钾素并不能被农作物直接吸收利用。白云母、正长石、黑云母、长石、伊利石和云母等矿石都是富含钾素的重要矿物，土壤中的钾素也主要储存于这些矿石当中以固定的形式存在。因为矿石中的钾素并不能被农作物直接吸收利用，因此农作物只能从土壤溶液中吸收溶解的钾素或从人为添加的化肥中获取钾素营养

(Sparks 和 Huang，1985)。大多数土壤中的全钾含量相对较高，但速效钾的含量相对较低。钾能够以多种形式存在于土壤中，包括矿物钾（超过土壤钾含量的 90%~98%）、非交换性钾（约占土壤钾含量的 1%~10%）、交换性钾和溶解性钾（或水溶性钾）。在这些不同形式的钾素状态中，有效（即水溶性钾）和可交换性钾是可以直接被农作物吸收和利用的，其余类型的钾素营养则几乎不能直接作用于农作物 (Sparks 和 Huang，1985)。农作物从土壤溶液中吸收的钾素形式主要是 K^+。由于 K^+ 是一种极易流失的营养元素，因此土壤溶液中钾素的含量变化很大，通常为 1~10 mg/kg。黏性土壤中含有大量的钾营养元素，因此有效钾（水溶性钾）与土壤中黏性土壤矿物含量之间往往存在正相互关系 (Sparks 和 Huang，1985)。农作物在集约栽培条件下，非交换性钾在植物营养中发挥着非常重要的作用，特别是在交换性钾明显低于农作物正常需求的情况下 (McLean 和 Watson，1985)。当土壤中可溶性钾和可交换性钾含量降低到低于农作物正常需求所需的最低阈值时，不可交换性钾则可以从黏土矿物层中释放出来供农作物生长所用 (Tributh 等，1987)。而在这种条件下，土壤中释放的钾素量则取决于含钾矿物的数量和类型，比如钾素来源于白云母、钾长石或蛭石等 (Steffens 和 Sparks，1997)。许多研究表明，在土壤溶液中释放不可交换性钾，可显著增加农作物吸收钾素的量 (Snapp 等，1995)。

除了施肥的不平衡之外，高产作物品种/杂交种的引进以及现代农业的逐步集约化生产正在以更快的速度消耗土壤中的钾储量 (Parmar 和 Sindhu，2013)。另外，土壤盐渍化也导致钾成为农作物生产过程中的主要制约因素之一。盐胁迫条件下，钠元素诱导的钾缺乏已经在多种农作物中被发现和报道 (Botella 等，1997)。据报道，Na^+ 离子影响农作物吸收 K^+ 的原因在于 Na^+ 能够通过干扰农作物根细胞质膜中的各种 K^+ 转运蛋白和选择性离子通道的正常功能，与 K^+ 竞争结合各种通道蛋白上的原本属于 K^+ 的结合位点，从而减少甚至是阻止农作物对 K^+ 的吸收 (Barea 等，2005；Colla 等，2008；Rawal 和 Kuligod，2014)。Na^+ 诱导的 K^+ 吸收减少归因于两种离子在细胞内的竞争性流动 (Cerda 等，1995)。农作物在盐渍化土壤中对 Na^+ 的吸收增加，从而导致农作物细胞内 Na^+/K^+ 比例增加，而 K^+ 是细胞中很多功能酶的激活剂，因此高比例的 Na^+/K^+ 破坏了农作物细胞中很多蛋白质的活性，从而影响了一系列生命活动所必需的代谢过程 (Tester 和 Davenport，2003)。如上所述，K^+ 在农作物生理代谢过程中有许多非常重要的作用，比如激活一系列细胞发挥正常功能的酶。高含量的 Na^+ 或高 Na^+/K^+ 比值不仅可以影响 K^+ 在激活酶过程中发挥的作用，而且还会破坏细胞质中各种酶促反应的过程 (Bhandal 和 Malik，1988)。因此，保持足够水平的 K^+ 对农作物在盐渍化土地上的生存至关重要 (Botella 等，1997)。为盐胁迫下的植物提供足够的钾素营养能够部分减轻 Na$^+$ 对农作物生长的负面影响，一旦农作物吸收了足够钾素，则会重新建立细胞中的 Na$^+$/K$^+$ 平衡（Bach Allen 和 Cunningham，1983）。

目前，有多种途径能够实现含钾矿物中钾素的增溶和释放，但由于土壤微生物具有易用性和低成本的特点，因此可用来实现含钾矿物中钾素的增溶和释放，作为农作物营养的良好补充（Etesami 等，2002、2017）。微生物在促进农作物根系周围土壤中钾矿石的溶解过程中的作用已经有很多文献报道（Etesamietal，2017；Nabiolahy 等，2006）。土壤钾矿石在钾溶解过程中发生的与矿物风化作用有关的生化过程主要发生在有土壤微生物存在的情况下，其进程受到土壤微生物生命活动的影响。有很多研究报道了农作物根系周围的土壤微生物在钾素溶解过程中的作用，并且农作物根系的很多分泌物质都会促进根际环境中钾矿石的风化作用（Arocena 等，2012；Wuet，2008）。

在众多的土壤微生物中，丛枝菌根真菌和细菌对植物钾素吸收的有效性具有重要意义。据报道，很多有益的土壤细菌都是溶钾细菌（potassium-solubilizing bacteria，KSB），主要包括氧化亚铁硫杆菌（*Acidithiobacillus ferrooxidans*）、霍氏肠杆菌（*Enterobacter hormaechei*）、节杆菌属（*Arthrobacter* sp.）、假单胞菌属（*Pseudomonas* spp.）、铜绿假单胞菌（*Pseudomonas aeruginosa*）、类芽孢杆菌（*Paenibacillus* sp.）、黏杆菌属（*Pseudomonas mucilaginosus*）、常见假单胞菌（*Pseudomonas frequentans*）、葡聚糖聚菌属（*Pseudomonas glucanolyticus*）、嗜氨基氨基杆菌（*Aminobacter*）、鞘氨醇单胞菌属（*Sphingomonas*）、伯克霍尔德氏菌（*Burkholderia*）、黏液芽孢杆菌（*Bacillus mucilaginosus*）、土壤芽孢杆菌（*Bacillus edaphicus*）以及环状芽孢杆菌（*Bacillus circulans*）等。这些溶钾细菌能够通过各种不同的机制将钾矿石中不溶性的钾素溶解为可溶性钾。这些机制包括产生有机酸，比如琥珀酸、柠檬酸、葡萄糖酸、α-酮葡萄糖酸、草酸、乳酸、丙酸、乙醇酸、丙二酸、琥珀酸、延胡索酸和酒石酸等，这些有机酸电离产生的酸根阴离子能够增强与钾素结合的阳离子（如硅和铝等）的结合力，有助于钾素从钾矿石中溶解出来。溶钾菌等通过分泌有机酸对农作物根际微环境的土壤颗粒进行的酸解和络合作用、通过分泌多糖等物质进行的置换等反应都能有效地促进土壤颗粒钾素的溶解（Etesami 等，2017；Meena 等，2014；Parmar 和 Sindhu，2013；Uroz 等，2009）。溶钾细菌还通过增加农作物的生根系统和改善在不同生长阶段与宿主农作物之间的共生互利关系，增加农作物根系周围环境中不溶性钾化合物的溶解，直接促进农作物的生长（Etesami 等，2017）。Johnston 和 Krauss（1998）研究表明，如果农作物在土壤中根的密度比较低，则土壤溶液中可供农作物直接吸收利用的钾素浓度应该更高，这样农作物才能够吸收足够的供生长发育所需的钾素营养。另外，具有吲哚-3-乙酸和1-氨基环丙烷-1-羧酸脱氨酶合成能力的溶钾菌，也能够在营养缺乏时通过增加农作物的根系生长，增加农作物对钾素的吸收和利用（Etesami 等，2017）。一些研究结果也表明，这类溶钾细菌能够对农作物的生长具有促进作用（Badr 等，2006；Basak 和 Biswas，2009、2010；Han 和 Lee，2006；Nadeem 等，2007；Sheng，2005；Sheng 和 He，2006）。Nadeem 等（2007）通过研究指出，接种能够合成 1-氨基环丙烷-1-羧酸脱氨酶的细菌的农作物比未接种该类细菌的农作物在盐胁迫条件下能够更好地生长，并且农作物细胞内具有更高的 K^+/Na^+ 比例和叶绿素含量。Mayak 等（2004）还报道了在盐胁迫条件下接种皮氏无色杆菌 RV8（*Achromobacter piechaudii* ARV8）后，番茄对磷素和钾素的吸收都有所增加。

除了溶钾菌之外，丛枝菌根真菌还可以通过增加农作物根系的表面积（增加根系吸收营养的表面积）来促进农作物对营养元素的吸收，从而提高农作物的生长能力，并提高最终产量（Meena 等，2014）。摩西管柄囊霉（*Funneliformis mosseae*）、土曲霉（*Aspergillus terreus*）、黑曲霉（*Aspergillus niger*）、青霉（*Penicillium sp.*）等真菌和根瘤菌（*Rhizoglomus*）等细菌都能够通过产生有机酸/有机酸阴离子（如柠檬酸、苹果酸和草酸）等方式或通过呼吸作用释放 CO_2 与农作物根系微环境中的水形成碳酸，进而释放 H^+，通过酸化农作物根系周围的土壤颗粒来提高土壤溶液中的 K^+ 水平，将不溶性含钾矿物（如钾长石和钾铝硅酸盐中的钾素）溶解出来（Meena 等，2014；Prajapati 等，2012；Sangeeth 等，2012；Sieverding 等，2015；Wu 等，2005）。这些丛枝菌根真菌还能够通过生长的菌丝体增加钾素在农作物根系周围的分布范围，从而提高农作物吸收钾素的范围（Rashid 等，2016）。据报道，丛枝菌根真菌在盐胁迫条件下也能为农作物的生长提供 K^+（Giri 等，2007；Mohammad 等，2003；Ojala 等，1983；Porcel 等，2012）。例如，Giri 等（2007）、Colla 等（2008）以及 Zuccarini 和 Okurowska（2008）的研究均表明，在盐胁迫条件下，与丛枝菌根真菌共生的农作物其根和茎组织中的 K^+ 浓度远高于未与丛枝菌根真菌共生的处理。丛枝菌根真菌促进农作物对 K^+ 吸收，并提高其在细胞内的积累程度，有助于保持农作物细胞或组织中较高的 K^+/Na^+ 比值，从而防止盐胁迫条件下各种酶促反应过程的中断和蛋白质合成受到抑制（Porcel 等，2012）。以上研究表明，在盐渍化土壤中应用溶钾微生物作为生物肥料进行农业生产改良，能够有效减少化学肥料的投入和使用，对保护生态环境有益，也符合农业可持续发展的要求。然而目前，对耐盐溶钾微生物的研究还很少，因此，从盐生植物和盐渍化环境中分离和鉴定抗盐溶钾微生物，对未来研究溶解微生物对含钾矿物增溶能力和利用溶钾微生物进行生物肥料的生产等方面具有非常重要的意义。

5.6 植物生长促生菌和丛枝菌根真菌介导的盐胁迫下微量元素利用有效性的提高

在盐渍化土壤中，农作物对大多数微量元素利用的有效性都很差，因此，生长在这种类型土壤中的农作物其生长往往会受到严重的抑制（Yousfi 等，2007）。

铁（Fe）是一种各种生物生长发育所必需的微量元素。这种微量元素也是许多酶的组成部分，这些酶在农作物中参与多种生化过程，包括光合作用、呼吸作用以及氮的固定等多种生理生化活动（Kobayashi 和 Nishizawa，2012）。在能够进行固氮作用的农作物中，缺铁会降低固氮结节的数量和质量，尤其是会降低固氮过程中血红蛋白的含量、类杆菌的数量以及固氮酶的活性等（Garcia，2015；Tang，1990）。一般而言，农作物对铁元素的吸收和利用往往是以亚铁（Fe^{2+}）的形式，然而在自然界中，铁元素主要以 Fe^{3+} 的形式存在。铁被氧化成 Fe^{3+} 后在土壤溶液中极易与一些阴离子形成不溶性化合物，这些化合物中的铁元素即被固定，并不能被农作物吸收利用（Ma，2005）。而这种现象在盐渍化土壤中尤其突出。一般来说，在世界范围内石灰土和盐渍化土壤中农作物对铁素吸收利用的有效性非常低（Rabhi 等，2007）。

包括植物生长促生菌在内的多种微生物可以通过各种机制增加农作物对多种微量营养素的可用性，例如通过降低土壤 pH 值和产生螯合剂等（Miransari，2013）。植物生长促生菌能够在铁素缺乏的情况下产生螯合剂，这种螯合剂被称为铁载体。铁载体是一种分子量较小、但对 Fe^{3+} 具有高亲和力的螯合化合物，能够十分有效地吸附 Fe^{3+}。而通过铁载体固定的 Fe^{3+} 则可以有效地被农作物吸收和利用（Kloeper 等，1980）。农作物对铁载体固定的 Fe^{3+} 的利用是通过配体交换作用实现的。细菌铁载体固定的 Fe^{3+} 在配体交换反应中与农作物的铁载体相互作用，最终转移到农作物的铁载体上，从而能够供给农作物吸收和利用（Latour 等，2009）。植物生长促生菌产生的铁载体能够给呈现缺铁症状的番茄提供铁元素，以保证植株的正常生长（Grobelak 和 Hiller，2017；Pii 等，2015；Radzki 等，2013；Scavino 和 Pedraza，2013）。值得注意的是，从盐生植物包括盐角草属植物（*Salicornia brachiate*）、碱蓬（*Aster tripolium* L.）和牧豆树属植物（*Prosopis strombulifera*）等中分离出的耐盐植物生长促生菌，例如苏拉氏短杆菌属（*Brachybacterium saurashtrense* sp.）、黏质沙雷氏菌属（*Serratia marcescens*）、不动杆菌（*Acinetobacter*）、假单胞菌属（*Pseudomonas* sp.）、施氏假单胞菌（*Pseudomonas stutzeri*）、恶臭假单胞菌（*Pseudomonas putida*）、微杆菌属（*Microbacterium*）、链霉菌属（*Streptomyces* sp.）、赭褐链霉菌（*Streptomyces* umbrinus)、灰平链霉菌（*Streptomyces griseoplanus*）、脱叶链霉菌（*Streptomyces exfoliatus*）、亲和素链霉菌（*Streptomyces avidinii*）、欧洲疮痂链霉菌（*Streptomyces europaeiscabiei*）、芽孢杆菌属（*Bacillus*）、蜡状芽孢杆菌属（*Bacillus cereus*）、短小芽孢杆菌（*Bacillus pumilus*）、简单芽孢杆菌（*Bacillus simplex*）、蕈状芽孢杆菌（*Bacillus mycoides*）、普城沙雷菌（*Serratia plymuthica*）、微杆菌属（*Microbacterium* sp.）、氧化微杆菌（*Microbacterium oxydans*）、红球菌属（*Rhodococcus* sp.）、红城红球菌（*Rhodococcus erythropolis*）、藤黄微球菌（*Micrococcus luteus*）、诺卡氏菌属（*Nocardia* sp.）、根癌土壤杆菌（*Agrobacterium tumefaciens*）、刘志恒氏菌（*Zhihengliuella* sp.）、短杆菌属（*Brachybacterium* sp.）、溶藻弧菌属（*Vibrio alginolyticus*）、乳酪短杆菌属（*Brevibacterium casei*）、阪崎克罗诺杆菌（*Cronobacter sakazakii*）、类产碱假单胞菌属（*Pseudomonas pseudoalcaligenes*）、血杆菌（*Haererehalobacter* sp.）、盐单胞菌属（*Halomonas* sp.）、巴西酸酯菌（*Brasilate*）和星耳酵母（*Asterotremella tripolium*）等，都具有非常显著的合成和分泌铁载体的能力，因此这些微生物可用于盐渍化土壤的改良（Etesami 和 Beattie，2018；Gontia，2011；Jha 等，2012；Sgroy，2009；Szymańska 等，2016）。这就意味着从盐生植物中分离出的具有抗盐性的、能产生铁载体的植物生长促生菌，在增加其他农作物必需微量元素（如锌、锰、铜等）可用性方面也有可能具有非常好的效果，但这仍需要通过大量细致的研究工作进行证实（Etesami 和 Beattie，2018）。

目前，有很多研究结果表明，部分植物生长促生菌能提高农作物对土壤中锰素利用的有效性。这些菌类包括芽孢杆菌（*Bacillus*）、假单胞菌（*Pseudomonas*）和地杆菌属（*Geobacter*）等。这些植物生长促生菌能够将土壤中的 Mn^{4+} 还原为 Mn^{2+}，Mn^{2+} 是锰素在农作物代谢过程中的常态形式（Osorio Vega，2007）。这些植物生长促生菌对土壤中锰素利用有效性提高的原因在于它们能够对农作物的生长和根系分泌物产生影响（Etesami 和 Maheshwari，2018）。细菌活动引起农作物根系分泌物增加，这些分泌物则大多是一些碳水化合物和有机酸类物质。这些碳水化合物通过分解作用以及有机酸电离作用产生的电子和质子，参与 MnO_2 中 Mn^{4+} 的还原作用，将其还原为 Mn^{2+}，从而供给农作物吸收和利用。此外，植物生长促生菌还能够产生不同类型的螯合剂，如酚类化合物和有机酸。这些螯合剂类化合物能够与锰、铁和其他微量元素形成复杂的可溶性化合物，防止这些微量元素在土壤溶液中沉淀。而微生物影响土壤中农作物对铜和锌利用的有效性与之类似，也是通过合成各种螯合类化合物（如酚类化合物和羧酸等）实现的；或与铁类似，通过间接影响农作物的生长和增加根系分泌物来实现（Iqbal 等，2010）。

只有在缺乏营养的生态系统中，丛枝菌根真菌在增加农作物对营养元素吸收方面起到的重要作用才能够体现得较为明显（Cardoso 和 Kuyper，2006）。有研究指出，丛枝菌根真菌能够增强农作物对锌、铜和铁等微量元素的吸收（Al-Karaki，2000；Al-Karaki 和 Clark，1998；Marschner 和 Dell，1994）。真菌的菌丝系统能够从生长环境中吸取氮、磷、钾、锌、铜、铁等营养物质，这些营养物质并不完全都被真菌所利用，有一部分营养物质能够通过各种途径被传递到农作物体内，供给农作物的生长和发育（Cardoso 和 Kuyper，2006）。Clark 和 Zeto（1996）的研究结果表明，在玉米生长的土壤中添加丛枝菌根真菌，能够增加玉米芽中铁元素的含量。此外，与丛枝菌根真菌共生的高粱植株内铁元素的含量也比未施加丛枝菌根真菌的高粱要高很多（Caris 等，1998）。由于锌元素在土壤中的迁移速率较低，导致其土壤中的分布呈现圈点状态，严重阻碍农作物对锌元素的吸收和利用，因此农作物对锌元素利用的有效性受到了很大的限制。同时，在土壤颗粒表面的吸附作用是影响土壤中锌元素生物利用有效性的重要决定因素。因此，与其他农作物所需的营养元素相比，通过丛枝菌根真菌在农作物根际定殖作用从而增加真菌对土壤颗粒表面的锌元素的吸收就变得尤其重要（Hajiboland 等，2009）。由于丛枝菌根真菌菌丝的生长可以增加其吸收微量元素的表面积，因此丛枝菌根真菌在农作物根部的定殖作用就相当于增加了农作物根系单位长度中锌元素的吸收量，这对提高农作物对锌元素的吸收可利用非常重要（Gao 等，2007）。通过以上研究表明，耐盐微生物可以通过合成和分泌有机酸和螯合剂类化合物，并通过合成吲哚-3-乙酸和1-氨基环丙烷-1-羧酸脱氨酶等物质来促进植物的生根系统从而增加盐胁迫条件下植物对必需微量营养素的利用效率（Etesami 等，2015a、2015b；Etesami 和 Maheshwari，2018）。

5.7 结论

耕作土壤中的水和盐分是干旱和半干旱地区农业生产中遇到的制约农作物生长的最重要的农业问题。利用土壤耐盐微生物缓解或者提高农作物对盐胁迫的耐受程度是应对这种压力的策略之一。这是因为在受盐胁迫影响的耕作土壤中，过量施用化肥会导致土壤中的盐分含量成倍增加，并且连续施用化肥也会导致土壤理化性质的下降，为生态和农业的可持续发展带来挑战。因此使用耐盐植物生长促生微生物能够减少盐渍化土壤中农作物生产过程中的化肥投入，并且这类微生物还具有促进农作物生长的功效。但是，在大田条件下将植物生长促生细菌作为生物肥料进行使用仍需要更多细致和深入的研究。在大田条件下施用植物生长促生菌或丛枝菌根真菌时，需要更好地了解耐盐细菌与丛枝菌根真菌和农作物之间的相互作用。阐明耐盐植物促生微生物在盐胁迫下提高农作物植株对微量元素和大量营养素利用效率的机理,将有助于开发新型的、成本低廉的工业产品,并最终提高盐渍化土壤的肥力和农作物生产能力。

[致谢]

感谢德黑兰大学为这项研究提供了必要的设施和资金。

参考文献

Abdelmoumen H, El Idrissi MM, 2009. Germination, growth and nodulation of Trigonella foenum graecum (Fenu Greek) under salt stress. Afr J Biotechnol 8: 2489-2496.

Adesemoye AO, Kloepper JW, 2009. Plant-microbes interactions in enhanced fertilizer-use efficiency. Appl Microbiol Biotechnol 85: 1-12.

Ahmad P, 2013. Oxidative damage to plants, antioxidant networks and signaling. Academic, Elsevier, San Diego.

Ahmed W, Shahroona B, Zahir ZA, Arshad M, 2004. Inoculation with ACC-deaminase containing rhizobacteria for improving growth and yield of wheat. Pak J Agric Sci 41: 119.

Albacete A et al., 2008. Hormonal changes in relation to biomass partitioning and shoot growth impairment in salinized tomato (*Solanum lycopersicum* L.) plants. J Exp Bot 59: 4119-4131.

Aliasgharzadeh N, Rastin SN, Towfighi H, Alizadeh A, 2001. Occurrence of *Arbuscular mycorrhizal* fungi in saline soils of the Tabriz Plain of Iran in relation to some physical and chemical properties of soil. Mycorrhiza 11: 119-122.

Al-Karaki GN, 2000. Growth of mycorrhizal tomato and mineral acquisition under salt stress. Mycorrhiza 10: 51-54.

Al-Karaki GN, 2006. Nursery inoculation of tomato with *Arbuscular mycorrhizal* fungi and subsequent performance under irrigation with saline water. Sci Horticult 109: 1-7.

Al-Karaki GN, Clark RB, 1998. Growth, mineral acquisition, and water use by mycorrhizal wheat grown under water stress. J Plant Nutr 21: 263-276.

America SSSo, 2001. Glossary of soil science terms. Soil Science Society of America, Madison.

Andrade G, Linderman RG, Bethlenfalvay GJ, 1998. Bacterial associations with the mycorrhizosphere and hyphosphere of the *Arbuscular mycorrhizal* fungus *Glomus mosseae*. Plant Soil 202: 79-87.

Arocena JM, Velde B, Robertson SJ, 2012. Weathering of biotite in the presence of *Arbuscular mycorrhizae* in selected agricultural crops. Appl Clay Sci 64: 12-17.

Asghari HR, 2012. Vesicular-arbuscular (VA) mycorrhizae improve salinity tolerance in pre-inoculation subterranean clover (*Trifolium subterraneum*) seedlings. Int J Plant Prod 2: 243-256.

Ashraf M, Hasnain S, Berge O, Mahmood T, 2004. Inoculating wheat seedlings with exopolysaccharide-producing bacteria restricts sodium uptake and stimulates plant growth under salt stress. Biol Fertil Soils 40: 157-162.

Aydi S, Sassi S, Abdelly C, 2008. Growth, nitrogenfixation and ion distribution in *Medicago truncatula* subjected to salt stress. Plant Soil 312: 59.

Bach Allen E, Cunningham GL, 1983. Effects of vesicular-*Arbuscular mycorrhizae* on Distichlis spicata under three salinity levels. New Phytol 93: 227-236.

Badr MA, Shafei AM, Sharaf El-Deen SH, 2006. The dissolution of K and P-bearing minerals by silicate dissolving bacteria and their effect on sorghum growth. Res J Agric Biol Sci 2: 5-11.

Banerjee S, Palit R, Sengupta C, Standing D, 2010. Stress induced phosphate solubilization by *Arthrobacter* sp. and *Bacillus* sp. isolated from tomato rhizosphere. Aust J Crop Sci 4: 378.

Banik P, Midya A, Sarkar BK, Ghose SS, 2006. Wheat and chickpea intercropping systems in an additive series experiment: advantages and weed smothering. Eur J Agron 24: 325-332.

Barassi CA, Ayrault G, Creus CM, Sueldo RJ, Sobrero MT, 2006. Seed inoculation with *Azospirillum mitigates* NaCl effects on lettuce. Sci Horticult 109: 8 - 14. https: //doi. org/10. 1016/j. scienta. 2006. 02. 025.

Barea JM, Azcón R, Azcón-Aguilar C, 2005. Interactions between mycorrhizal fungi and bacteria to improve plant nutrient cycling and soil structure. In: Microorganisms in soils: roles in genesis and functions. Springer, pp. 195-212.

Barrow JR, Osuna P, 2002. Phosphorus solubilization and uptake by dark septate fungi in four-wing saltbush, *Atriplex canescens* (Pursh) Nutt. J Arid Environ 51: 449-459.

Basak BB, Biswas DR, 2009. Influence of potassium solubilizing microorganism (*Bacillus mucilaginosus*) and waste mica on potassium uptake dynamics by sudan grass (*Sorghum vulgare* Pers.) grown under two Alfisols. Plant Soil 317: 235-255.

Basak BB, Biswas DR, 2010. Co - inoculation of potassium solubilizing and nitrogenfixing bacteria on solubilization of waste mica and their effect on growth promotion and nutrient acquisition by a forage crop. Biol Fertil Soils 46: 641-648.

Bashan Y, Moreno M, Troyo E, 2000. Growth promotion of the seawater-irrigated oilseed halophyte *Salicornia bigelovii* inoculated with mangrove rhizosphere bacteria and halotolerant *Azospirillum* spp. Biol Fertil Soils 32: 265-272.

Belimov AA, Dodd IC, Hontzeas N, Theobald JC, Safronova VI, Davies WJ, 2009. Rhizosphere bacteria containing 1 - aminocyclopropane - 1 - carboxylate deaminase increase yield of plants grown in drying soil via both local and systemic hormone signalling. New Phytol 181: 413-423. https: //doi. org/10. 1111/j. 1469-8137. 2008. 02657. x.

Bhandal IS, Malik CP, 1988. Potassium estimation, uptake, and its role in the physiology and metabolism of flowering plants. In: International review of cytology, vol 110. Elsevier, pp. 205-254.

Bian G, Zhang Y, Qin S, Xing K, Xie H, Jiang J, 2011. Isolation and biodiversity of heavy metal tolerant endophytic bacteria from halotolerant plant species located in coastal shoal of Nantong. Wei Sheng Wu Xue Bao=Acta Microbiol Sin 51: 1538-1547.

Botella MA, Martinez V, Pardines J, Cerdá A, 1997. Salinity induced potassium deficiency in maize plants. J Plant Physiol 150: 200 - 205. https: //doi. org/10. 1016/S0176 - 1617 (97) 80203-9.

Cantrell IC, Linderman RG, 2001. Preinoculation of lettuce and onion with VA mycorrhizal fungi reduces deleterious effects of soil salinity. Plant Soil 233: 269-281.

Cardoso IM, Kuyper TW, 2006. Mycorrhizas and tropical soil fertility. Agric Ecosyst Environ 116: 72-84.

Caris C, Hördt W, Hawkins HJ, Römheld V, George E, 1998. Studies of iron transport by *Arbuscular mycorrhizal* hyphae from soil to peanut and sorghum plants. Mycorrhiza 8: 35-39.

Carmen B, Roberto D, 2011. Soil bacteria support and protect plants against abiotic stresses. In: Shanker A (ed) Abiotic stress in plants mechanisms and adaptations. Pub InTech, Rijeka, pp. 143-170.

Cerda A, Pardines J, Botella MA, Martinez V, 1995. Effect of potassium on growth, water relations, and the inorganic and organic solute contents for two maize cultivars grown under saline conditions. J Plant Nutr 18: 839-851.

Chakraborty U, Chakraborty B, Basnet M, 2006. Plant growth promotion and induction of resistance in *Camellia sinensis* by *Bacillus megaterium*. J Basic Microbiol 46: 186-195. https://doi.org/10.1002/jobm.200510050.

Chakraborty D et al., 2008. Effect of mulching on soil and plant water status, and the growth and yield of wheat (*Triticum aestivum* L.) in a semi-arid environment. Agric Water Manag 95: 1323-1334.

Chakraborty U, Chakraborty B, Dey P, Chakraborty AP, 2015. Role of microorganisms in alleviation of abiotic stresses for sustainable agriculture abiotic stresses in crop plants. CABI, Wallingford, p. 232.

Chalk PM, Souza RF, Urquiaga S, Alves BJR, Boddey RM, 2006. The role of arbuscular mycorrhiza in legume symbiotic performance. Soil Biol Biochem 38: 2944-2951.

Chalk PM, Alves BJR, Boddey RM, Urquiaga S, 2010. Integrated effects of abiotic stresses on inoculant performance, legume growth and symbiotic dependence estimated by 15 N dilution. Plant Soil 328: 1-16.

Chen JH, 2006. The combined use of chemical and organic fertilizers and/or biofertilizer for crop growth and soil fertility. Citeseer, p. 20.

Chen YP, Rekha PD, Arun AB, Shen FT, Lai WA, Young CC, 2006. Phosphate solubilizing bacteria from subtropical soil and their tricalcium phosphate solubilizing abilities. Appl Soil Ecol 34: 33-41.

Cheng SX, He ZD, Zhen L, Wei T, Zhang DM, Wei JL, Xiang QK, 2010. Effects of N, P, and K fertilizer application on cotton growing in saline soil in yellow river delta. Acta Agron Sin 36: 1698-1706.

Cherif-Silini H, Silini A, Ghoul M, Yahiaoui B, Arif F, 2013. Solubilization of phosphate by the *Bacillus* under salt stress and in the presence of osmoprotectant compounds. Afr J Microbiol Res 7: 4562-4571.

Choudhary D, 2012. Microbial rescue to plant under habitat-imposed abiotic and biotic stresses. Appl Microbiol Biotechnol 96: 1137-1155.

Choudhary DK, Kasotia A, Jain S, Vaishnav A, Kumari S, Sharma KP, Varma A, 2015. Bacterial-mediated tolerance and resistance to plants under abiotic and biotic stresses. J Plant Growth Regul 1-25.

Clark RB, Zeto SK, 1996. Mineral acquisition by mycorrhizal maize grown on acid and alkaline soil. Soil Biol Biochem 28: 1495-1503.

Colla G, Rouphael Y, Cardarelli M, Tullio M, Rivera CM, Rea E, 2008. Alleviation of salt stress by *Arbuscular mycorrhizal* in zucchini plants grown at low and high phosphorus concentration. Biol Fertil Soils 44: 501-509.

Contreras-Cornejo HA, Macías-Rodríguez L, Cortés-Penagos C, López-Bucio J, 2009. *Trichoderma virens*, a plant beneficial fungus, enhances biomass production and promotes lateral root growth through an auxin – dependent mechanism in *Arabidopsis*. Plant Physiol 149: 1579-1592.

Cortina C, Culiáñez-Macià FA, 2005. Tomato abiotic stress enhanced tolerance by trehalose biosynthesis. Plant Sci 169: 75-82.

Cozzolino V, Di Meo V, Piccolo A, 2013. Impact of *Arbuscular mycorrhizal* fungi applications on maize production and soil phosphorus availability. J Geochem Explor 129: 40-44.

Dahmardeh M, Ghanbari A, Syasar B, Ramroudi M, 2009. Effect of intercropping maize (*Zea mays* L.) with cow pea (*Vigna unguiculata* L.) on green forage yield and quality evaluation. Asian J Plant Sci 8: 235.

Damodaran T et al., 2014. Rhizosphere and endophytic bacteria for induction of salt tolerance in gladiolus grown in sodic soils. J Plant Interact 9: 577-584.

Davies WJ, Zhang J, Yang J, Dodd IC, 2011. Novel crop science to improve yield and resource use efficiency in water-limited agriculture. J Agric Sci 149: 123-131.

del Amor Francisco M, Cuadra-Crespo P, 2012. Plant growth-promoting bacteria as a tool to improve salinity tolerance in sweet pepper. Funct Plant Biol 39: 82-90.

Delavechia C, Hampp E, Fabra A, Castro S, 2003. Influence of pH and calcium on the growth, polysaccharide production and symbiotic association of *Sinorhizobium meliloti* SEMIA 116 with alfalfa roots. Biol Fertil Soils 38: 110-114.

Delgado-García M, De la Garza-Rodríguez I, Cruz-Hernández MA, Balagurusamy N, Aguilar C, Rodríguez-Herrera R, 2013. Characterization and selection of halophilic microorganisms isolated from Mexican soils. J Agric Biol Sci 8: 457-464.

Diby P, Bharathkumar S, Sudha N, 2005a. Osmotolerance in biocontrol strain of *Pseudomonas pseudoalcaligenes* MSP-538: a study using osmolyte, protein and gene expression profiling. Ann Microbiol 55: 243-247.

Diby P, Sarma YR, Srinivasan V, Anandaraj M, 2005b. *Pseudomonas fluorescens* mediated vigour in black pepper (*piper nigrum* L.) under green house cultivation. Ann Microbiol 55: 171-174.

Dimkpa C, Weinand T, Asch F, 2009. Plant-rhizobacteria interactions alleviate abiotic stress conditions. Plant Cell Environ 32: 1682-1694. https: //doi. org/10. 1111/j. 1365-3040. 2009. 02028. x.

Dolkar D, Dolkar P, Angmo S, Chaurasia OP, Stobdan T, 2018. Stress tolerance and plant growth promotion potential of *Enterobacter ludwigii* PS1 isolated from *Seabuckthorn* rhizosphere. Biocatal Agric Biotechnol. https: //doi. org/10. 1016/j. bcab. 2018. 04. 012.

Dulormne M, Musseau O, Muller F, Toribio A, Bâ A, 2010. Effects of NaCl on growth, water status, N_2 fixation, and ion distribution in *Pterocarpus officinalis* seedlings. Plant Soil 327: 23-34.

Dunlap JR, Binzel ML, 1996. NaCl reduces indole-3-acetic acid levels in the roots of tomato plants independent of stress-induced abscisic acid. Plant Physiol 112: 379-384.

Duzan HM, Zhou X, Souleimanov A, Smith DL, 2004. Perception of *Bradyrhizobium japonicum* Nod factor by soybean [*Glycine max* (L.) Merr.] root hairs under abiotic stress conditions. J Exp Bot 55: 2641-2646.

Egamberdieva D, Kucharova Z, 2009. Selection for root colonising bacteria stimulating wheat growth in saline soils. Biol Fertil Soils 45: 563-571.

Egamberdieva D, Kamilova F, Validov S, Gafurova L, Kucharova Z, Lugtenberg B, 2008. High incidence of plant growth-stimulating bacteria associated with the rhizosphere of wheat grown on salinated soil in Uzbekistan. Environ Microbiol 10: 1-9. https://doi.org/10.1111/j.1462-2920.2007.01424.x.

Egamberdieva D, Berg G, Lindström K, Räsänen L, 2013. Alleviation of salt stress of symbiotic *Galega officinalis* L. (goat's rue) by co-inoculation of Rhizobium with root-colonizing *Pseudomonas*. Plant Soil 369: 453-465. https://doi.org/10.1007/s11104-013-1586-3.

Egamberdiyeva D, 2005. Plant-growth-promoting rhizobacteria isolated from a Calcisol in a semiarid region of Uzbekistan: biochemical characterization and effectiveness. J Plant Nutr Soil Sci 168: 94-99.

Elbein AD, Pan YT, Pastuszak I, Carroll D, 2003. New insights on trehalose: a multifunctional molecule. Glycobiology 13: 17R-27R.

El-Tarabily KA, Youssef T, 2010. Enhancement of morphological, anatomical and physiological characteristics of seedlings of the mangrove *Avicennia marina* inoculated with a native phosphate-solubilizing isolate of *Oceanobacillus picturae* under greenhouse conditions. Plant Soil 332: 147-162.

Esquivel-Cote R, Ramírez-Gama RM, Tsuzuki-Reyes G, Orozco-Segovia A, Huante P, 2010. *Azospirillum lipoferum* strain AZm5 containing 1-aminocyclopropane-1-carboxylic acid deaminase improves early growth of tomato seedlings under nitrogen deficiency. Plant Soil 337: 65-75. https://doi.org/10.1007/s11104-010-0499-7.

Etesami H, 2018. Can interaction between silicon and plant growth promoting rhizobacteria benefit in alleviating abiotic and biotic stresses in crop plants? Agric Ecosyst Environ 253: 98-112. https://doi.org/10.1016/j.agee.2017.11.007.

Etesami H, Alikhani HA, 2016a. Co-inoculation with endophytic and rhizosphere bacteria allows reduced application rates of N-fertilizer for rice plant. Rhizosphere 2: 5-12.

Etesami H, Alikhani HA, 2016b. Rhizosphere and endorhiza of oilseed rape (*Brassica napus* L.) plant harbor bacteria with multifaceted beneficial effects. Biol Control 94: 11-24. https://doi.org/10.1016/j.biocontrol.2015.12.003.

Etesami H, Beattie GA, 2017. Plant-microbe interactions in adaptation of agricultural crops to abiotic stress conditions. In: Probiotics and plant health. Springer, pp. 163-200.

Etesami H, Beattie G, 2018. Mining halophytes for plant growth-promoting halotolerant bacteria to enhance the salinity tolerance of non-halophytic crops. Front Microbiol 9: 148. https://doi.org/10.3389/fmicb.2018.00148.

Etesami H, Maheshwari DK, 2018. Use of plant growth promoting rhizobacteria (PGPRs) with multiple plant growth promoting traits in stress agriculture: action mechanisms and future prospects. Ecotoxicol Environ Saf 156: 225-246. https://doi.org/10.1016/j.ecoenv.2018.03.013.

Etesami H, Alikhani HA, Hosseini HM, 2015a. Indole – 3 – acetic acid (IAA) production trait, a useful screening to select endophytic and rhizosphere competent bacteria for rice growth promoting agents. MethodsX 2: 72-78.

Etesami H, Alikhani HA, Mirseyed Hosseini H, 2015b. Indole-3-acetic acid and 1-aminocyclopropane- 1 – carboxylate deaminase: bacterial traits required in rhizosphere, rhizoplane and/or endophytic competence by beneficial bacteria. In: Maheshwari DK (ed) Bacterial metabolites in sustainable agroecosystem. Springer, Cham, pp. 183-258. https://doi.org/10.1007/978-3-319-24654-3_ 8.

Etesami H, Emami S, Alikhani HA, 2017. Potassium solubilizing bacteria (KSB): mechanisms, promotion of plant growth, and future prospects – a review. J Soil Sci Plant Nutr 17: 897-911.

Evelin H, Kapoor R, Giri B, 2009. *Arbuscular mycorrhizal* fungi in alleviation of salt stress: a review. Ann Bot 104: 1263-1280.

Fageria NK, Moreira A, 2011. 4) The role of mineral nutrition on root growth of crop plants. Adv Agron 110: 251-331.

FAO, 2005. Available on URL: http://www.fao.org.

FAO, 2008. Land and plant nutrition management service. http://www.fao.org/ag/agl/agll/spush.

Farías-Rodríguez R, Mellor RB, Arias C, Peña-Cabriales JJ, 1998. The accumulation of trehalose in nodules of several cultivars of common bean (*Phaseolus vulgaris*) and its correlation with resistance to drought stress. Physiol Plant 102: 353-359.

Feigin A, 1985. Fertilization management of crops irrigated with saline water. Plant Soil 89: 285-299. https://doi.org/10.1007/BF02182248.

Feng G, Zhang F, Li X, Tian C, Tang C, Rengel Z, 2002. Improved tolerance of maize plants to salt stress by *Arbuscular mycorrhiza* is related to higher accumulation of soluble sugars in roots. Mycorrhiza 12: 185-190.

Fernandez O, Béthencourt L, Quero A, Sangwan RS, Clément C, 2010. Trehalose and plant stress responses: friend or foe? Trends Plant Sci 15: 409-417.

Flowers TJ, 2004. Improving crop salt tolerance. J Exp Bot 55: 307-319.

Founoune H, Duponnois R, Bâ AM, El Bouami F, 2002. Influence of the dual arbuscular endomycorrhizal/ectomycorrhizal symbiosis on the growth of *Acacia holosericea* (A. Cunn. ex G. Don) in glasshouse conditions. Ann For Sci 59: 93-98.

Franco-Correa M, Quintana A, Duque C, Suarez C, Rodríguez MX, Barea JM, 2010. Evaluation of actinomycete strains for key traits related with plant growth promotion and mycorrhiza helping activities. Appl Soil Ecol 45: 209-217.

Franzini VI, Azcon R, Mendes FL, Aroca R, 2010. Interactions between *Glomus* species and *Rhizobium* strains affect the nutritional physiology of drought-stressed legume hosts. J Plant Physiol 167: 614-619.

Fu Q, Liu C, Ding N, Lin Y, Guo B, 2010. Ameliorative effects of inoculation with the plant growth-promoting rhizobacterium *Pseudomonas* sp. DW1 on growth of eggplant (*Solanum melongena* L.) seedlings under salt stress. Agric Water Manag 97: 1994-2000. https://doi.org/10.1016/j.agwat.2010.02.003.

Gamalero E, Glick BR, 2011. Mechanisms used by plant growth-promoting bacteria. In: Bacteria in agrobiology: plant nutrient management. Springer, pp. 17-46.

Gao X, Kuyper TW, Zou C, Zhang F, Hoffland E, 2007. Mycorrhizal responsiveness of aerobic rice genotypes is negatively correlated with their zinc uptake when nonmycorrhizal. Plant Soil 290: 283-291.

Garcia NS, Fu F, Sedwick PN, Hutchins DA, 2015. Iron deficiency increases growth and nitrogen-fixation rates of phosphorus-deficient marine cyanobacteria. ISME J 9: 238-245. https://doi.org/10.1038/ismej.2014.104.

Garg N, Chandel S, 2011. The effects of salinity on nitrogenfixation and trehalose metabolism in mycorrhizal *Cajanus cajan* (L.) mill sp. plants. J Plant Growth Regul 30: 490-503.

Garg N, Manchanda G, 2008. Effect of *Arbuscular mycorrhizal* inoculation on salt-induced nodule senescence in *Cajanus cajan* (pigeonpea). J Plant Growth Regul 27: 115.

Garg N, Manchanda G, 2009. Role of *Arbuscular mycorrhizae* in the alleviation of ionic, osmotic and oxidative stresses induced by salinity in *Cajanus cajan* (L.) Mill sp. (*pigeonpea*). J Agron Crop Sci 195: 110-123.

Giri B, Mukerji KG, 2004. Mycorrhizal inoculant alleviates salt stress in *Sesbania aegyptiaca* and *Sesbania grandiflora* under field conditions: evidence for reduced sodium and improved magnesium uptake. Mycorrhiza 14: 307-312.

Giri B, Kapoor R, Mukerji KG, 2003. Influence of *Arbuscular mycorrhizal* fungi and salinity on growth, biomass, and mineral nutrition of *Acacia auriculiformis*. Biol Fertil Soils 38: 170-175.

Giri B, Kapoor R, Mukerji KG, 2007. Improved tolerance of Acacia nilotica to salt stress by *Arbuscular mycorrhiza*, *Glomus fasciculatum* may be partly related to elevated K/Na ratios in root and shoot tissues. Microb Ecol 54: 753-760.

Glick BR, 2005. Modulation of plant ethylene levels by the bacterial enzyme ACC deaminase. FEMS Microbiol Lett 251: 1-7.

Glick BR, 2012. Plant growth-promoting bacteria: mechanisms and applications. Scientifica 2012: 963401.

Glick BR, 2014. Bacteria with ACC deaminase can promote plant growth and help to feed the world. Microbiol Res 169: 30-39.

Gontia I, Kavita K, Schmid M, Hartmann A, Jha B, 2011. *Brachybacterium saurashtrense* sp. nov., a halotolerant root-associated bacterium with plant growth-promoting potential. Int J Syst Evol Microbiol 61: 2799-2804.

Gosling P, Hodge A, Goodlass G, Bending GD, 2006. *Arbuscular mycorrhizal* fungi and organic farming. Agric Ecosyst Environ 113: 17-35.

Goss MJ, De Varennes A, 2002. Soil disturbance reduces the efficacy of mycorrhizal associations for early soybean growth and N_2 fixation. Soil Biol Biochem 34: 1167-1173.

Grant CA, Flaten DN, Tomasiewicz DJ, Sheppard SC, 2001. The importance of early season phosphorus nutrition. Can J Plant Sci 81: 211-224.

Gray EJ, Smith DL, 2005. Intracellular and extracellular PGPR: commonalities and distinctions in the plant-bacterium signaling processes. Soil Biol Biochem 37: 395-412. https://doi.org/10.1016/j.soilbio.2004.08.030.

Greaves JE, 1922. Influence of salts on bacterial activities of soil. Bot Gaz 73: 161-180.

Grobelak A, Hiller J, 2017. Bacterial siderophores promote plant growth: screening of catechol and hydroxamate siderophores. Int J Phytoremediation 19: 825-833.

Grover M, Ali SZ, Sandhya V, Rasul A, Venkateswarlu B, 2011. Role of microorganisms in adaptation of agriculture crops to abiotic stresses. World J Microbiol Biotechnol 27: 1231-1240.

Gulati A, Sharma N, Vyas P, Sood S, Rahi P, Pathania V, Prasad R, 2010. Organic acid production and plant growth promotion as a function of phosphate solubilization by *Acinetobacter rhizosphaerae* strain BIHB 723 isolated from the cold deserts of the trans-Himalayas. Arch Microbiol 192: 975-983.

Gyaneshwar P, Kumar GN, Parekh LJ, Poole PS, 2002. Role of soil microorganisms in improving P nutrition of plants. In: Food security in nutrient-stressed environments: exploiting plants' genetic capabilities. Springer, pp. 133-143.

Hajiboland R, Aliasgharzad N, Barzeghar R, 2009. Influence of *Arbuscular mycorrhizal* fungi on uptake of Zn and P by two contrasting rice genotypes. Plant Soil Environ 55: 93-100.

Hajiboland R, Aliasgharzadeh N, Laiegh SF, Poschenrieder C, 2010. Colonization with *Arbuscular mycorrhizal* fungi improves salinity tolerance of tomato (*Solanum lycopersicum* L.) plants. Plant Soil 331: 313-327.

Halvorson HO, Keynan A, Kornberg HL, 1990. Utilization of calcium phosphates for microbial growth at alkaline pH. Soil Biol Biochem 22: 887-890.

Hamdia MAES, Shaddad MAK, Doaa MM, 2004. Mechanisms of salt tolerance and interactive effects of *Azospirillum brasilense* inoculation on maize cultivars grown under salt stress conditions. Plant Growth Regul 44: 165-174. https://doi.org/10.1023/B:GROW.0000049414.03099.9b.

Hamilton CE, Bever JD, Labbé J, Yang X, Yin H, 2016. Mitigating climate change through managing constructed-microbial communities in agriculture. Agric Ecosyst Environ 216: 304-308.

Han HS, Lee KD, 2006. Effect of co-inoculation with phosphate and potassium solubilizing bacteria on mineral uptake and growth of pepper and cucumber. Plant Soil Environ 52: 130.

Hingole SS, Pathak AP, 2016. Saline soil microbiome: a rich source of halotolerant PGPR. J Crop Sci Biotechnol 19: 231-239.

Hirrel MC, Gerdemann JW, 1980. Improved growth of onion and bell pepper in saline soils by two vesicular-*Arbuscular mycorrhizal* fungi. 1. Soil Sci Soc Am J 44: 654-655.

Illmer P, Schinner F, 1992. Solubilization of inorganic phosphates by microorganisms isolated from forest soils. Soil Biol Biochem 24: 389-395.

Iniguez AL, Dong Y, Carter HD, Ahmer BM, Stone JM, Triplett EW, 2005. Regulation of enteric endophytic bacterial colonization by plant defenses. Mol Plant-Microbe Interact MPMI 18: 169-178. https://doi.org/10.1094/mpmi-18-0169.

Iqbal Hussain M, Naeem Asghar H, Javed Akhtar M, Arshad M, 2013. Impact of phosphate solubilizing bacteria on growth and yield of maize. Soil Environ 32: 71-78.

Iqbal U, Jamil N, Ali I, Hasnain S, 2010. Effect of zinc-phosphate-solubilizing bacterial isolates on growth of *Vigna radiata*. Ann Microbiol 60: 243-248.

Jamil A, Riaz S, Ashraf M, Foolad MR, 2011. Gene expression profiling of plants under salt stress. Crit Rev Plant Sci 30: 435-458. https://doi.org/10.1080/07352689.2011.605739.

Jebara S, Drevon JJ, Jebara M, 2010. Modulation of symbiotic efficiency and nodular antioxidant enzyme activities in two *Phaseolus vulgaris* genotypes under salinity. Acta Physiol Plant 32: 925-932.

Jha B, Gontia I, Hartmann A, 2012. The roots of the halophyte *Salicornia brachiata* are a source of new halotolerant diazotrophic bacteria with plant growth-promoting potential. Plant Soil 356: 265-277.

Joe MM, Devaraj S, Benson A, Sa T, 2016. Isolation of phosphate solubilizing endophytic bacteria from *Phyllanthus amarus* Schum & Thonn: evaluation of plant growth promotion and antioxidant activity under salt stress. J Appl Res Med Aromat Plants 3: 71-77.

Johnston AE, Krauss A, 1998. The essential role of potassium in diverse cropping systems: future research needs and benefits. In: Essential role of potassium in diverse cropping systems. International Potash Institute, Basel, pp. 101-120.

Johri JK, Surange S, Nautiyal CS, 1999. Occurrence of salt, pH, and temperature-tolerant, phosphate-solubilizing bacteria in alkaline soils. Curr Microbiol 39: 89-93.

Jones DL, Oburger E, 2011. Solubilization of phosphorus by soil microorganisms. In: Phosphorus in action. Springer, pp. 169-198.

Karlidag H, Yildirim E, Turan M, Pehluvan M, Donmez F, 2013. Plant growth-promoting rhizobacteria mitigate deleterious effects of salt stress on strawberry plants (*Fragaria×ananassa*). Hortscience 48: 563-567.

Katznelson H, Peterson EA, Rouatt JW, 1962. Phosphate-dissolving microorganisms on seed and in the root zone of plants. Can J Bot 40: 1181-1186.

Kaya C, Ashraf M, Sonmez O, Aydemir S, Tuna AL, Cullu MA, 2009a.The influence of Arbuscular mycorrhizal colonisation on key growth parameters and fruit yield of pepper plants grown at high salinity. Sci Horticult 121: 1-6.

Kaya C, Tuna AL, Yokaş I, 2009b.The role of plant hormones in plants under salinity stress. In: Salinity and water stress. Springer, pp. 45-50.

Kaymak HC, Guvenc I, Yarali F, Donmez MF, 2009. The effects of bio-priming with PGPR on germination of radish (*Raphanus sativus* L.) seeds under saline conditions. Turk J Agric For 33: 173-179.

Khadri M, Tejera NA, Lluch C, 2006. Alleviation of salt stress in common bean (*Phaseolus vulgaris*) by exogenous abscisic acid supply. J Plant Growth Regul 25: 110-119.

Khan MS, Zaidi A, Wani PA, 2007. Role of phosphate-solubilizing microorganisms in sustainable agriculture—a review. Agron Sustain Dev 27: 29-43.

Khan MS, Zaidi A, Wani PA, Ahemad M, Oves M, 2009. Functional diversity among plant growth-promoting rhizobacteria: current status. In: Microbial strategies for crop improvement. Springer, pp. 105-132.

Khan AL, Hamayun M, Kang SM, Kim YH, Jung HY, Lee JH, Lee IJ, 2012. Endophytic fungal association via gibberellins and indole acetic acid can improve plant growth under abiotic stress: an example of *Paecilomyces formosus* LHL10. BMC Microbiol 12: 1.

Kim J, Rees DC, 1994. Nitrogenase and biological nitrogenfixation. Biochemistry 33: 389-397. https://doi.org/10.1021/bi00168a001.

Kim KY, Jordan D, McDonald GA, 1997. Effect of phosphate-solubilizing bacteria and vesicular-Arbuscular mycorrhizae on tomato growth and soil microbial activity. Biol Fertil Soils 26: 79-87.

Kloepper JW, Leong J, Teintze M, Schroth MN, 1980. Enhanced plant growth by siderophores produced by plant growth-promoting rhizobacteria. Nature 286: 885-886.

Kobayashi T, Nishizawa NK, 2012. Iron uptake, translocation, and regulation in higher plants. Annu Rev Plant Biol 63: 131-152.

Koide RT, Kabir Z, 2000. Extraradical hyphae of the mycorrhizal fungus *Glomus intraradices* can hydrolyse organic phosphate. New Phytol 148: 511-517.

Krüger M, Krüger C, Walker C, Stockinger H, Schüßler A, 2012. Phylogenetic reference data for systematics and phylotaxonomy of *Arbuscular mycorrhizal* fungi from phylum to species level. New Phytol 193: 970-984.

Kucey RMN, 1983. Phosphate-solubilizing bacteria and fungi in various cultivated and virgin Alberta soils. Can J Soil Sci 63: 671-678.

Kucey RMN, Janzen HH, Leggett ME, 1989. Microbially mediated increases in plant-available phosphorus. In: Advances in agronomy, vol 42. Elsevier, pp. 199-228.

Kumar A, Sharma S, Mishra S, 2010. Influence of *Arbuscular mycorrhizal* (AM) fungi and salinity on seedling growth, solute accumulation, and mycorrhizal dependency of *Jatropha curcas* L. J Plant Growth Regul 29: 297-306.

Latour X, Delorme S, Mirleau P, Lemanceau P, 2009. Identification of traits implicated in the rhizosphere competence of *fluorescent pseudomonads*: description of a strategy based on population and model strain studies. In: Sustainable agriculture. Springer, pp. 285-296.

Long RL et al., 2008. Seed persistence in thefield may be predicted by laboratory-controlled aging. Weed Sci 56: 523-528.

López M, Lluch C, 2008. Nitrogenfixation is synchronized with carbon metabolism in *Lotus japonicus* and *Medicago truncatula* nodules under salt stress. J Plant Interact 3: 137-144.

López M, Tejera NA, Iribarne C, Lluch C, Herrera-Cervera JA, 2008. Trehalose and trehalase in root nodules of *Medicago truncatula* and *Phaseolus vulgaris* in response to salt stress. Physiol Plant 134: 575-582.

Ma JF, 2005. Plant root responses to three abundant soil minerals: silicon, aluminum and iron. Crit Rev Plant Sci 24: 267-281.

Ma Y, Galinski EA, Grant WD, Oren A, Ventosa A, 2010. Halophiles 2010: life in saline environments. Appl Environ Microbiol 76: 6971-6981.

Mapelli F et al., 2013. Potential for plant growth promotion of rhizobacteria associated with *Salicornia* growing in Tunisian hypersaline soils. BioMed Res Int 2013: 248078.

Marasco R et al., 2012. A drought resistance-promoting microbiome is selected by root system under desert farming. PLoS One 7: e48479.

Margesin R, Schinner F, 2001. Potential of halotolerant and halophilic microorganisms for biotechnology. Extremophiles 5: 73-83.

Marschner H, 1995. Mineral nutrition of higher plants, 2nd edn. Academic Press, London.

Marschner H, Dell B, 1994. Nutrient uptake in mycorrhizal symbiosis. Plant Soil 159: 89-102.

Mayak S, Tirosh T, Glick BR, 2004a. Plant growth-promoting bacteria confer resistance in tomato plants to salt stress. Plant Physiol Biochem 42: 565-572.

Mayak S, Tirosh T, Glick BR, 2004b. Plant growth-promoting bacteria that confer resistance to water stress in tomatoes and peppers. Plant Sci 166: 525-530.

McLean EO, Watson ME, 1985. Soil measurements of plant-available potassium. In: Potassium in agriculture, pp. 277-308.

McMillen BG, Juniper S, Abbott LK, 1998. Inhibition of hyphal growth of avesicular-*Arbuscular mycorrhizal* fungus in soil containing sodium chloride limits the spread of infection from spores. Soil Biol Biochem 30: 1639-1646.

Meena VS, Maurya BR, Verma JP, 2014. Does a rhizospheric microorganism enhance K^+ availability in agricultural soils? Microbiol Res 169: 337-347.

Meena VS, Maurya BR, Verma JP, Aeron A, Kumar A, Kim K, Bajpai VK, 2015. Potassium solubilizing rhizobacteria (KSR): isolation, identification, and K-release dynamics from waste mica. Ecol Eng 81: 340-347.

Mensah JK, Ihenyen J, 2009. Effects of salinity on germination, seedling establishment and yield of three genotypes of mung bean (*Vigna mungo* L. Hepper) in Edo State, Nigeria. Niger Ann Nat Sci 8: 17-24.

Miller SH, Browne P, Prigent-Combaret C, Combes-Meynet E, Morrissey JP, O'Gara F, 2010. Biochemical and genomic comparison of inorganic phosphate solubilization in *Pseudomonas* species. Environ Microbiol Rep 2: 403-411.

Mills HAJJ et al., 1996. Plant analysis handbook II: a practical preparation, analysis, and interpretation guide. Potash and Phosphate Institute.

Milošević NA, Marinković JB, Tintor BB, 2012. Mitigating abiotic stress in crop plants by microorganisms. Zbornik Matice Srpske Za Prirodne Nauke 17-26.

Miransari M, 2013. Soil microbes and the availability of soil nutrients. Acta Physiol Plant 35: 3075-3084.

Miransari M, Smith D, 2008. Using signal molecule genistein to alleviate the stress of suboptimal root zone temperature on soybean-*Bradyrhizobium* symbiosis under different soil textures. J Plant Interact 3: 287-295.

Mohammad MJ, Malkawi HI, Shibli R, 2003. Effects of *Arbuscular mycorrhizal* fungi and phosphorus fertilization on growth and nutrient uptake of barley grown on soils with different levels of salts. J Plant Nutr 26: 125-137.

Moradi A, Tahmourespour A, Hoodaji M, Khors F, 2011. Effect of salinity on free living-diazotroph and total bacterial populations of two saline soils. Afr J Microbiol Res 5: 144-148.

Müller J, Boller T, Wiemken A, 2001. Trehalose becomes the most abundant non-structural carbohydrate during senescence of soybean nodules. J Exp Bot 52: 943-947.

Munns R, 2005. Genes and salt tolerance: bringing them together. New Phytol 167: 645-663.

Munns R, Tester M, 2008. Mechanisms of salinity tolerance. Annu Rev Plant Biol 59: 651-681. https://doi.org/10.1146/annurev.arplant.59.032607.092911.

Nabiollahy K, Khormali F, Bazargan K, Ayoubi S, 2006. Forms of K as a function of clay mineralogy and soil development. Clay Miner 41: 739-749.

Nadeem SM, Zahir ZA, Naveed M, Arshad M, 2007. Preliminary investigations on inducing salt tolerance in maize through inoculation with rhizobacteria containing ACC deaminase activity. Can J Microbiol 53: 1141-1149.

Nadeem SM, Zahir ZA, Naveed M, Arshad M, 2009. Rhizobacteria containing ACC-deaminase confer salt tolerance in maize grown on salt-affectedfields. Can J Microbiol 55: 1302-1309.

Nadeem SM, Zahir ZA, Naveed M, Nawaz S, 2013. Mitigation of salinity-induced negative impact on the growth and yield of wheat by plant growth-promoting rhizobacteria in naturally saline conditions. Ann Microbiol 63: 225-232.

Navarro JM, Botella MA, Cerdá A, Martinez V, 2001. Phosphorus uptake and translocation in salt-stressed melon plants. J Plant Physiol 158: 375-381.

Naz I, Bano A, Ul-Hassan T, 2009. Isolation of phytohormones producing plant growth promoting rhizobacteria from weeds growing in Khewra salt range, Pakistan and their implication in providing salt tolerance to *Glycine max* L. Afr J Biotechnol 8: 5762-5768.

Ocón A, Hampp R, Requena N, 2007. Trehalose turnover during abiotic stress in *Arbuscular mycorrhizal* fungi. New Phytol 174: 879-891.

Ojala JC, Jarrell WM, Menge JA, Johnson ELV, 1983. Influence of mycorrhizal fungi on the mineral nutrition and yield of onion in saline soil. 1. Agron J 75: 255-259.

Orhan F, Gulluce M, 2015. Isolation and characterization of salt-tolerant bacterial strains in salt-affected soils of Erzurum, Turkey. Geomicrobiol J 32: 521-529.

Osorio Vega NW, 2007. A review on beneficial effects of rhizosphere bacteria on soil nutrient availability and plant nutrient uptake. Rev Fac Nac Agron Medellin 60: 3621-3643.

Owen D, Williams AP, Griffith GW, Withers PJA, 2015. Use of commercial bio-inoculants to increase agricultural production through improved phosphorus acquisition. Appl Soil Ecol 86: 41-54.

Parida AK, Das AB, 2005. Salt tolerance and salinity effects on plants: a review. Ecotoxicol Environ Saf 60: 324-349. https://doi.org/10.1016/j.ecoenv.2004.06.010.

Parmar P, Sindhu SS, 2013. Potassium solubilization by rhizosphere bacteria: influence of nutritional and environmental conditions. J Microbiol Res 3: 25-31.

Patel AD, Jadeja H, Pandey AN, 2010. Effect of salinization of soil on growth, water status and nutrient accumulation in seedlings of *Acacia auriculiformis* (Fabaceae). J Plant Nutr 33: 914-932.

Patreze CM, Cordeiro L, 2004. Nitrogen-fixing and vesicular-*Arbuscular mycorrhizal* symbioses in some tropical legume trees of tribe Mimoseae. For Ecol Manag 196: 275-285.

Paul D, 2012. Osmotic stress adaptations in rhizobacteria. J Basic Microbiol 52: 1-10.

Paul D, 2013. Osmotic stress adaptations in rhizobacteria. J Basic Microbiol 53: 101-110. https://doi.org/10.1002/jobm.201100288.

Paul D, Lade H, 2014. Plant-growth-promoting rhizobacteria to improve crop growth in saline soils: a review. Agron Sustain Dev 34: 737-752.

Paul D, Nair S, 2008. Stress adaptations in a Plant Growth Promoting Rhizobacterium (PGPR) with increasing salinity in the coastal agricultural soils. J Basic Microbiol 48: 378-384. https://doi.org/10.1002/jobm.200700365.

Paul D, Sarma YR, 2006. Plant growth promoting rhizobacteria (PGPR) -mediated root proliferation in black pepper (*Piper nigrum* L.) as evidenced through GS Root software. Arch Phytopathol Plant Protect 39: 311-314.

Penfold C, 2001. 32. In: May R (ed) Phosphorus management in broadacre organic farming systems, p. 179.

Pereira PAA, Bliss FA, 1989. Selection of common bean (*Phaseolus vulgaris* L.) for N_2 fixation at different levels of available phosphorus under field and environmentally-controlled conditions. Plant Soil 115: 75-82.

Pereira SIA, Castro PML, 2014. Phosphate-solubilizing rhizobacteria enhance *Zea mays* growth in agricultural P-deficient soils. Ecol Eng 73: 526-535.

Peters NK, Crist-Estes DK, 1989. Nodule formation is stimulated by the ethylene inhibitor aminoethoxyvinylglycine. Plant Physiol 91: 690-693.

Peterson RL, Massicotte HB, Melville LH, 2004. Mycorrhizas: anatomy and cell biology. NRC Research Press, Ottawa.

Pfetffer CM, Bloss HE, 1988. Growth and nutrition of guayule (*Parthenium argentatum*) in a saline soil as influenced by vesicular-arbuscular mycorrhiza and phosphorus fertilization. New Phytol 108: 315-321.

Piccoli P, Travaglia C, Cohen A, Sosa L, Cornejo P, Masuelli R, Bottini R, 2011. An endophytic bacterium isolated from roots of the halophyte *Prosopis strombulifera* produces ABA, IAA, gibberellins A1 and A3 and jasmonic acid in chemically-defined culture medium. Plant Growth Regul 64: 207-210.

Pii Y, Penn A, Terzano R, Crecchio C, Mimmo T, Cesco S, 2015. Plant-microorganism-soil interactions influence the Fe availability in the rhizosphere of cucumber plants. Plant Physiol Biochem 87: 45-52.

Pond EC, Menge JA, Jarrell WM, 1984. Improved growth of tomato in salinized soil by vesicular-*Arbuscular mycorrhizal* fungi collected from saline soils. Mycologia 76: 74-84.

Poonguzhali S, Madhaiyan M, Sa T, 2008. Isolation and identification of phosphate solubilizing bacteria from Chinese cabbage and their effect on growth and phosphorus utilization of plants. J Microbiol Biotechnol 18: 773-777.

Porcel R, Aroca R, Ruiz-Lozano JM, 2012. Salinity stress alleviation using *Arbuscular mycorrhizal* fungi. A review. Agron Sustain Dev 32: 181-200.

Porras-Soriano A, Soriano-Martín ML, Porras-Piedra A, Azcón R, 2009. *Arbuscular mycorrhizal* fungi increased growth, nutrient uptake and tolerance to salinity in olive trees under nursery conditions. J Plant Physiol 166: 1350-1359.

Poss JA, Pond E, Menge JA, Jarrell WM, 1985. Effect of salinity on mycorrhizal onion and tomato in soil with and without additional phosphate. Plant Soil 88: 307-319.

Prajapati K, Sharma MC, Modi HA, 2012. Isolation of two *potassium solubilizing* fungi from ceramic industry soils. Life Sci Leafl 5: 71-75.

Rabhi M, Barhoumi Z, Ksouri R, Abdelly C, Gharsalli M, 2007. Interactive effects of salinity and iron deficiency in *Medicago ciliaris*. C R Biol 330: 779-788.

Rabie G, Almadini AM, 2005. Role of bioinoculants in development of salt-tolerance of *Vicia faba* plants under salinity stress. Afr J Biotechnol 4: 210-222.

Radzki W, Mañero FJG, Algar E, García JAL, García-Villaraco A, Solano BR, 2013. Bacterial siderophores efficiently provide iron to iron-starved tomato plants in hydroponics culture. Antonie Van Leeuwenhoek 104: 321-330.

Ramadoss D, Lakkineni VK, Bose P, Ali S, Annapurna K, 2013. Mitigation of salt stress in wheat seedlings by halotolerant bacteria isolated from saline habitats. SpringerPlus 2: 1-6.

Rashid MI, Mujawar LH, Shahzad T, Almeelbi T, Ismail IMI, Oves M, 2016. Bacteria and fungi can contribute to nutrients bioavailability and aggregate formation in degraded soils. Microbiol Res 183: 26-41.

Rawal R, Kuligod VB, 2014. Influence of graded doses of nitrogen on nutrient uptake and grain yield of maize (*Zea mays*) under varying levels of soil salinity. Karnataka J Agric Sci 27: 22-24..

Rawat R, Tewari L, 2011. Effect of abiotic stress on phosphate solubilization by biocontrol fungus *Trichoderma* sp. Curr Microbiol 62: 1521-1526.

Rehman A, Nautiyal CS, 2002. Effect of drought on the growth and survival of the stress-tolerant bacterium *Rhizobium* sp. NBRI2505 sesbania and its drought-sensitive transposon Tn5 mutant. Curr Microbiol 45: 368-377.

Richardson AE, 2001. Prospects for using soil microorganisms to improve the acquisition of phosphorus by plants. Funct Plant Biol 28: 897-906.

Richardson AE, Simpson RJ, 2011. Soil microorganisms mediating phosphorus availability update on microbial phosphorus. Plant Physiol 156: 989-996.

Rinaldelli E, Mancuso S, 1996. Response of young mycorrhizal and non-mycorrhizal plants of olive tree (*Olea europaea* L.) to saline conditions. I. Short-term electrophysiological and long-term vegetative salt effects. Adv Hortic Sci 10: 126-134.

Rogers ME, Grieve CM, Shannon MC, 2003. Plant growth and ion relations in lucerne (*Medicago sativa* L.) in response to the combined effects of NaCl and P. Plant Soil 253: 187-194.

Rojas-Tapias D, Moreno-Galván A, Pardo-Díaz S, Obando M, Rivera D, Bonilla R, 2012. Effect of inoculation with plant growth-promoting bacteria (PGPB) on amelioration of saline stress in maize (*Zea mays*). Appl Soil Ecol 61: 264-272.

Roychoudhury P, Kaushik BD, 1989. Solubilization of Mussoorie rock phosphate by cyanobacteria. Curr Sci 58: 569-570.

Ruiz-Lozano JM, 2003. *Arbuscular mycorrhizal* symbiosis and alleviation of osmotic stress. New perspectives for molecular studies. Mycorrhiza 13: 309-317.

Ruppel S, Franken P, Witzel K, 2013. Properties of the halophyte microbiome and their implications for plant salt tolerance. Funct Plant Biol 40: 940-951.

Saber MSM, Zanati MR, 1984. Effectiveness of inoculation with silicate bacteria in relation to the potassium content of plants using the intensive cropping technique. Agric Res Rev.

Sadeghi A, Karimi E, Dahaji PA, Javid MG, Dalvand Y, Askari H, 2012. Plant growth promoting activity of an auxin and siderophore producing isolate of *Streptomyces* under saline soil conditions. World J Microbiol Biotechnol 28: 1503-1509. https://doi.org/10.1007/s11274-011-0952-7.

Sagoe CI, Ando T, Kouno K, Nagaoka T, 1998. Relative importance of protons and solution calcium concentration in phosphate rock dissolution by organic acids. Soil Sci Plant Nutr 44: 617-625.

Saleem M, Arshad M, Hussain S, Bhatti AS, 2007. Perspective of plant growth promoting rhizobacteria (PGPR) containing ACC deaminase in stress agriculture. J Ind Microbiol Biotechnol 34: 635-648.

Sánchez-Porro C, Rafael R, Soto-Ramírez N, Márquez MC, Montalvo-Rodríguez R, Ventosa A, 2009. Description of *Kushneria aurantia* gen. nov., sp. nov., a novel member of the family Halomonadaceae, and a proposal for reclassification of *Halomonas marisflavi* as *Kushneria marisflavi* comb. nov., of *Halomonas indalinina* as *Kushneria indalinina* comb. nov. and of *Halomonas avicenniae* as *Kushneria avicenniae* comb. nov. Int J Syst Evol Microbiol 59: 397-405.

Sangeeth KP, Bhai RS, Srinivasan V, 2012. *Paenibacillus glucanolyticus*, a promising potassium solubilizing bacterium isolated from black pepper (*Piper nigrum* L.) rhizosphere. J Spices Aromat Crop 21: 118-124.

Sannazzaro AI, Echeverría M, Albertó EO, Ruiz OA, Menéndez AB, 2007. Modulation of polyamine balance in *Lotus glaber* by salinity and *Arbuscular mycorrhiza*. Plant Physiol Biochem 45: 39-46.

Scavino AF, Pedraza RO, 2013. The role of siderophores in plant growth-promoting bacteria. In: Bacteria in agrobiology: crop productivity. Springer, pp. 265-285.

Sengupta A, Chaudhuri S, 2002. *Arbuscular mycorrhizal* relations of mangrove plant community at the Ganges river estuary in India. Mycorrhiza 12: 169-174.

Sgroy V, Cassán F, Masciarelli O, Del Papa MF, Lagares A, Luna V, 2009. Isolation and characterization of endophytic plant growth-promoting (PGPB) or stress homeostasis-regulating (PSHB) bacteria associated to the halophyte *Prosopis strombulifera*. Appl Microbiol Biotechnol 85: 371-381.

Sha Valli Khan PS, Nagamallaiah GV, Dhanunjay Rao M, Sergeant K, Hausman JF, 2014. Chapter 2: Abiotic stress tolerance in plants: insights from proteomics. In: Emerging technologies and management of crop stress tolerance. Academic, San Diego, pp. 23-68. https://doi.org/10.1016/B978-0-12-800875-1.00002-8.

Shamseldin A, Werner D, 2004. Selection of competitive strains of *Rhizobium* nodulating *Phaseolus vulgaris* and adapted to environmental conditions in Egypt, using the gus-reporter gene technique. World J Microbiol Biotechnol 20: 377-382.

Shamseldin A, Werner D, 2005. High salt and high pH tolerance of new isolated *Rhizobium etli* strains from Egyptian soils. Curr Microbiol 50: 11-16.

Shannon MC, Grieve CM, 1998. Tolerance of vegetable crops to salinity. Sci Horticult 78: 5-38.

Sharifi M, Ghorbanli M, Ebrahimzadeh H, 2007. Improved growth of salinity-stressed soybean after inoculation with salt pre-treated *mycorrhizal* fungi. J Plant Physiol 164: 1144-1151.

Sharma SB, Sayyed RZ, Trivedi MH, Gobi TA, 2013. Phosphate solubilizing microbes: sustainable approach for managing phosphorus deficiency in agricultural soils. SpringerPlus 2: 587.

Sharma S, Kulkarni J, Jha B, 2016. Halotolerant rhizobacteria promote growth and enhance salinity tolerance in peanut. Front Microbiol 7: 1600.

Shedova E, Lipasova V, Velikodvorskaya G, Ovadis M, Chernin L, Khmel I, 2008. *Arbuscular mycorrhizal* relations of mangrove plant community at the Ganges river estuary in India. Mycorrhiza 53: 110-114.

Sheng XF, 2005. Growth promotion and increased potassium uptake of cotton and rape by a potassium releasing strain of *Bacillus edaphicus*. Soil Biol Biochem 37: 1918-1922.

Sheng XF, He LY, 2006. Solubilization of potassium-bearing minerals by a wild-type strain of *Bacillus edaphicus* and its mutants and increased potassium uptake by wheat. Can J Microbiol 52: 66-72.

Sheng M, Tang M, Chen H, Yang B, Zhang F, Huang Y, 2008. Influence of *Arbuscular mycorrhizae* on photosynthesis and water status of maize plants under salt stress. Mycorrhiza 18: 287-296.

Shrivastava P, Kumar R, 2015. Soil salinity: a serious environmental issue and plant growth promoting bacteria as one of the tools for its alleviation. Saudi J Biol Sci 22: 123-131.

Shukla PS, Agarwal PK, Jha B, 2012. Improved salinity tolerance of *Arachis hypogaea* (L.) by the interaction of halotolerant plant-growth-promoting *rhizobacteria*. J Plant Growth Regul 31: 195-206.

Siddikee MA, Chauhan PS, Anandham R, Han GH, Sa T, 2010. Isolation, characterization, and use for plant growth promotion under salt stress, of ACC deaminase-producing halotolerant bacteria derived from coastal soil. J Microbiol Biotechnol 20: 1577-1584.

Siddikee MA, Glick BR, Chauhan PS, Jong Yim W, Sa T, 2011. Enhancement of growth and salt tolerance of red pepper seedlings (*Capsicum annuum* L.) by regulating stress ethylene synthesis with halotolerant bacteria containing 1-aminocyclopropane-1-carboxylic acid deaminase activity. Plant Physiol Biochem 49: 427-434.

Siddiqui MH, Mohammad F, Khan MN, 2009. Morphological and physio-biochemical characterization of *Brassica juncea* L. Czern. & Coss. genotypes under salt stress. J Plant Interact 4: 67-80.

Sieverding E, da Silva GA, Berndt R, Oehl F, 2015. Rhizoglomus, a new genus of the Glomeraceae. Mycotaxon 129: 373-386.

Singh RP, Jha PN, 2016. A halotolerant bacterium *Bacillus licheniformis* HSW-16 augments induced systemic tolerance to salt stress in wheat plant (*Triticum aestivum*). Front Plant Sci 7: 1890.

Smith SE, Read DJ, 2010. Mycorrhizal symbiosis. Academic.

Smith SE, Jakobsen I, Grønlund M, Smith FA, 2011. Roles of *Arbuscular mycorrhizas* in plant phosphorus nutrition: interactions between pathways of phosphorus uptake in *Arbuscular mycorrhizal* roots have important implications for understanding and manipulating plant phosphorus acquisition. Plant Physiol 156: 1050-1057.

Snapp S, Koide R, Lynch J, 1995. Exploitation of localized phosphorus-patches by common bean roots. Plant Soil 177: 211-218.

Soleimani M, Akbar S, Hajabbasi MA, 2011. Enhancing phytoremediation efficiency in response to environmental pollution stress. Plants Environ 23: 10-14.

Sparks DL, Huang PM, 1985. Physical chemistry of soil potassium. In: Potassium in agriculture, pp. 201-276.

Srinivasan R, Yandigeri MS, Kashyap S, Alagawadi AR, 2012. Effect of salt on survival and P-solubilization potential of phosphate solubilizing microorganisms from salt affected soils. Saudi J Biol Sci 19: 427-434.

Steffens D, Sparks DL, 1997. Kinetics of nonexchangeable ammonium release from soils. Soil Sci Soc Am J 61: 455-462.

Stevenson FJ, Cole MA, 1999. Cycles of soils: carbon, nitrogen, phosphorus, sulfur, micronutrients. Wiley, Hoboken.

Sugawara M, Okazaki S, Nukui N, Ezura H, Mitsui H, Minamisawa K, 2006. Rhizobitoxine modulates plant-microbe interactions by ethylene inhibition. Biotechnol Adv 24: 382-388. htps://doi.org/10.1016/j.biotechadv.2006.01.004.

Szymańska S, Płociniczak T, Piotrowska-Seget Z, Złoch M, Ruppel S, Hrynkiewicz K, 2016. Metabolic potential and community structure of endophytic and rhizosphere bacteria associated with the roots of the halophyte *Aster tripolium* L. Microbiol Res 182: 68-79.

Tang C, Robson AD, Dilworth MJ, 1990. A split-root experiment shows that iron is required for nodule initiation in *Lupinus angustifolius* L. New Phytol 115: 61-67.

Teng S, Liu Y, Zhao L, 2010. Isolation, identification and characterization of ACC deaminase-containing endophytic bacteria from halophyte *Suaeda salsa*. Wei Sheng Wu Xue Bao=ActaMicrobiol Sin 50: 1503-1509.

Tester M, Davenport R, 2003. Na^+ tolerance and Na^+ transport in higher plants. Ann Bot 91: 503-527.

Theunis M, 2005. IAA biosynthesis in rhizobia and its potential role in symbiosis. PhD thesis, Universiteit Antwerpen.

Tiwari S, Singh P, Tiwari R, Meena KK, Yandigeri M, Singh DP, Arora DK, 2011. Salt-tolerant rhizobacteria-mediated induced tolerance in wheat (*Triticum aestivum*) and chemical diversity in rhizosphere enhance plant growth. Biol Fertil Soils 47: 907.

Toljander JF, Lindahl BD, Paul LR, Elfstrand M, Finlay RD, 2007. Influence of *Arbuscular mycorrhizal* mycelial exudates on soil bacterial growth and community structure. FEMS Microbiol Ecol 61: 295-304.

Tributh HV, Boguslawski EV, Lieres AV, Steffens D, Mengel K, 1987. Effect of potassium removal by crops on transformation of illitic clay minerals. Soil Sci 143: 404-409.

Upadhyay SK, Singh DP, 2015. Effect of salt-tolerant plant growth-promoting rhizobacteria on wheat plants and soil health in a saline environment. Plant Biol 17: 288-293.

Upadhyay S, Singh D, Saikia R, 2009. Genetic diversity of plant growth promoting rhizobacteria isolated from rhizospheric soil of wheat under saline condition. Curr Microbiol 59: 489-496. https://doi.org/10.1007/s00284-009-9464-1.

Uroz S, Calvaruso C, Turpault MP, Frey-Klett P, 2009. Mineral weathering by bacteria: ecology, actors and mechanisms. Trends Microbiol 17: 378-387.

van Hoorn JW, Katerji N, Hamdy A, Mastrorilli M, 2001. Effect of salinity on yield and nitrogen uptake of four grain legumes and on biological nitrogen contribution from the soil. Agric Water Manag 51: 87-98. https://doi.org/10.1016/S0378-3774(01)00114-7.

Van Vuuren DP, Bouwman AF, Beusen AHW, 2010. Phosphorus demand for the 1970-2100 period: a scenario analysis of resource depletion. Glob Environ Chang 20: 428-439.

Vassilev N, Fenice M, Federici F, 1996. Rock phosphate solubilization with gluconic acid produced by immobilized *Penicillium variabile* P16. Biotechnol Tech 10: 585-588.

Venkateswarlu B, Rao AV, Raina P, 1984. Evaluation of phosphorus solubilisation by microorganisms isolated from Aridisols. J Indian Soc Soil Sci 32: 273-277.

Venkateswarlu B, Desai S, Prasad YG, 2008. Agriculturally important microorganisms for stressed ecosystems: challenges in technology development and application. In: Khachatourians GG, Arora DK, Rajendran TP, Srivastava AK (eds) Agriculturally important, Microorganisms, vol 1. Academic World, Bhopal, pp. 225-246.

Veresoglou SD, Shaw LJ, Sen R, 2011. *Glomus intraradices* and *Gigaspora margarita*, *Arbuscular mycorrhizal* associations differentially affect nitrogen and potassium nutrition of *Plantago lanceolata* in a low fertility dune soil. Plant Soil 340: 481-490.

Vreeland RH, 1987. Mechanisms of halotolerance in microorganisms. CRC Crit Rev Microbiol 14: 311-356.

Waller F et al., 2005. The endophytic fungus *Piriformospora indica* reprograms barley to salt-stress tolerance, disease resistance, and higher yield. Proc Natl Acad Sci U S A 102: 13386-13391.

Wang Q, Dodd IC, Belimov AA, Jiang F, 2016. Rhizosphere bacteria containing. 1-aminocyclopropane-1-carboxylate deaminase increase growth and photosynthesis of pea plants under salt stress by limiting Na^+ accumulation. Funct Plant Biol 43: 161-172.

Werner D, 1992. Symbiosis of plants and microbes. Chapman & Hall, London.

Whitelaw MA, Harden TJ, Bender GL, 1997. Plant growth promotion of wheat inoculated with *Penicillium radicum* sp. nov. Soil Res 35: 291-300.

Wild A, 1988. Plant nutrients in soil: phosphate. In: Russell's soil conditions and plant growth.

Williams A, Ridgway HJ, Norton DA, 2013. Different *Arbuscular mycorrhizae* and competition with an exotic grass affect the growth of *Podocarpus cunninghamii* Colenso cuttings. New For 44: 183-195.

Wu SC, Cao ZH, Li ZG, Cheung KC, Wong MH, 2005. Effects of biofertilizer containing N-fixer, P and K solubilizers and AM fungi on maize growth: a greenhouse trial. Geoderma 125: 155-166.

Wu QS, Xia RX, Zou YN, 2008. Improved soil structure and citrus growth after inoculation with three *Arbuscular mycorrhizal* fungi under drought stress. Eur J Soil Biol 44: 122-128.

Wu QS, Zou YN, He XH, 2010. Contributions of *Arbuscular mycorrhizal* fungi to growth, photosynthesis, root morphology and ionic balance of citrus seedlings under salt stress. Acta Physiol Plant 32: 297-304.

Wu Z, Yue H, Lu J, Li C, 2012. Characterization of rhizobacterial strain Rs-2 with ACC deaminase activity and its performance in promoting cotton growth under salinity stress. World J Microbiol Biotechnol 28: 2383-2393.

Xie H, Pasternak JJ, Glick BR, 1996. Isolation and characterization of mutants of the plant growth-promoting rhizobacterium *Pseudomonas putida* GR12-2 that overproduce indoleacetic acid. Curr Microbiol 32: 67-71.

Yang J, Kloepper JW, Ryu CM, 2009. *Rhizosphere bacteria* help plants tolerate abiotic stress. Trends Plant Sci 14: 1-4.

Yao L, Wu Z, Zheng Y, Kaleem I, Li C, 2010. Growth promotion and protection against salt stress by *Pseudomonas putida* Rs-198 on cotton. Eur J Soil Biol 46: 49-54.

Yasmin H, Bano A, 2011. Isolation and characterization of phosphate solubilizing bacteria from rhizosphere soil of weeds of Khewra salt range and Attock. Pak J Bot 43: 1663-1668.

Yildirim E, Taylor AG, 2005. Effect of biological treatments on growth of bean plans under salt stress. Ann Rep Bean Improv Coop 48: 176-177.

Yoon JH, Choi SH, Lee KC, Kho YH, Kang KH, Park YH, 2001. *Halomonas marisflavae* sp. nov., a halophilic bacterium isolated from the Yellow Sea in Korea. Int J Syst Evol Microbiol 51: 1171-1177.

Yousfi S, Mahmoudi H, Abdelly C, Gharsalli M, 2007. Effect of salt on physiological responses of barley to iron deficiency. Plant Physiol Biochem 45: 309-314.

Zacarías JJJ, Altamirano-Hernández J, Cabriales JJP, 2004. Nitrogenase activity and trehalose content of nodules of drought-stressed common beans infected with effective (Fix+) and ineffective (Fix) rhizobia. Soil Biol Biochem 36: 1975-1981.

Zhang L, Fan J, Ding X, He X, Zhang F, Feng G, 2014. Hyphosphere interactions between an *Arbuscular mycorrhizal* fungus and a phosphate solubilizing bacterium promote phytate mineralization in soil. Soil Biol Biochem 74: 177−183.

Zhou N, Zhao S, Tian CY, 2017. Effect of halotolerant rhizobacteria isolated from halophytes on the growth of sugar beet (*Beta vulgaris* L.) under salt stress. FEMS Microbiol Lett 364. https://doi.org/10.1093/femsle/fnx091.

Zhu F, Qu L, Hong X, Sun X, 2011. Isolation and characterization of a phosphate−solubilizing halophilic bacterium *Kushneria* sp. YCWA18 from Daqiao Saltern on the coast of Yellow Sea of China. Evid Based Complement Alternat Med 2011.

Zuccarini P, Okurowska P, 2008. Effects of mycorrhizal colonization and fertilization on growth and photosynthesis of sweet basil under salt stress. J Plant Nutr 31: 497−513.

Zwetsloot MJ, Lehmann J, Bauerle T, Vanek S, Hestrin R, Nigussie A, 2016. Phosphorus availabilityfrom bone char in a P−fixing soil influenced by root−mycorrhizae−biochar interactions. Plant Soil 408: 95−105.

6 盐生植物内生细菌：如何帮助植物缓解盐胁迫危害

Ignacio D. Rodríguez-Llorente, Eloisa Pajuelo,
Salvadora Navarro-Torre, Jennifer Mesa-Marín, Miguel A. Caviedes

[摘要]
  土壤盐渍化是造成全球农作物减产的主要因素之一。开发以生物系统为基础的盐渍化农业解决方案，比如培育相关农作物品种使其可以在遭受到盐分和微生物富集的盐渍化土壤中良好生长，这样的农业措施正在引起人们的关注。盐生植物是一类能够在高盐浓度环境中正常生存和繁殖的植物种类，它们也是开展植物对盐胁迫环境适应性研究的良好模式植物。植物的生长和环境适应能力（特别是在逆境条件下）受到根际和内生微生物的高度影响。盐生植物根际微生物群需要适应土壤中的高浓度盐分，还能在高浓度盐分存在的条件下促进植物的生长。从生长在盐渍化土壤环境中的盐生植物中分离得到的内生细菌可以通过改变植物激素种类和含量、农作物对营养元素的吸收，以及通过调节1-氨基环丙烷-1-羧酸脱氨酶活性、磷酸盐增溶、氮的固定、吲哚-3-乙酸、脱落酸、铁载体和挥发物的产生等在内的多种不同机制，减轻农作物在遭受盐胁迫危害时细胞内产生的活性氧物质对细胞的损伤作用，从而在植株水平上缓解农作物受到的盐渍化胁迫，并促进其生长。本章综述了从盐生植物中分离出的主要内生菌菌属，综述并讨论了其在缓解农作物遭受到的盐胁迫危害中是如何发挥作用的。

[关键词]
  内生细菌；盐生植物；盐胁迫；植物生长促生菌

I. D. Rodríguez-Llorente (✉), E. Pajuelo, S. Navarro-Torre, J. Mesa-Marín, M. A. Caviedes, 西班牙塞维利亚大学药学院微生物学和寄生虫学系, E-mail: I. D. Rodríguez-Llorente: irodri@us.es

6.1 前言

土壤盐渍化是一个全球性问题，目前估计可能影响到 10 亿~100 亿 hm^2 的土地，并且还在以每年 15% 的速度增加（Yensen 和 Biel，2006）。据估计，可能多达 50% 的灌溉土地受到盐分或碱度的影响（Pitman 和 Lauchli，2002）。作物生长环境问题是影响世界各地作物减产的主要因素之一，因为盐分严重影响植物的生长（Gerhardt 等，2017）。为了能够养活世界上不断增长的人口，必须对受盐胁迫影响的生态系统进行植被恢复和补救。

淋溶（lixiviation）是降低土壤盐分的一种传统方式，但是它需要大量的水，还需要添加有机化合物和化学产品，这种方式投入成本高，还会使土壤贫瘠（Jesus 等，2015）。而涉及植物和相关微生物的生物系统修复方式在盐渍土修复中越来越受到关注（Qin 等，2016）。

植物耐盐的生理和分子机制已经被广泛研究和报道（Deinlein 等，2014；Gupta 和 Huang，2014；Mickelbart 等，2015；Zhu，2016）。这些机制已经被用于制定通过植物来缓解作物盐分胁迫的策略（Ismail 和 Horie，2017）。这种策略的很大一部分是建立在转基因植物的基础上的，通过在植物中过表达与植物盐适应相关的基因来实现（Roy 等，2014）。尽管这些方法仍在发展中，但有受到几个方面限制，包括：①需要大量的时间和努力（Coleman Derr 和 Tringe，2014）；②突变株有时不稳定（Jewell 等，2010）；③有很多重要的四倍体和六倍体作物不适合用这项分子技术（Kumar 等，2015）；④在很大程度上没有得到公众的接受，而且这种技术在未来能否取得公众认可仍然是不确定的（Fedoroff 等，2010）。此外，盐胁迫在本质上经常与其他非生物胁迫（如碱胁迫）联系在一起（Bui 等，2014），并存在有机和金属污染物污染造成的胁迫影响。

由于所有这些限制因素，可选择的生态策略，比如能够很好地在受盐分胁迫和与这些植物相关的微生物影响的土壤中生长的植物或品种，在治理土壤盐渍化的过程中越来越受到重视。在这种情况下，盐生植物能够在盐浓度超过 200 mmol/L NaCl 的环境中生存和繁殖（Flowers 和 Colmer，2008），是研究植物对盐环境的适应和制定提高植物耐盐性策略的极好模型（Shabala，2013）。

人们普遍认为，植物的生长和适应，特别是在胁迫条件下，受到定殖在植物根际和植物内生微生物（即植物微生物群）的影响（Bhattacharyya 和 Jha，2012）。这些微生物大多具有促进植物生长（plant growth-promoting，PGP）的特性，通过各自直接或间接的机制促进植物生长。虽然很难区分根际微生物和内生微生物（大多数根际微生物在根表面定殖，甚至渗透到根皮层），本章将重点讨论内生菌，特别是从盐生植物中分离出来的内生菌帮助植物缓解盐胁迫的机制。

6.2 盐生植物及其相关微生物

与人类微生物研究相比，植物微生物的研究仍处于起步阶段（Bai 等，2015）。在过去 10 年中，对于盐生植物微生物组群只有很少的研究，并且大多数研究仅仅是描述性的结果，试图对微生物物种进行分类和计数（Ruppel 等，2013）。近年来，人们的研究聚焦于从盐生植物根际分离微生物（特别是细菌），并对其植物促生特性（Plant growth-promoting，PGP）和对 NaCl 的耐性进行了筛选。比如从生长在沙特阿拉伯的海马齿属植物（*Sesuvium verrucosum*）（El Awady 等，2015）、生长在西班牙的互花米草（*Spartina maritima*）和蝎节木属植物（*Arthrocnemum Macrostachum*）（Mesa 等，2015a；Navarro Torre 等，2016），以及从土耳其的咸猪毛菜（*Salsola grandis*）（Kataoka 等，2017）中分离出了植物生长促进细菌。也有从盐生植物中分离内生微生物的例子，比如从生长在印度的臂状盐角草（*Salicornia brachiata*）、绒毛戴星草（*Sphaeranthus indicus*）、扫帚花属植物（*Cressa cretica*）和裸地碱蓬（*Suaeda nudiflora*）植物（Arora 等，2014）；生长于葡萄牙的合滨藜属植物（*Halimione portulacoides*）（Fidalgo 等，2016），以及生长在西班牙的互花米草（*Spartina maritima*）和蝎节木属植物（*Arthrocnemum Macrostachum*）（Mesa 等，2015b；Navarro Torre 等，2016；Navarro Torre 等，2017a）。

正如所期望的，从盐生植物微生物群中分离出的大多数细菌都是耐盐菌或嗜盐菌，即它们能够在高浓度 NaCl 或有 NaCl 的条件下生长。这些微生物是如何影响盐生植物适应盐渍化环境的，这也正在引起科学界的兴趣。对这些耐盐微生物的研究兴趣正在日益增长，这也是最近分离出大量具有促进植物生长特性的新微生物物种（表 6.1）的原因。对这些新微生物物种的研究大多需要对它们进行完整的基因组测序，为研究促进植物生长的机制提供良好的基因组资源。在这一背景下，盐生植物盐地碱蓬（*Suaeda salsa*）的全菌群研究表明，存在大量的与盐胁迫驯化和营养增溶过程相关的基因（Yuan 等，2016）。盐生植物微生物群必须适应土壤盐分，这一优势可用于设计盐生植物修复用细菌菌剂或维持盐渍土壤中的作物生产。从这个意义上讲，最近出版的一份关于从盐生植物中分离的根际细菌的详尽修订说明以及它们是如何帮助植物处理盐分胁迫的示例（Etesami 和 Beattie，2018）则很有意义。

表 6.1 近年从盐生植物微生物群中分离到的具有植物生长促生特性的新细菌种类

盐生植物	微生物新种	位置	参考文献
互花米草（Spartina maritima）	斯巴达海洋单胞菌（Marinomonas spartinae）	内生菌	Lucena 等（2016）
	斯巴达弧菌（Vibrio spartinae）	内生菌	Lucena 等（2017）
	根际微泡菌（Microbulbifer rhizosphaerae）	根际菌	Camacho 等（2016a）
蝎节木属植物（Arthrocnemum macrostachyum）	盐拉布伦茨氏菌（Labrenzia salina）	根际菌	Camacho 等（2016b）
	帕鲁斯特弧菌（Vibrio palustris）	内生菌	Lucena 等（2017）
	盐考克氏菌（Kocuria salina）	根际菌	Camacho 等（2017）
合滨藜属植物（Halimione portulacoides）	王祖农氏内生菌（Zunongwangia endophytica）	内生菌	Fidalgo 等（2017a）
	交替红色杆菌（Altererythrobacter halimionae）	内生菌	Fidalgo 等（2017b）
	植物内生交替红色杆菌（Altererythrobacter endophyticus）		
	科雷亚糖孢子菌（Saccharospirillum correiae）	内生菌	Fidalgo 等（2017c）
车前属植物（Plantago winteri）	普拉蒂萨尔西屈曲杆菌（Ancylobacter pratisalsi）	根际菌	Suarez 等（2017）
柽柳（Tamarix chinensis）	塔毛利盐藻（Salinicola tamaricis）	内生菌	Zhao 等（2017）
补血草（Limonium sinense）	嗜盐谷氨酸杆菌（Glutamicibacter halophytocola）	内生菌	Feng 等（2017）
高枝假木贼（Anabasis elatior）	内生阿尔塔米拉洞单胞菌（Aurantimonas endophytica）	内生菌	Liu 等（2016）
碱蓬属植物（Suaeda maritime）	碱蓬马特尔氏菌（Martelella suaedae）	内生菌	Chung 等（2016）
补血草（Limonium tetragonum）	马特尔氏菌（Martelella limonii）	内生菌	Chung 等（2016）

6.3 盐生植物-微生物互作与非生物胁迫

盐生植物已经进化成能够耐受高盐浓度的植物，多种不同的耐盐机制也得到了相应的研究（Kumari 等，2015a，2015b；Shabala，2013）。这些机制可能不限于 Na^+ 和 Cl^-，并且可能对其他有毒离子和化合物也具有耐受性。事实上，一些盐生植物已经显示出它们能够耐受和（或）积累高浓度的有毒污染物。例如，互花米草属（*Spartina*）的一些物种就能够耐受并在组织中积累一些有机和无机污染物（Mackova 等，2006；Redondo-Gómez 等，2011）。特别是密花米草（*Spartina densiflora*）和互花米草（*Spartina maritima*）能够在它们的组织中（主要是在根中）通过生物积累作用积累高浓度的重金属（Redondo-Gómez，2013）。类似地，蝎节木属植物（*Arthrocnemum macrostachyum*）能够大量积累镉的能力已经有相关报道（Redondo-Gómez 等，2010）。

与盐生植物相关的微生物可以通过降低植物的应激反应、产生污染物的解毒作用或影响污染物的生物利用度等方式来提高植物的修复能力（Mittler 等，2004）。最近的几项研究报道了利用盐生植物微生物减轻污染物对盐生植物的毒害作用。例如，从密花米草（*Spartina densiflora*）和互花米草（*Spartina maritima*）根际分离出的微生物能够提高植物在盐胁迫和金属胁迫下的生长（Mateos Naranjo 等，2015；Mesa 等，2015c；Paredes-Páliz 等，2017）；而从互花米草（*Spartina maritima*）中分离出的内生菌也能够改善生长在遭受污染的沼泽土壤中的植物生长（Mesa 等，2015b）。类似地，从蝎节木属植物（*Arthrocnemum macrostachum*）中分离出的内生菌和根际细菌都促进了重金属污染沼泽土壤中的植物生长（Navarro Torre 等，2017b）。盐生植物，比如灯心草（*Juncus acutus*）、盐节木（*Halocnemum strobilaceum*）和三棱水葱（*Scirpus triqueter*），以及与他们相关的微生物（根际微生物和内生细菌），已被证明是对有机污染物如石油污染的土壤中的碳氢化合物修复的有力工具（Al-Mailem 等，2010；Liu 等，2011；Syranidou 等，2017；Zhao 等，2013）。根据这些研究结果，盐生植物-微生物群的组合被认为是用来修复受污染和退化沼泽土壤的有效工具（Zhao 等，2013）。

6.4 从盐生植物中分离的内生微生物如何帮助植物缓解盐胁迫?

植物共生微生物对植物适应盐胁迫至关重要（Munns 和 Gilliham，2015；Tkacz 和 Poole，2015）。一般来说，诱导植物对不同非生物胁迫产生的耐受性被称为系统性诱导耐逆性（induced systemic tolerance，IST）（Yang 等，2009）。植物生长促生细菌通过改变植物激素状态（Vurukonda 等，2016）、吸收营养元素（Gerhardt 等，2017）和（或）调节活性氧的产生（Gururani 等，2013）等来诱导植物产生对盐胁迫和其他非生物胁迫的耐受性。一些微生物的特性也因此包括在这些过程中，比如 1-氨基环丙烷-1-羧酸-脱氨酶活性、磷酸盐增溶能力、氮固定能力和激素的产生，特别是吲哚-3-乙酸和脱落酸、铁载体以及相关挥发物等（图6.1）（Farag 等，2013；Kumari 等，2015；Nadeem 等，2016）。

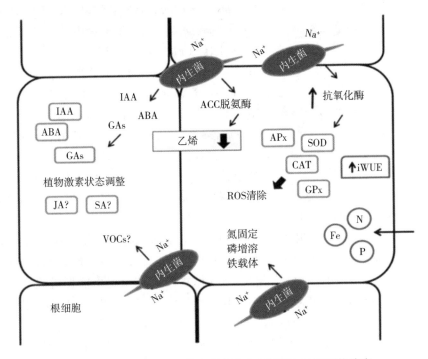

图6.1 从盐生微生物群落中分离的内生细菌有助于改善盐度升高条件下植物生长的机制概述

内生菌能产生1-氨基环丙烷-1-羧酸脱氨酶和植物激素来调节植物激素的状态。产1-氨基环丙烷-1-羧酸脱氨酶的内生菌减少了盐胁迫引起的过量乙烯生成。内生菌具有增强植物抗氧化系统和清除活性氧的能力。内生菌通过多种机制促进植物对养分的吸收,并能积累或结合Na^+。ABA—脱落酸;Apx—抗坏血酸过氧化物酶;CAT—过氧化氢酶;GA—赤霉素;GPx—愈创木酚过氧化物酶;IAA—吲哚乙酸;iWUE—水资源利用效率;JA—茉莉酸;ROS—活性氧;SA—水杨酸;SOD—超氧化物歧化酶;VOCs—挥发性有机物。

6.4.1 内生菌和植物激素状况

遭受盐胁迫的植物会产生高浓度的乙烯,从而延缓植物根系的发育(Mahajan和Tuteja,2005)。植物内生菌能够产生1-氨基环丙烷-1-羧酸脱氨酶,这种酶能够催化乙烯前体1-氨基环丙烷-1-羧酸转化为氨和α-酮丁酸酯,因此可以降低植物中的乙烯水平,从而克服盐胁迫所导致的生长抑制(Senthilkumar等,2009)。此外,这些内生菌还可以为植物提供氮源(氨)(Hardoim等,2008)。目前已经从合滨藜属植物(*Halimione portulacoides*)、补血草(*Limonium* sinense)、牧豆树属植物（*Prosopis strombulifera*）、互花米草（*Spartina maritima*）和盐角草属植物（*Salicornia europaea*）等盐生植物中分离到了能够产生 1-氨基环丙烷-1-羧酸脱氨酶的内生菌（Fidalgo 等，2016；Mesa 等，2015b；Qin 等，2014；Sgroy 等，2009；Zhao 等，2016）。然而，这些微生物在盐胁迫下对植物生长的直接影响还没有得到充分地研究。从补血草（*Limonium sinense*）中分离出的节杆菌属（*Arthrobacter*）、芽孢杆菌属（*Bacillus*）、居白蚁菌属（*Isoptericola*）和链霉菌属（*Streptomyces*）的能够产 1-氨基环丙烷-1-羧酸脱氨酶的内生菌在盐胁迫下具有促进植物生长的能力（Qin 等，2014）。有趣的是，对植物生长促进作用最好的菌株其吲哚-3-乙酸的含量非常低或几乎无，并且不能溶解磷酸盐，这表明这些内生菌缓解盐胁迫的能力可能与其产生 1-氨基环丙烷-1-羧酸脱氨酶的能力有关（Qin 等，2014）。类似地，从盐角草属植物（*Salicornia europaea*）中分离出的属于节杆菌属（*Arthrobacter*）、芽孢杆菌属（*Bacillus*）、游动球菌（*Planococcus*）和贪噬菌属（*Variovorax*）的具有 1-氨基环丙烷-1-羧酸脱氨酶活性的植物内生菌在 NaCl 浓度增加的情况下能促进离体幼苗生长；但这些菌株也能溶解磷酸盐或（和）产生吲哚-3-乙酸，这些都有利于促进盐角草属植物（*Salicornia europaea*）的生长（Zhao 等，2016）。

植物激素的分泌，特别是吲哚-3-乙酸的体外分泌，很可能是植物生长促生菌最常见的特征（Duca 等，2014）。吲哚-3-乙酸直接参与根系的产生和生长（Birnbaum，2016），因此产生吲哚-3-乙酸的内生菌可以通过刺激根系增殖来提高植物的耐盐性。已经从多种盐生植物中分离出能够分泌吲哚-3-乙酸的内生菌，包括牧豆树属植物（*Prosopis strombulifera*）、蝎节木属植物（*Arthrocnemum macrostachyum*）、互花米草（*Spartina maritima*）、盐角草（*Salicornia europaea*）、合滨藜属（*Halimione portulacoides*）和补骨脂（*Psoralea corylifolia*）（Fidalgo 等，2016；Mesa 等，2015b；Zhao 等，2016；Navarro Torre 等，2017a；NavarroTorre 等，2017b；Sgroy 等，2009；Sorty 等，2016）。从补骨脂（*Psoralea corylifolia*）中分离出的芽孢杆菌属（*Bacillus*）、海杆菌属（*Marinobacterium*）和中华根瘤菌属（*Sinorhizobium*）的能够产吲哚-3-乙酸的内生菌，即便通过单独接种的方式，也能够促进盐胁迫下小麦幼苗的生长（Sorty 等，2016）。最近，有研究报道了接种从蝎节木属植物（*Arthrocnemum macrostachyum*）中分离出的内生菌菌群能够减轻高盐度对植物生长和生理特性的影响（Navarro Torre 等，2017a）。在这项研究中，内生菌菌群主要由属于芽孢杆菌属（*Bacillus*）和纤细芽孢杆菌属（*Gracilibacillus*）能够产吲哚-3-乙酸的内生菌组成；然而，对植物生长的促进作用和植物耐盐性的增加直接归因于内生菌吲哚-3-乙酸的产生，因为内生菌还有一些其他的促生特征，比如磷酸盐的增溶能力和铁载体的产生。

植物脱落酸是一种重要的植物响应胁迫的应激激素，通过脱落酸产生和（或）植物生长促生菌的代谢来调节植物脱落酸的合成，也可能有助于减轻盐胁迫对植物的危害（Gerhardt 等，2017）。然而，对从盐生植物中分离出的植物内生菌产生或代谢脱落酸的能力研究甚少。Sgroy 等（2009）报道了从盐生植物牧豆树属植物（*Prosopis strombulifera*）中分离出 7 株具有体外产生脱落酸能力的内生细菌。这些菌株属于无色杆菌属（*Achromobacter*）、芽孢杆菌属（*Bacillus*）、短杆菌属（*Brevibacterium*）、赖氨酸芽孢杆菌（*Lysinibacillus*）和假单胞菌属（*Pseudomonas*）。这些菌株中的大多数也能够产生赤霉素，赤霉素是一种积极调节细胞分裂和伸长、下胚轴和茎生长以及叶和根分生组织大小的植物激素（Martinez 等，2016；Wang 等，2015）。尽管赤霉素信号传导是在逆境条件下抑制植物生长的关键因素（Martinez 等，2016），但由内生菌产生赤霉素的研究很少有报道。在植物中也有其他植物激素参与抵抗非生物胁迫中，比如茉莉酸和水杨酸（Ahmad 等，2016），它们也能够由一些内生菌合成（Chen 等，2014；Forchetti 等，2007），但直到现在还没有从盐生植物中分离出来的内生植物能够产生这两种激素的报道。

6.4.2 植物内生菌与养分吸收

植物生长促生菌能促进植物对养分的吸收。具有溶解不溶性磷酸盐、钾和锌，能够固定大气中的氮元素和（或）释放铁载体以清除离子等能力的内生菌，在非生物胁迫下能够促进植物生长（Santoyo 等，2016）。从盐生植物中分离出来的内生菌的这些特征有很多研究，但在盐胁迫下用选定的细菌接种植物并不多（Sgroy 等，2009）。从蝎节木属（*Arthrocnemum macrostachyum*）中分离出的芽孢杆菌属（*Bacillus*）和纤细芽孢杆菌属（*Gracilibacillus*）内生菌具有溶解磷酸盐、固定大气氮素和产生铁载体的能力（Navarro-Torre 等，2017a）。含有其中 3 种内生菌的细菌群能够在高盐浓度下促进植物生长，同时也能增加植物积累 NaCl 的水平（Navarro Torre 等，2017a）。Mesa 等（2015b）还报道了包括微球菌属（*Micrococcus*）、盐藻属（*Salinicola*）和弧菌属（*Vibrio*）在内的内生菌联合体是如何帮助减轻互花米草（*Spartina maritima*）在受到重金属和 NaCl 胁迫的盐沼土壤中对生长的影响的。这些内生菌在体外结合了吲哚-3-乙酸和铁载体的产生、磷酸盐的溶解和固氮作用。有趣的是，接种了内生菌的互花米草（*Spartina maritima*）能够提高 15% 的水分利用效率（Water use efficiency，WUE），水分利用效率是植物在胁迫条件下如何管理水分的一个重要指标（Tardieu，2012）。类似的结果是使用从蝎节木属（*Arthrocnemum macrostachyum*）中分离出的内生菌群作为接种剂（Navarro Torre 等，2017b）。在这项研究中，芽孢杆菌属（*Bacillus*）、盐单胞菌（*Halomonas*）、克锡勒氏菌（*Kushneria*）和微球菌属（*Micrococcus*）的内生菌改善了生长在被重金属污染的盐沼土壤中的蝎节木属（*Arthrocnemum macrostachyum*）的生长。这个内生菌联合体能够产生 IAA 和铁载体，并且具有溶解磷酸盐和固定大气中氮元素的能力。

6.4.3 内生菌和活性氧水平

在高盐度下生长的植物中通常可以观察到活性氧的增加（Miller 等，2010）。内生菌可以通过调节活性氧的产生来提高植物对盐和其他非生物胁迫的耐受性。由芽孢杆菌属（*Bacillus*）、盐单胞菌（*Halomonas*）、克锡勒氏菌（*Kushneria*）和微球菌属（*Micrococcus*）的内生菌构成的菌群能够诱导抗坏血酸过氧化物酶、过氧化氢酶、愈创木酚过氧化物酶和超氧化物歧化酶等抗氧化酶活性的变化（Navarro Torre 等，2017b）。这些酶有助于维持体内对高水平活性氧的平衡（Mesnoua 等，2016）。接种了这个内生菌群的植物蝎节木属（*Arthrocnemum macrostachyum*）其在重金属污染的盐渍沼泽土壤中抗坏血酸过氧化物酶、谷胱甘肽过氧化物酶和超氧化物歧化酶的活性都有所增强，表明内生菌有助于降低植物积累金属和 NaCl 所引起的氧化胁迫反应。但是，利用内生菌和 NaCl 作为单一非生物胁迫的分析未在文献中发现。

6.4.4 内生菌和其他耐盐机制

其他可能与植物生长促生菌赋予的耐盐性相关的机制已通过根际细菌进行了部分研究。例如，接种根瘤菌能够诱导水通道蛋白相关基因的上调表达（Gond 等，2015；Qin 等，2016；Vurukonda 等，2016），这与根系导水率直接相关，并决定着根系吸水能力（Moshelin 等，2015）。还有一些根际植物生长促生菌能够释放挥发性有机化合物（Volatile organic compounds，VOCs），能够诱导植物发生生理变化（Liu 和 Zhang，2015）。然而，这些机制在植物内生菌诱导的耐盐性中的作用至今尚未被研究。

存在于盐生植物体内的内生菌必须适应植物组织中的盐浓度。如前所述，在本章节中提到的从盐生植物中分离出的内生菌大多是嗜盐菌或耐盐菌。这些细菌能够在高浓度 NaCl 存在的情况下，依赖不同的生长机制正常生长，这些机制通常包括有机溶质的合成或积累，以提供细胞质与周围介质的渗透平衡（Ventosa 等，1998）。由于嗜盐菌和耐盐菌能够在其细胞质内或细胞表面积累 Na^+，因此可以推测内生菌的 Na^+ 吸收和（或）积累，也有助于提高植物在盐胁迫环境中的耐性和生长。

6.5 结论和展望

存在于盐生植物体内的内生菌必须适应盐胁迫，还能够有助于植物在盐渍土壤中的适应和生长。越来越多的文献报道了从盐生植物中分离和鉴定出的具有植物生长促进特性的内生菌。尽管与植物激素调节、养分吸收和活性氧产生等相关的内生细菌特性可能在诱导植物耐盐性方面具有协同效应，但迄今为止，尚未直接证明诱导耐盐性所涉及的具体分子机制。研究盐生植物的完整微生物组和比较最近从这些植物中分离出来的内生菌的基因组序列，为揭示不同植物生长、促进特性对植物耐受性的影响提供了基因组学工具。例如，构建影响具体性状产生的基因敲除突变体，并接种这些突变体的植物，将有助于确定每个性状或这些性状在植物耐盐性方面的作用。这些研究可为选择和设计生物菌剂以提高盐渍土作物产量提供有用的信息。

[致谢]

感谢塞维利亚大学为这项研究提供了必要的设施。

参考文献

Ahmad P, Rasool S, Gul A, Sheikh SA, Akram NA, Ashraf M, Kazi AM, Gucel S, 2016. Jasmonates: multifunctional roles in stress tolerance. Front Plant Sci 7: 813.

Al-Mailem D, Sorkhoh N, Marafie M, Al-Awadhi H, Eliyas M, Radwan S, 2010. Oil phytoremediation potential of hypersaline coasts of the Arabian Gulf using rhizosphere technology. Bioresour Technol 101: 5786-5792.

Arora S, Patel PN, Vanza MJ, Rao GG, 2014. Isolation and characterization of endophytic bacteria colonizing halophyte and other salt tolerant plant species from coastal Gujarat. Afr J Microbiol Res 8: 1779-1788.

Bai Y, Müller DB, Srinivas G, Garrido-Oter R, Potthoff E, Rott M, Dombrowski N, Münch PC, Spaepen S, Remus-Emsermann M, Hüttel B, McHardy AC, Vorholt JA, Schulze-Lefert P, 2015. Functional overlap of the *Arabidopsis* leaf and root microbiota. Nat Res 528: 364-369.

Bhattacharyya PN, Jha DK, 2012. Plant growth-promoting rhizobacteria (PGPR): emergence in agriculture. World J Microbiol Biotechnol 28: 1327-1350.

Birnbaum KD, 2016. How many ways are there to make a root? Curr Opin Plant Biol 34: 61-67.

Bui EN, Thornhill A, Miller JT, 2014. Salt and alkaline tolerance are linked in Acacia. Biol

Lett 10: 20140278.

Camacho M, Montero-Calasanz M, Redondo-Gómez S, Rodríguez-Llorente I, Schumann P, Klenk HP, 2016a. *Microbulbifer rhizosphaerae* sp. nov., isolated from the rhizosphere of the halophyte *Arthrocnemum macrostachyum*. Int J Syst Evol Microbiol 66: 1844-1850.

Camacho M, Redondo-Gómez S, Rodríguez-Llorente I, Rohde M, Spröer C, Schumann P, Klenk HP, Montero-Calasanz MD, 2016b. *Labrenzia salina* sp. nov., isolated from the rhizosphere of the halophyte *Arthrocnemum macrostachyum*. Int J Syst Evol Microbiol 66: 5173-5180.

Camacho M, Redondo-Gomez S, Rodríguez-Llorente I, Rohde M, Spröer C, Schumann P, Klenk HP, Montero-Calasanz MDC, 2017. *Kocuria salina* sp. nov., an actinobacterium isolated from the rhizosphere of the halophyte *Arthrocnemum macrostachyum* and emended description of *Kocuria turfanensis*. Int J Syst Evol Microbiol 67: 5006-5012.

Chen Y, Fan JB, Du L, Xu H, Zhang QH, He YQ, 2014. The application of phosphate solubilizing endophyte *Pantoea dispersa* triggers the microbial community in red acidic soil. Appl Soil Ecol 84: 235-244.

Chung EJ, Hwang JM, Kim KH, Jeon CO, Chung YR, 2016. *Martelella suaedae* sp. nov. and *Martelella limonii* sp. nov., isolated from the root of halophytes. Int J Syst Evol Microbiol 66: 3917-3922.

Coleman-Derr D, Tringe SG, 2014. Building the crops of tomorrow: advantages of symbiont-based approaches to improving abiotic stress tolerance. Front Microbiol 5: 283.

Deinlein U, Stephan AB, Horie T, Luo W, Xu G, Schroeder JI, 2014. Plant salt tolerance mechanisms. Trends Plant Sci 19: 371-379.

Duca D, Lorv J, Patten CL, Rose D, Glick BR, 2014. Indole-3-acetic acid in plant-microbe interactions. Ant Leeuw 106: 85-125.

El-Awady MAM, Hassan MM, Al-Sodany YM, 2015. Isolation and characterization of salt tolerant endophytic and rhizospheric plant growth promoting bacteria (PGPB) associated with the halophyte plant (*Sesuvium verrucosum*) grown in KSA. Int J Appl Sci Biotechnol 33: 552-560.

Etesami H, Beattie GA, 2018. Mining halophytes for plant growth-promoting halotolerant bacteria to enhance the salinity tolerance of non-halophytic crops. Front Microbiol 9: 148.

Farag MA, Zhang H, Ryu CM, 2013. Dynamic chemical communication between plants and bacteria through airborne signals: induced resistance by bacterial volatiles. J Chem Ecol 39: 1007-1018.

Fedoroff NV, Battisti DS, Beachy RN, Cooper PJ, Fischhoff DA, Hodges CN, Knauf VC, Lobell D, Mazur BJ, Molden D, Reynolds MP, Ronald PC, Rosegrant MW, Sanchez PA, Vonshak A, Zhu JK, 2010. Radically rethinking agriculture for the 21st century. Science 327: 833-834.

Feng WW, Wang TT, Bai JL, Ding P, Xing K, Jiang JH, Peng X, Qin S, 2017. *Glutamicibacter halophytocola* sp. nov., an endophytic actinomycete isolated from the roots of a coastal halophyte, *Limonium sinense*. Int J Syst Evol Microbiol 67: 1120-1125.

Fidalgo C, Henriques I, Rocha J, Tacao M, Alves A, 2016. Culturable endophytic bacteria from the salt marsh plant *Halimione portulacoides*: phylogenetic diversity, functional characterization, and influence of metal (loid) contamination. Environ Sci Pollut Res 23:

10200-11024.

Fidalgo C, Martins R, Proença DN, Morais PV, Alves A, Henriques I, 2017a. *Zunongwangia endophytica* sp. nov., an endophyte isolated from the salt marsh plant, *Halimione portulacoides*, and emended description of the genus *Zunongwangia*. Int J Syst Evol Microbiol 67: 3004-3009.

Fidalgo C, Rocha J, Martins R, Proença DN, Morais PV, Henriques I, Alves A, 2017b. *Altererythrobacter halimionae* sp. nov. and *Altererythrobacter endophyticus* sp. nov., two endophytes from the salt marsh plant *Halimione portulacoides*. Int J Syst Evol Microbiol 67: 3057-3062.

Fidalgo C, Rocha J, Proença DN, Morais PV, Alves A, Henriques I, 2017c. *Saccharospirillum correiae* sp. nov., an endophytic bacterium isolated from the halophyte *Halimione portulacoides*. Int J Syst Evol Microbiol 67: 2026-2030.

Flowers TJ, Colmer TD, 2008. Salinity tolerance in halophytes. New Phytol 179: 945-963.

Forchetti G, Masciarelli O, Alemano S, Alvarez D, Abdala G, 2007. Endophytic bacteria in sunflower (*Helianthus annuus* L.): isolation, characterization, and production of jasmonates and abscisic acid in culture medium. Appl Microbiol Biotechnol 76: 1145-1152.

Gerhardt KE, Macneill GJ, Gerwing PD, Greenberg BM, 2017. Phytoremediation of salt-impacted soils and use of plant growth-promoting rhizobacteria (PGPR) to enhance phytoremediation. Phytoremediation. https://doi.org/10.1007/978-3-319-52381-1.

Gond SK, Torres MS, Bergen MS, Helsel Z, White JF, 2015. Induction of salt tolerance and up-regulation of aquaporin genes in tropical corn by rhizobacterium *Pantoea agglomerans*. Lett Appl Microbiol 60: 392-399.

Gupta B, Huang B, 2014. Mechanism of salinity tolerance in plants: physiological, biochemical, and molecular characterization. Int J Genomics 2014: 701596.

Gururani MA, Upadhyaya CP, Baskar V, Venkatesh J, Nookaraju A, Park SW, 2013. Plant growthpromoting rhizobacteria enhance abiotic stress tolerance in *Solanum tuberosum* through inducing changes in the expression of ROS-scavenging enzymes and improved photosynthetic performance. J Plant Growth Regul 32: 245-258.

Hardoim PR, Overbeek LSV, Elsas JDV, 2008. Properties of bacterial endophytes and their proposed role in plant growth. Trends Microbiol 16: 463-471.

Ismail AM, Horie T, 2017. Genomics, physiology, and molecular breeding approaches for improving salt tolerance. Annu Rev Plant Biol 68: 405-434.

Jesús JM, Danko AS, Fiúza A, Borges MT, 2015. Phytoremediation of salt-affected soils: a review of processes, applicability, and the impact of climate change. Environ Sci Pollut Res 22: 6511-6525.

Jewell MC, Campbell BC, Godwin ID, 2010. Transgenic plants for abiotic stress resistance. In: Kole C, Michler C, Abbott AG, Hall TC (eds) Transgenic crop plants. Springer, Berlin, pp. 67-132.

Kataoka R, Güneri E, Turgay OC, Yaprak AE, Sevilir B, Başköse I, 2017. Sodium-resistant plant growth-promoting rhizobacteria isolated from a halophyte, *Salsola grandis*, in saline-alkaline soils of Turkey. Eur J Soil Sci 6: 216-225.

Kumar M, Choi JY, Kumari N, Pareek A, Kim SR, 2015. Molecular breeding in *Brassica* for salt tolerance: importance of microsatellite (SSR) markers for molecular breeding in *Brassica*.

Front Plant Sci 6: 688.

Kumari S, Vaishnav A, Jain S, Varma A, Choudhary DK, 2015a. Bacterial-mediated induction of systemic tolerance to salinity with expression of stress alleviating enzymes in soybean (*Glycine max* L. Merrill). J Plant Growth Regul 34: 558-573.

Kumari A, Das P, Parida AK, Agarwal PK, 2015b. Proteomics, metabolomics, and ionomics perspectives of salinity tolerance in halophytes. Front Plant Sci 6: 537.

Liu XM, Zhang H, 2015. The effects of bacterial volatile emissions on plant abiotic stress tolerance. Front Plant Sci 6: 774.

Liu X, Wang Z, Zhang X, Wang J, Xu G, Cao Z, Zhong C, Su P, 2011. Degradation of dieseloriginated pollutants in wetlands by Scirpus triqueter and microorganisms. Ecotoxcol Environ Safe 74: 1967-1972.

Liu BB, Wang HF, Li QL, Zhou XK, Zhang YG, Xiao M, Li QQ, Zhang W, Li WJ, 2016. *Aurantimonas endophytica* sp. nov., a novel endophytic bacterium isolated from roots of Anabasis elatior (C. A. Mey.) Schischk. Int J Syst Evol Microbiol 66: 4112-4117.

Lucena T, Mesa J, Rodriguez-Llorente ID, Pajuelo E, Caviedes MÁ, Ruvira MA, Pujalte MJ, 2016. *Marinomonas spartinae* sp. nov., a novel species with plant-beneficial properties. Int J Syst Evol Microbiol 66: 1686-1691.

Lucena T, Arahal DR, Ruvira MA, Navarro-Torre S, Mesa J, Pajuelo E, Rodriguez-Llorente ID, Rodrigo-Torres L, Piñar MJ, Pujalte MJ, 2017. *Vibrio palustris* sp. nov. and *Vibrio spartinae* sp. nov., two novel members of the Gazogenes clade, isolated from salt-marsh plants (*Arthrocnemum macrostachyum* and *Spartina maritima*). Int J Syst Evol Microbiol 67: 3506-3512.

Mackova M, Dowling D, Macek T, 2006. Phytoremediation and rhizoremediation. Springer, Dordrecht. (Focus on Biotechnology). Available at: https://books.google.es/books?Id=mph14ko78hYC.

Mahajan S, Tuteja N, 2005. Cold, salinity and drought stresses: an overview. Arch Biochem Biophys 444: 139-158.

Martínez C, Espinosa-Ruiz A, Prat S, 2016. Gibberellins and plant vegetative growth. Annu Plant Rev 49: 285-322.

Mateos-Naranjo E, Mesa J, Pajuelo E, Perez-Martin A, Caviedes MA, Rodríguez-Llorente ID, 2015. Deciphering the role of plant growth-promoting rhizobacteria in the tolerance of the invasive cordgrass *Spartina densiflora* to physicochemical properties of salt-marsh soils. Plant Soil 394: 45-55.

Mesa J, Mateos-Naranjo E, Caviedes MA, Redondo-Gómez S, Pajuelo E, Rodríguez-Llorente ID, 2015a. Scouting contaminated estuaries: heavy metal resistant and plant growth promoting rhizobacteria in the native metal rhizoaccumulator *Spartina maritima*. Mar Pollut Bull 90: 150-159.

Mesa J, Mateos-Naranjo E, Caviedes MA, Redondo-Gomez S, Pajuelo E, Rodriguez-Llorente ID, 2015b. Endophytic cultivable bacteria of the metal bioaccumulator *Spartina maritima* improve plant growth but not metal uptake in polluted marshes soils. Front Microbiol 6: 1450.

Mesa J, Rodriguez-Llorente ID, Pajuelo E, Piedras JM, Caviedes MA, Redondo-Gómez S, MateosNaranjo E, 2015c. Moving closer towards restoration of contaminated estuaries: bioaugmentation with autochthonous rhizobacteria improves metal rhizoaccumulation in native *Spartina maritima*. J Hazard Mater 300: 263-271.

Mesnoua M, Mateos-Naranjo E, Barcia-Piedras JM, Perez-Romero JA, Lotmani B, RedondoGomez S, 2016. Physiological and biochemical mechanisms preventing Cd-toxicity in the hyperaccumulator *Atriplex halimus* L. Plant Physiol Biochem 106: 30-38.

Mickelbart MV, Hasegawa PM, Bailey-Serres J, 2015. Genetic mechanisms of abiotic stress tolerance that translate to crop yield stability. Nat Rev Genet 16: 237-251.

Miller G, Suzuki N, Ciftci-Yilmaz S, Mittler R, 2010. Reactive oxygen species homeostasis and signalling during drought and salinity stresses. Plant Cell Environ 33: 453-467.

Mittler R, Vanderauwera S, Gollery M, Van Breusegem F, 2004. Reactive oxygen gene network of plants. Trends Plant Sci 9: 490-498.

Moshelion M, Halperin O, Wallach R, Oren R, Way DA, 2015. Role of aquaporins in determining transpiration and photosynthesis in water-stressed plants: crop water use efficiency, growth and yield. Plant Cell Environ 38: 1785-1793.

Munns R, Gilliham M, 2015. Salinity tolerance of crops- what is the cost? New Phytol 208: 668-673.

Nadeem SM, Ahmad M, Naveed M, Imran M, Zahir ZA, Crowley DE, 2016. Relationship between in vitro characterization and comparative efficacy of plant growth-promoting rhizobacteria for improving cucumber salt tolerance. Arch Microbiol 198: 379-387.

Navarro-Torre S, Mateos-Naranjo E, Caviedes MA, Pajuelo E, Rodríguez-Llorente ID, 2016. Isolation of plant-growth-promoting and metal-resistant cultivable bacteria from *Arthrocnemum macrostachyum* in the Odiel marshes with potential use in phytoremediation. Mar Pollut Bull 110: 133-142.

Navarro-Torre S, Barcia-Piedras JM, Mateos-Naranjo E, Redondo-Gómez S, Camacho M, Caviedes MA, Pajuelo E, Rodríguez-Llorente ID, 2017a. Assessing the role of endophytic bacteria in the halophyte *Arthrocnemum macrostachyum* salt tolerance. Plant Biol 19: 249-256.

Navarro-Torre S, Barcia-Piedras JM, Caviedes MA, Pajuelo E, Redondo-Gómez S, RodríguezLlorente ID, Mateos-Naranjo E, 2017b. Bioaugmentation with bacteria selected from the microbiome enhances *Arthrocnemum macrostachyum* metal accumulation and tolerance. Mar Pollut Bull 117: 340-347.

Paredes-Páliz KI, Mateos-Naranjo E, Doukkali B, Caviedes MA, Redondo-Gómez S, RodríguezLlorente ID, Pajuelo E, 2017. Modulation of *Spartina densiflora* plant growth and metal accumulation upon selective inoculation treatments: a comparison of gram negative and gram positive rhizobacteria. Mar Pollut Bull 125: 77-85.

Pitman MG, Läuchli A, 2002. Global impacts of salinity and agricultural ecosystem. In: Läuchli A, Lüttge U (eds) Salinity: environment-plants-molecules. Kluwer Academic, Dordrecht, pp. 3-20.

Qin S, Zhang YJ, Yuan B, Xu PY, Xing K, Wang J, Jiang JH, 2014. Isolation of ACC deaminase-producing habitat-adapted symbiotic bacteria associated with halophyte *Limonium sinense* (Girard) Kuntze and evaluating their plant growth-promoting activity under salt stress. Plant Soil 374: 753-766.

Qin Y, Druzhinina IS, Pan X, Yuan Z, 2016. Microbially mediated plant salt tolerance and microbiome-based solutions for saline agriculture. Biotechnol Adv 34: 1245-1259.

Redondo-Gómez S, 2013. Bioaccumulation of heavy metals inSpartina. Funct Plant Biol 40: 913-921.

Redondo-Gómez S, Mateos-Naranjo E, Andrades-Moreno L, 2010. Accumulation and tolerance characteristics of cadmium in a halophytic Cd-hyperaccumulator, *Arthrocnemum macrostachyum*. J Hazard Mater 184: 299-307.

Redondo-Gómez S, Andrades-Moreno L, Parra R, Valera-Burgos J, Real M, Mateos-Naranjo E, Cox L, Cornejo J, 2011. *Spartina densiflora* demonstrates high tolerance to phenanthrene in soil and reduces it concentration. Mar Pollut Bull 62: 1800-1808.

Roy SJ, Negrão S, Tester M, 2014. Salt resistant crop plant. Curr Opin Biotechnol 26: 115-124.

Ruppel S, Franken P, Witzel K, 2013. Properties of the halophyte microbiome and their implications for plant salt tolerance. Funct Plant Biol 40: 940-951.

Santoyo G, Moreno-Hagelsieb G, Orozco-Mosqueda MC, Glick B, 2016. Plant growth-promoting bacterial endophytes. Microbiol Res 183: 92-99.

Senthilkumar M, Swarnalakshmi K, Govindasamy V, Lee YK, Annapurna K, 2009. Biocontrol potential of soybean bacterial endophytes against charcoal root fungus, *Rhizoctonia bataticola*. Curr Microbiol 58: 288-293.

Sgroy V, Cassán F, Masciarelli O, Del Papa MF, Lagares A, Luna V, 2009. Isolation and characterization of endophytic plant growth-promoting (PGPB) or stress homeostasis-regulating (PSHB) bacteria associated to the halophyte *Prosopis strombulifera*. Appl Microbiol Biotechnol 85: 371-381.

Shabala S, 2013. Learning from halophytes: physiological basis and strategies to improve abiotic stress tolerance in crops. Ann Bot 112: 1209-1221.

Sorty AM, Meena KK, Choudhary K, Bitla UM, Minhas PS, Krishnani KK, 2016. Effect of plant growth promoting bacteria associated with halophytic weed (*Psoralea corylifolia* L) on germination and seedling growth of wheat under saline conditions. Appl Biochem Biotechnol 180: 872-882.

Suarez C, Ratering S, Schäfer J, Schnell S, 2017. *Ancylobacter pratisalsi* sp. nov. with plant growth promotion abilities from the rhizosphere of *Plantago winteri* Wirtg. Int J Syst Evol Microbiol 67: 4500-4506.

Syranidou E, Christofilopoulos S, Politi M, Weyens N, Venieri D, Vangronsveld J, Kalogerakis N, 2017. Bisphenol-a removal by the halophyte *Juncus acutus* in a phytoremediation pilot: characterization and potential role of the endophytic community. J Hazard Mater 32: 350-358.

Tardieu F, 2012. Any trait or trait related allele can confer drought tolerance: just design the right drought scenario. J Exp Bot 63: 25-31.

Tkacz A, Poole P, 2015. Role of root microbiota in plant productivity. J Exp Bot 66: 2167-2175.

Ventosa A, Nieto JJ, Oren A, 1998. Biology of moderately halophilic aerobic bacteria. Microbiol Mol Biol Rev 62: 504-544.

Vurukonda SSKP, Vardharajula S, Shrivastava M, SkZ A, 2016. Enhancement of drought stress tolerance in crops by plant growth promoting rhizobacteria. Microbiol Res 184: 13-24.

Wang GL, Que F, Xu ZS, Wang F, Xiong AS, 2015. Exogenous gibberellin altered morphology, anatomic and transcriptional regulatory networks of hormones in carrot root and shoot. BMC Plant Biol 15: 290.

Yang J, Kloepper JW, Ryu CM, 2009. Rhizosphere bacteria help plants tolerate abiotic stress. Trends Plant Sci 14: 1-4.

Yensen NP, Biel KY, 2006. Soil remediation via salt-conduction and the hypotheses of halosynthesis and photoprotection ecophysiology of high salinity tolerant plants. In: Khan MA, Weber DJ (eds) Task for vegetation science, vol 34. Springer, Dordrecht, pp. 313-344.

Yuan Z, Druzhinina IS, Labbé J, Redman R, Qin Y, Rodriguez R, Zhang C, Tuskan GA, Lin F, 2016. Specialized microbiome of a halophyte and its role in helping non-host plants to withstand salinity. Sci Rep 6: 32467.

Zhao S, Zhou N, Wang L, Tian CY, 2013. Halophyte-endophyte coupling: a promising bioremediation system for oil-contaminated soil in Northwest China. Environ Sci Technol 47: 11938-11939.

Zhao S, Zhou N, Zhao ZY, Zhang K, Wu GH, Tian CY, 2016. Isolation of endophytic plant growth-promoting bacteria associated with the halophyte *Salicornia europaea* and evaluation of their promoting activity under salt stress. Curr Microbiol 73: 574-581.

Zhao GY, Zhao LY, Xia ZJ, Zhu JL, Liu D, Liu CY, Chen XL, Zhang YZ, Zhang XY, Dai MX, 2017. *Salinicola tamaricis* sp. nov., a heavy-metal-tolerant, endophytic bacterium isolated from the halophyte *Tamarix chinensis* Lour. Int J Syst Evol Microbiol 67: 1813-1819.

Zhu JK, 2016. Abiotic stress signaling and responses in plants. Cell 167: 313-324.

7 盐胁迫条件下嗜盐细菌对水稻品种生化特性的影响

Mehvish Riaz Khattak, Sami Ullah Jan, Ijaz Malook,
Sehrish Riaz Khattak, Nazneen Akhtar,
Sehresh Khan, Muhammad Jamil

[摘要]

土壤盐渍化是一个十分严重的生态问题，盐渍化土壤影响包括水稻在内的谷类作物的产量。目前，有很多研究已经报道了从盐渍化土壤中分离鉴定出了多种耐盐细菌，这些细菌菌株已经被开发成盐渍化土壤的生物修复剂，但是嗜盐细菌的农业应用却受到了限制。在本项研究中，研究了两株耐盐芽孢杆菌NCCP-71 和 NCCP-77（Bacillus strains NCCP-71 和 Bacillus strains NCCP-77）在不同浓度盐胁迫（0 mmol/L、50 mmol/L、100 mmol/L 和 150 mmol/L NaCl）环境下对两个水稻品种 NIAB-IR-9 和 KSK-282 在生理生化方面的影响。研究结果表明，在高浓度盐胁迫下，水稻植株内 Na^+ 含量增加，K^+ 和 Ca^{2+} 含量降低；而接种芽孢杆菌后，水稻幼苗中 Na^+ 含量降低，K^+ 含量增加。芽孢杆菌 NCCP-71 可促使两个品种植株中的 Ca^{2+} 含量增加。但在 50 mmol/L 浓度的 NaCl 处理下，接种芽孢杆菌 NCCP-77 的处理则可降低水稻品种 KSK-282 和 NIAB-IR-9 植株中 Ca^{2+} 的含量。水稻在遭受盐胁迫危害下，其植株中叶绿素的含量不断降低，但是接种芽孢杆菌 NCCP-71 和 NCCP-77 后却能使因盐胁迫而降低的植物光合色素的含量（叶绿素a、叶绿素b和类胡萝卜素）显著增加。另外，随着盐胁迫的加剧水稻植株中蛋白质和氮的含量显著降低。接种芽孢杆菌 NCCP-71 菌株能够明显提高两个水稻品种中蛋白质和氮的含量，而接种芽孢杆菌 NCCP-77 菌株的处理则只提高了 50 mmol/L 浓度 NaCl 处理下水稻植株的蛋白质和氮含量。在盐胁迫条件下两种耐盐芽孢杆菌能够通过改变水稻的生理生化反应，从而对接种水稻的耐盐性指标造成很大的影响。

M. R. Khattak, S. U. Jan, I. Malook, S. R. Khattak, N. Akhtar, S. Khan, M. Jamil (✉)，巴基斯坦科哈特科技大学生物技术与基因工程系，E-mail: M. Jamil: dr.jamil@kust.edu.pk

[关键词]
嗜盐菌；生化特性；水稻；盐胁迫

7.1 前言

植物最易受气候变化的影响。主要作物的产量受到不同胁迫的影响（Grover 等，2011）。盐胁迫是当今农业生产中遇到的主要障碍之一。由于灌溉、沙漠化和不合理的施肥，耕作土壤中的盐分含量一直呈现增加的状态（Bacilio 等，2004）。盐胁迫会导致植物内稳态、离子平衡和渗透势等方面发生变化，抑制细胞分裂和伸长生长等，从而影响植物正常发育（Princie 等，2007）。盐胁迫通过降低发芽率影响植物生长的生理参数。而非生物胁迫能够导致植物这些生长参数的改变（Kaymak 等，2009）。植物依靠碳水化合物调节渗透势，具有植物毒性离子的积累，会导致植物从土壤中吸收水分的能力下降，从而导致光合作用速率降低和气孔关闭（Yildirim 等，2011）。

为了缓解这一问题，研究人员使用并简化了很多不同的技术，其中大多数是使用蔗糖（Siringam 等，2012）、磷（Naheed 等，2007）、激素（Rafique 等，2011；Afzal 等，2006）、植物源性烟雾溶液（Jamil 等，2013；Malook 等，2014；Malook 等，2017）以及植物生长促生菌（Bhattacharyya 和 Jha，2012；Ashrafuzzaman 等，2009；Kaymak 等，2008；Shah 等，2017；Khan 等，2017）。

定殖在植物根部并能促进萌发生长的自由生活的微生物称为植物生长促生菌（plant growth-promoting rhizobacteria，PGPR）（Pallai，2005）。植物生长促生菌通过刺激植物生长激素（如吲哚乙酸、细胞分裂素和赤霉素）的生物合成和促进磷元素的最大溶解和吸收，以及通过生产富铁载体增加植物根系对氮和铁的利用这两种机制促进植物生长（Mallesh，2008）。间接机制包括对病原菌产生抗生素、真菌细胞分裂酶的生物合成、脯氨酸含量增加和甜菜碱的形成，从而增强对病原体和生物胁迫的整体系统性抗性（Pallai，2005）。根际细菌也经常被用作生物杀虫剂、生物肥料、植物刺激剂和根状茎调节剂（Antoun 和 Prévost，2005）。据报道，有许多细菌菌株能够促进植物的生长，包括假单胞菌（*Pseudomonas*）、固氮螺菌属菌（*Azospirillum*）、固氮菌（*Azotobacter*）、肠杆菌（*Enterobacter*）、克雷伯氏菌属（*Klebsiella* sp.）、产碱杆菌属（*Alcaligenes*）、伯克霍尔德氏菌（*Burkholderia*）、节杆菌属（*Arthrobacter*）、沙雷氏菌属（*Serratia*）、根瘤菌（*Rhizobium*）（Saharan 和 Nehra，2011）。芽孢杆菌（*Bacillus*）具有植物生长促生活性，尤其对植物病原菌具有抗性。这些微生物能够产生内生孢子，还能够产生许多可以用于农业用途的有益化合物（Nihorimbere 等，2010）。芽孢杆菌在各种非生物胁迫条件下（包括盐和重金属胁迫）能够通过增加有益化合物（如丁二醇和丙酮），促进植物生长（Ryu 等，2003）。从根际分离到的植物生长促生菌中，最常见的是芽孢杆菌属（*Bacillus* spp.）微生物。大豆种子接种芽孢杆菌菌株后，其结瘤能力和生长情况都得到了增强（Gururani 等，2013）。

土壤盐渍化能够抑制植物生长，包括用作主食的水稻。很多不同的技术被用来减少或修复受盐渍化影响的土壤，但这些方法通常成本较高，并且会改变土壤的结构和质地。因此，人们更倾向于采用环境友好型和更加经济性的方法来解决这一问题。很多不同的细菌种类被用来减少土壤中的有机或无机污染物。从植物根际分离到的植物生长促生菌菌株具有修复土壤污染的潜力。然而目前，利用嗜盐细菌修复盐渍土的资料还很少。因此，本研究旨在探讨嗜盐芽孢杆菌 NCCP-71 和 NCCP-77（*Bacillus* strains NCCP-71 和 *Bacillus* strains NCCP-77）在不同盐分条件下对水稻品种生化特性影响中的作用。

7.2 植物材料

水稻品种种子（编号：NIAB-IR-9 和 KSK-282）来自巴基斯坦伊斯兰堡国家农业研究中心（NARC）。种子表面用 3.5% 次氯酸钠溶液（v/v）消毒后，用蒸馏水冲洗干净。

7.3 细菌菌株

嗜盐细菌菌株取自保存在巴基斯坦科哈特科技大学植物和微生物生物技术实验室中从卡拉克盐矿中分离出的菌株（Roohi 等，2014）。根据 Roohi 等（2012）的方法，对嗜盐芽孢杆菌 NCCP-71 和 NCCP-77 进行了培养。菌株在恒温 37℃ 的条件下在振荡培养箱（Wise Cube，WIS-20R）中连续培养 24h。根据 Rani 和 Arundhathi Reddy（2012）的标准方案，使用添加 5%（w/v）Na^+ 含量的琼脂培养基计算菌落形成单位（colony-forming unit，CFU），在接种前菌株浓度需要达到 10^8 CFU/mL。CFU 的计算公式如下。

$$CFU = （菌落数目 \times 稀释倍数）/接种体积$$

7.3.1 沙培试验

将 NIAB-IR-9 和 KS-282 两个品种 10 日龄的幼苗移入温室内装有约 1 kg 洗好的干沙的塑料盆中（塑料盆底部有一个孔，用布覆盖）。幼苗种在沙下 2～3 cm 深的地方，幼苗彼此之间间隔几厘米的距离。所有的塑料盆浇灌霍格兰溶液

(Hoagland 和 Arnon, 1950), pH 值为 5~6, 生长 24 d。

7.3.2 盐胁迫和 PGPR 处理

24 d 后,将不同浓度的 NaCl (50 mmol/L、100 mmol/L 和 150 mmol/L,包括对照 (NaCl 浓度为 0 mmol/L) 与霍格兰溶液混合,浇灌植株 1 周。为了研究嗜盐芽孢杆菌 NCCP-71 和 NCCP-77 在盐胁迫条件下对植物生长的影响,两个菌株各取 10 mL 菌液,同时混合于霍格兰溶液中浇灌植物 (Guo 等, 2011; Nabti 等, 2010)。在各个处理中随机取 3 份植物样品 (新鲜叶片),粉碎后放在离心管中置于-40℃条件下保存进行分子测定。其余剩下的材料在 80℃的培养箱中干燥 48 h,然后处理成粉末以进行不同的生化试验。

7.3.3 生化参数

7.3.3.1 离子分析 (Na^+、K^+、Ca^{2+})

使用火焰光度计 (Jenway-PFP7) 根据 Awan 和 Salim (1997) 的标准方案对各种离子进行测定。材料预处理按照如下方法进行:植物材料干燥后与 H_2SO_4 和 H_2O_2 (2:1, v/v) 混合后在热板上一起消化。消化液用 20 mL 蒸馏水稀释后经滤纸过滤后进行离子分析。

7.3.3.2 光合色素的测定

光合色素 (叶绿素 a、叶绿素 b 和类胡萝卜素) 的测定,取 25 mg 植物材料和 25 mg MgO 置于 10 mL 试管中。然后向试管中加入 5 mL 甲醇后在室温下使用斡旋振荡器振荡。然后将混合均匀的混合物在摇床上以 250 r/min 的速度振荡 3 h。之后,所有样品在室温下以 6 000 r/min 的速度离心 7 min。以甲醇为空白,在 470 nm、653 nm 和 666 nm 的紫外可见分光光度计上测定所有植物提取物的吸光度。

7.3.3.3 氮和蛋白质含量测定

干燥的植物材料放在消化瓶中,加入硫酸后在 400℃条件下消化,直到溶液变成淡黄色。然后将混合物用蒸馏水稀释至 20 mL,然后加入 50% (w/v) 的氢氧化钠溶液反应 4 s。在 250 mL 的容量瓶中,取 50 mL 含有甲基红为指示剂的 4% 的硼酸溶液,当溶液颜色变为无色时蒸馏终止。然后用 0.1 mol/L 的硫酸溶液滴定混合物,直到粉红色出现 (Peter 和 Young, 1980)。总蛋白和有机氮含量计算公式如下。

总蛋白 (g/g) = (样品体积-空白体积×0.1)/样品干重×1.400 7

总氮（g/g）=（样品体积-空白体积×0.1）/样品干重×6.25

7.3.4 离子含量

未接种微生物的两个品种中 Na^+ 含量随着盐浓度的增加而增加，而钾和钙离子含量则随着盐浓度增加而降低（图 7.1 和图 7.2）。嗜盐菌的接种使水稻植株离子含量发生变化。用嗜盐菌株 NCCP-71 和 NCCP-77 接种植物样品，两个品种的 Na^+ 含量降低，K^+ 含量增加（图 7.1 和图 7.2）。接种 NCCP-71 以后，两个品种的钙离子含量都有所增加（图 7.1 和图 7.2）。水稻品种 NIAB-IR-9 接种 NCCP-77 后钙离子含量增加，但在 100 mmol/L NaCl 处理条件下除外（图 7.1C）；而品种 KSK-282 接种 NCCP-77 后钙离子含量降低（图 7.2C），这表明

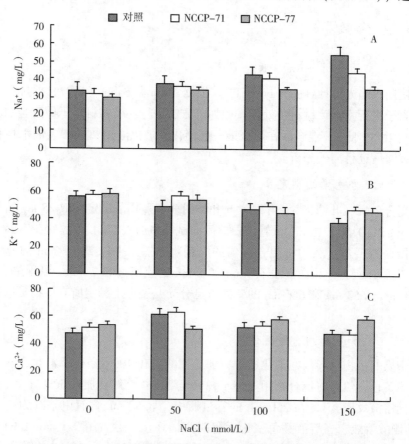

图 7.1 不同 NaCl 浓度胁迫条件下植物生长促生菌对 NIAB-IR-9 中 Na^+（A）、K^+（B）、Ca^{2+}（C）离子含量的影响

NCCP-71 具有更好的缓解植物盐胁迫的潜力。

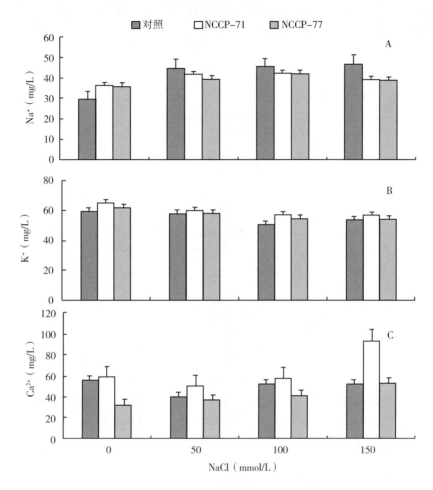

图 7.2　不同 NaCl 浓度胁迫条件下植物生长促生菌对 KSK-28 中 Na^+（A）、K^+（B）、Ca^{2+}（C）离子含量的影响

7.3.5　光合色素

两个品种的叶绿素 a、叶绿素 b 和类胡萝卜素的浓度在不同盐浓度下都有所降低（图 7.3 和图 7.4）。然而微生物的接种改变了在盐胁迫条件下两个水稻品种光合色素（叶绿素 a、叶绿素 b 和类胡萝卜素）的含量。接种 NCCP-71 导致品种 NIAB-IR-9 中的色素浓度增加（图 7.3 A-C）；而接种 NCCP-77，除在 150 mmol/L NaCl 处理条件下，品种 NIAB-IR-9 中叶绿素 a 和叶绿素 b 的浓度均

有所增加（图7.3）。菌种 NCCP-77 对品种 NIAB-IR-9 中类胡萝卜素的含量无促进作用（图7.3C）。而给品种 KSK-282 接种 NCCP-71 和 NCCP-77 菌株后，对照组和 100 mmol/L NaCl 处理的样品中叶绿素 a、叶绿素 b 和类胡萝卜素浓度增加（图7.4）。

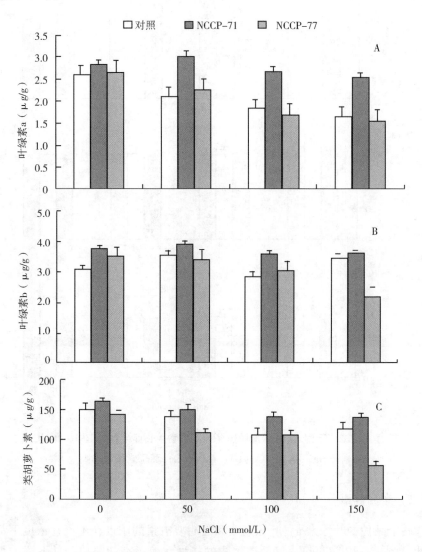

图7.3 不同 NaCl 浓度胁迫条件下植物生长促生菌对 NIAB-IR-9 光合色素中叶绿素 a（A）、叶绿素 b（B）和类胡萝卜素（C）含量的影响

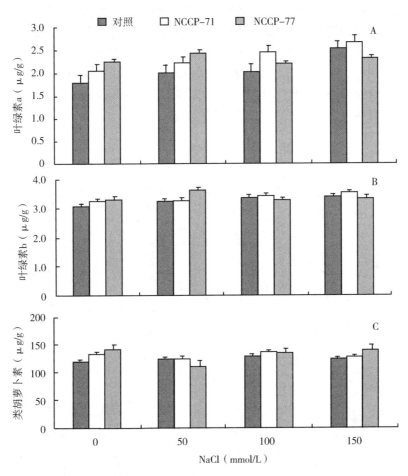

图 7.4 不同 NaCl 浓度胁迫条件下植物生长促生菌对 KSK-28 光合色素中
叶绿素 a (A)、叶绿素 b (B) 和类胡萝卜素 (C) 含量的影响

7.3.6 有机氮和总蛋白含量

盐浓度增加导致蛋白质和氮含量下降（图 7.5）。嗜盐菌的接种也引起了蛋白质和氮含量的变化。接种 NCCP-71 的 NIAB-IR-9 植株在各种盐浓度胁迫条件下蛋白质含量均较高。接种 NCCP-77 增加了 0 mmol/L 和 50 mmol/L NaCl 胁迫样品中的蛋白质含量（图 7.5A）。两种菌株的接种都增加了 NIABIR-9 中在 0 mmol/L 和 100 mmol/L NaCl 浓度胁迫下的氮含量（图 7.5B）。与接种 NCCP-77 相比，接种 NCCP-71 菌株的处理显示出更好的结果。在品种 KSK-282 上的试验表明，随着盐浓度的增加，总有机氮和蛋白质含量下降（图 7.6）。接种 NCCP-

71菌种的KSK-282植株在所有盐胁迫条件下都能提高蛋白质含量。而接种NCCP-77菌株只促进了0 mmol/L和50 mmol/L NaCl胁迫处理下样品蛋白质含量的增加（图7.6A）。接种NCCP-71的植株在所有盐浓度胁迫下都表现出氮含量的增加；而接种NCCP-77菌株只增加了0 mmol/L和50 mmol/L NaCl处理下植株样品的氮含量（图7.6B）。与NCCP-77相比，接种NCCP-71菌株的品种植株中氮和蛋白质的含量增加更多。

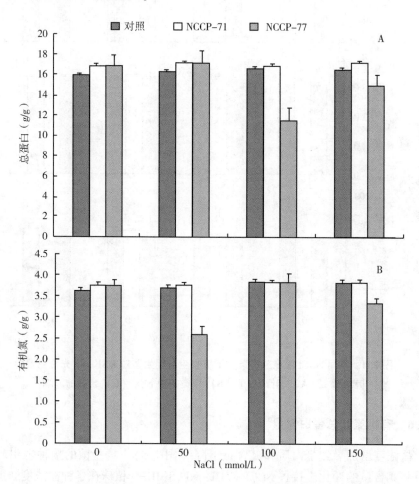

图7.5 不同NaCl浓度胁迫条件下植物生长促生菌对NIAB-IR-9中总蛋白（A）和有机氮（B）含量的影响

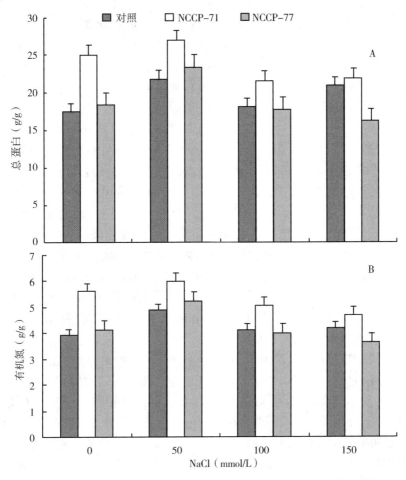

图7.6 不同 NaCl 浓度胁迫条件下植物生长促生菌对 KSK-28 中总蛋白（A）和总有机氮（B）含量的影响

7.4 讨论

从根际分离的微生物已被广泛应用于无机金属、有机石油废料和农药的修复以及盐渍土的复垦（Bose 等, 2008; Zeeb 等, 2006; Huang 等, 2004, 2005; Olson 等, 2008; Lunney 等, 2004; Qadir 等, 2007; Su 等, 2008）。从含盐生境中分离的细菌提高了这一过程的有效性。因此，本研究旨在探讨嗜盐芽孢杆菌（NCCP-71 和 NCCP-77）对生长在盐渍土环境下水稻生化特性的影响。在盐胁迫条件下，土壤中盐浓度的增加使植物中 Na^+ 离子的含量也相应增加，而 K^+ 和 Ca^{2+} 离子含量则降低（图 7.1 和图 7.2）。高浓度的 Na^+ 和较低的 K^+ 和 Ca^{2+} 含量可能是由于在胁迫条件下植物与外界直接接触的离子池有关。暴露在高盐浓度下的植物会引起离子和渗透胁迫（Niu 等，1995；Serrano 等，1999；Zhu 等，1998）。众所周知，疏水性静电失衡发生在细胞水平。离子浓度的调节对细胞酶在维持细胞蛋白质的过程中发挥最佳功能至关重要（Wyri Jones 和 Pollard，1983）。Wang 等（2012）报道在盐胁迫条件下，随着 NaCl 浓度的增加植物中 Na^+ 含量较高。Jamil 等（2012）在研究盐胁迫对 NIAB-IR-9 等 3 个水稻品种生理生化的影响的过程中也得到了类似的结果。用 NCCP-71 和 NCCP-77 接种水稻幼苗，能够降低植物中 Na^+ 离子含量，并增加 K^+ 和 Ca^{2+} 离子的含量（图 7.1 和图 7.2）。相对于未经接种的对照植株，两个菌株都能有效地降低接种植株中 Na^+ 的浓度，同时它们也能保持较高的 K^+ 和 Ca^{2+} 浓度。很多其他的研究也表明，接种植物生长促生细菌能够降低植物体内 Na^+ 离子含量，增加 K^+ 和 Ca^{2+} 含量（Nadeem 等，2006；Zahir 等，2009；Mishra 等，2011；Yildirim 等，2011）。

在水稻中，光合色素是关系到光合能力的重要性状（Teng 等，2004）。光合色素随着盐浓度的增加而减少（图 7.3 和图 7.4）。盐胁迫使叶绿素 a 含量下降的幅度大于叶绿素 b（图 7.3 和图 7.4），因为它们主要受到高盐胁迫的影响（Daiz 等，2002；Santos，2004）。光合色素含量的降低是导致水稻光合效率降低的原因（Moradi 和 Ismail，2007；ZhenHua 等，2012）。在盐胁迫条件下，接种 NCCP-71 显著增加了水稻幼苗光合色素含量；而接种 NCCP-77 增加了在 100 mmol/L NaCl 浓度处理下两个品种的色素含量（图 7.3 和图 7.4）。而这种含量的增加可能是由于细菌形成的丁二醇和丙酮等重要化合物，它们可以促进多种植物的生长（Ryu 等，2003）。不同的研究指出，很多促进植物生长的细菌能够通过提高光合色素含量的方式促进植物发育（Zahir 等，2009；Mishra 等，2011；Yildirim 等，2011；Nadeem 等，2006）。

盐浓度的增加也降低了蛋白质和氮的含量（图 7.5 和图 7.6）。有研究指出，高盐含量能够降低蛋白质含量，但却能增加多核糖体的含量（Jones，1996）。其他研究也有报道，盐胁迫条件下植物中蛋白质含量随 NaCl 浓度的增加而降低（Khan，1998；Azooz 等，2004；Dagar 等，2004）。在盐胁迫条件下，接种 PGPR 能够导致有机氮和蛋白质含量的增加（Nadeem 等，2006；Mishra 等，2011）。在本研究中，接种 NCCP-71 提高了两个品种幼苗在盐胁迫下的蛋白质和氮含量，而接种 NCCP-77 只提高了 50 mmol/L NaCl 浓度处理下水稻幼苗氮和蛋白质的含量（图 7.5 和图 7.6）。接种植物生长促生菌还能刺激植物增加氮和铁，从而增加植物的蛋白质含量（Mallesh，2008）。

7.5 结论

从卡拉克盐矿中分离出的嗜盐芽孢杆菌 NCCP-71 和 NCCP-77 通过接种到植物上能够有效降低盐胁迫的危害。这些促生菌导致离子含量的增加、光合色素的增加以及氮和蛋白质含量的增加。然而，菌株 NCCP-77 对一些盐浓度处理敏感。

参考文献

Afzal I, Basra SM, Farooq M, Nawaz A, 2006. Alleviation of salinity stress in spring wheat by hormonal priming with ABA, salicylic acid and ascorbic acid. Int J Agric Biol 8: 23-28.

Antoun H, Prévost D, 2005. Ecology of plant growth promoting rhizobacteria. In: PGPR: biocontrol and biofertilization. Springer, pp. 1-38.

Ashrafuzzaman M, Hossen FA, Ismail MR, Hoque A, Islam MZ, Shahidullah SM, Meon S, 2009. Efficiency of plant growth-promoting rhizobacteria (PGPR) for the enhancement of rice growth Afr J Biotechnol 8.

Awan JA, Salim UR, 1997. Food analysis manual. Pak Vet J Publ 5: 2-7.

Azooz MM, Shaddad MA, Abdel-Latef AA, 2004. The accumulation and compartmentation of proline in relation to salt tolerance of three sorghum cultivars. Indian J Plant Physiol 9: 1-8.

Bacilio M, Rodriguez H, Moreno M, Hernandez JP, Bashan Y, 2004. Mitigation of salt stress in wheat seedlings by a gfp-tagged *Azospirillum lipoferum*. Biol Fertil Soils 40: 188-193.

Bhattacharyya PN, Jha DK, 2012. Plant growth-promoting rhizobacteria (PGPR): emergence in agriculture. World J Microbiol Biotechnol 28: 1327-1350.

Bose S, Vedamati J, Rai V, Ramanathan AL, 2008. Metal uptake and transport by *Typha angustata* L. grown on metal contaminated waste amended soil: an implication of phytoremediation. Geoderma 145: 136-142.

Dagar JC, Bhagwan H, Kumar Y, 2004. Effect on growth performance and biochemical contents of *Salvadora persica* when irrigated with water of different salinity. Indian J Plant Physiol 9: 234-238.

Daiz MGQ, Bebing NN, Villegas MJ, Manuel MC, 2002. Cell biology, laboratory manual (fifth edition): genetics and molecular biology division. The University of the Philippines, Philippines, p. 47.

Grover M, Ali SZ, Sandhya V, Rasul A, Venkateswarlu B, 2011. Role of microorganisms in adaptation of agriculture crops to abiotic stresses. World J Microbiol Biotechnol 27: 1231-1240.

Guo J, Tang S, Ju X, Ding Y, Liao S, Song N, 2011. Effects of inoculation of a plant growth promoting rhizobacterium *Burkholderia* sp. D54 on plant growth and metal uptake by a hyperaccumulator *Sedum alfredii* Hance grown on multiple metal contaminated soil. World J Microbiol Biotechnol 27: 2835-2844.

Gururani MA, Upadhyaya CP, Baskar V, Venkatesh J, Nookaraju A, Park SW, 2013. Plant growth promoting rhizobacteria enhance abiotic stress tolerance in *Solanum tuberosum* through inducing changes in the expression of ROS-scavenging enzymes and improved photosynthetic performance. J Plant Growth Regul 32: 245-258.

Hoagland DR, Arnon DI, 1950. The water-culture method for growing plants without soil. Circ Calif Agric Exp Station 347: 32.

Huang XD, El-Alawi Y, Penrose DM, Glick BR, Greenberg BM, 2004. A multi-process phytoremediation system for removal of polycyclic aromatic hydrocarbons from contaminated soils. Environ Pollution 130 (3): 465-476.

Huang XD, El-Alawi Y, Gurska J, Glick BR, Greenberg BM, 2005. A multi-process phytoremediation system for decontamination of persistent total petroleum hydrocarbons (TPHs) from soils. Microchem J 81 (1): 139-147.

Jamil M, Bashir S, Anwar S, Bibi S, Bangash A, Ullah F, Rha ES, 2012. Effect of salinity on physiological and biochemical characteristics of different varieties of rice. Pak J Bot 44: 7-13.

Jamil M, Malook I, Parveen S, Naz T, Ali A, Jan SU, Rehman SU, 2013. Smoke priming, a potent protective agent against salinity: effect on proline accumulation, elemental uptake, pigmental attributes and protein banding patterns of rice (*Oryza Sativa*). J Stress Physiol & Biochem 9 (1): 169-183.

Jones GH, 1996. Plants and microclimate, 2nd edn. Cambridge, MA, pp. 72-108.

Kaymak HC, Yarali F, Guvenc I, Donmez MF, 2008. The effect of inoculation with plant growth rhizobacteria (PGPR) on root formation of mint (*Mentha piperita* L.) cuttings. Afr J Biotechnol 7: 4479-4483.

Kaymak HÇ, Güvenç IS, Yarali F, Dönmez MF, 2009. The effects of bio-priming with PGPR on germination of radish (*Raphanus sativus* L.) seeds under saline conditions. Turk J Agric For 33: 173-179.

Khan GS, 1998. Soil salinity/sodicity status in Pakistan. Soil Survey of Pakistan, Lahore 59

Khan N, Bano A, Babar MA, 2017. The root growth of wheat plants, the water conservation and fertility status of sandy soils influenced by plant growth promoting rhizobacteria. Symbiosis 72: 195-205.

Lichtenthaler HK, Wellburn AR, 1985. Determination of total carotenoids and chlorophyllA and B of leaf in different solvents. Biochem Soc Trans 11: 591-592.

Lunney AI, Zeeb BA, Reimer KJ, 2004. Uptake of weathered DDT in vascular plants: potential for phytoremediation. Environ Sci Technol 38 (22): 6147-6154.

Mallesh SB, 2008. Plant growth promoting rhizobacteria, their characterization and mechanisms in the suppression of soil bore pathogens of *Coleus* and *Ashwagandha*, P. HD thesis, University of agricultural sciences, Dharwad.

Malook I, Atlas A, Rehman SU, Wang W, Jamil M, 2014. Smoke alleviates adverse effects induced by stress on rice. Toxicol Environ Chem 96: 755-767.

Malook I, Shah G, Jan M, Shinwari KI, Aslam MM, Rehman S, Jamil M, 2017. Smoke priming regulates growth and the expression of myeloblastosis and zinc-finger genes in rice under salt stress. Arab J Sci Eng 42: 2207-2215. https://doi.org/10.1007/s13369-016-2378-x.

Mishra PK et al., 2011. Alleviation of cold stress in inoculated wheat (*Triticum aestivum* L.) seedlings with psychrotolerant pseudomonads from NW Himalayas. Arch Microbiol 193: 497-513.

Moradi F, Ismail AM, 2007. Responses of photosynthesis, chlorophyll fluorescence and ROS-scavenging systems to salt stress during seedling and reproductive stages in rice. Ann Bot 99: 1161-1173. https://doi.org/10.1093/aob/mcm052.

Nabti E, Sahnoune M, Ghoul M, Fischer D, Hofmann A, Rothballer M, Schmid M, Hartmann A, 2010. Restoration of growth of durum wheat (*Triticum durum* var. *waha*) under saline conditions due to inoculation with the rhizosphere bacterium *Azospirillum brasilense* NH and extracts of the marine alga *Ulva lactuca*. J Plant Growth Regul 29: 6-22.

Nadeem S, Zahir Z, Naveed M, Arshad M, Shahzad S, 2006. Variation in growth and ion uptake of maize due to inoculation with plant growth promoting rhizobacteria under salt stress. Soil Environ 25: 78-84.

Naheed G, Shahbaz M, Latif T, Rha ES, 2007. Alleviation of the adverse effects of salt stress on rice (*Oryza sativa* L.) by phosphorus applied through rooting medium: growth and gas exchange characteristics. Pak J Bot 39: 729-737.

Nihorimbere V, Ongena M, Cawoy H, Brostaux Y, Kakana P, Jourdan E, Thonart P, 2010. Beneficial effects of *Bacillus subtilis* on field-grown tomato in Burundi: reduction of local fusarium disease and growth promotion. Afr J Microbiol Res 4: 1135-1142.

Niu X, Bressan RA, Hasegawa PM, Pardo JM, 1995. Ion homeostasis in NaCl. Stress Environ Plant Physiol 109: 735.

Olson PE, Castro A, Joern M, DuTeau NM, Pilon-Smits E, Reardon KF, 2008. Effects of agronomic practices on phytoremediation of an aged PAH-contaminated soil. J Environ Quality 37 (4): 1439-1446.

Pallai R, 2005. Effect of plant growth-promoting rhizobacteria on Canola (*Brassica napus* L.) and lentil (*Lens culinaris* Medik.) plants. MSc thesis, University of Saskatchewan.

Peter LP, Young VR, 1980. Nutritional evaluation of proteins foods. United Nation University, Japan, p. 8.

Príncipe A, Alvarez F, Castro MG, Zachi L, Fischer SE, Mori GB, Jofré E, 2007. Biocontrol and PGPR features in native strains isolated from saline. Soils Argent Curr Microbiol 55: 314-322. https://doi.org/10.1007/s00284-006-0654-9.

Qadir M, Oster JD, Schubert S, Noble AD, Sahrawat KL, 2007. Phytoremediation of sodic and saline-sodic soils. In: Sparks DL (ed) Advan Agronomy, vol 96, pp. 197-247.

Rafique N, Raza SH, Qasim M, Iqbal N, 2011. Pre-sowing application of ascorbic acid and salicylic acid to seed of pumpkin and seedling response to salt. Pak J Bot 43 (6): 2677-2682.

Rani UM, Arundhathi Reddy G, 2012. Screening of rhizobacteria containing plant growth promoting (PGPR) traits in rhizosphere soils and their role in enhancing growth of pigeon pea. Afr J Biotechnol 11 (32): 8085-8091.

Roohi A, Ahmed I, Iqbal M, Jamil M, 2012. Preliminary isolation and characterization of halotolerant and halophilic bacteria from salt mines of Karak, Pakistan. Pak J Bot 44: 365-370.

Roohi A, Ahmed I, Khalid N, Iqbal M, Jamil M, 2014. Isolation and phylogenetic identification of halotolerant/halophilic bacteria from the salt mines of Karak, Pakistan. Int J Agric Biol 16: 564-570.

Ryu CM, Farag MA, Hu CH, Reddy MS, Wei HX, Pare PW, Kloepper JW, 2003. Bacterial volatiles promote growth in *Arabidopsis*. Proc Natl Acad Sci U S A 100: 4927-4932.

Saharan BS, Nehra V, 2011. Plant growth promoting rhizobacteria: a critical review. Life Sci Med Res 21: 1-30. LSMR-21.

Santos CV, 2004. Regulation of chlorophyll biosynthesis and degradation by salt stress in sun flower leaves. Sci Hortic 103: 93-99.

Serrano R et al., 1999. A glimpse of the mechanisms of ion homeostasis during salt stress. J Exp Bot 50: 1023-1036.

Shah G et al., 2017. Halophilic bacteria mediated phytoremediation of salt-affected soils cultivated with rice. J Geochem Explor 174: 59-65. https://doi.org/10.1016/j.gexplo.2016.03.011.

Siringam K, Juntawong N, Cha-um S, Boriboonkaset T, Kirdmanee C, 2012. Salt tolerance enhancement in indica rice (*Oryza sativa* L. spp. indica) seedlings using exogenous sucrose supplementation. Plant Omics 5: 52.

Su Y, Han FX, Chen J, Sridhar BBM, Monts DL, 2008. Phytoextraction and accumulation of mercury in three plant species: Indian mustard (*Brassica juncea*), beard grass (*Polypogon monospeliensis*), and Chinese brake fern (*Pteris vittata*). Int J Phytoremediation 10 (6): 547-560.

Teng S, Qian Q, Zeng D, Kunihiro Y, Fujimoto K, Huang D, Zhu L, 2004. QTL analysis of leaf photosynthetic rate and related physiological traits in rice (*Oryza sativa* L.). Euphytica 135: 1-7.

Wang H, Wu Z, Zhou Y, Han J, Shi D, 2012. Effects of salt stress on ion balance and nitrogen metabolism in rice. Plant Soil Environ 58: 62-67.

Wyn Jones R, Pollard A, 1983. Proteins, enzymes and inorganic ions. In: Encyclopedia of plant physiology new series.

Yildirim E, Turan M, Ekinci M, Dursun A, Cakmakci R, 2011. Plant growth promoting rhizobacteria ameliorate deleterious effect of salt stress on lettuce. Sci Res Essays 6: 4389-4396.

Zahir ZA, Ghani U, Naveed M, Nadeem SM, Asghar HN, 2009. Comparative effectiveness of *Pseudomonas* and *Serratia* sp. containing ACC-deaminase for improving growth and yield of wheat (*Triticum aestivum* L.) under salt-stressed conditions. Arch Microbiol 191: 415-424.

Zeeb BA, Amphlett JS, Rutter A, Reimer KJ, 2006. Potential for phytoremediation of polychlorinated biphenyl- (PCB-) contaminated soil. Int J Phytoremediation 8 (3): 199-221.

Zhang ZH, Liu Q, Song HX, Rong XM, Ismail AM, 2012. Responses of different rice (*Oryza sativa* L.) genotypes to salt stress and relation to carbohydrate metabolism and chloro phyll content African. J Agric Res 7: 19-27.

Zhu JK, Liu J, Xiong L, 1998. Genetic analysis of salt tolerance in *Arabidopsis*: evidence for a critical role of potassium nutrition. Plant Cell 10: 1181-1191.

8 耐盐细菌海水发酵生产鼠李糖脂及其在草莓白粉病防治中的应用

Xiangsheng Zhang, Boping Tang

[摘要]

生物表面活性剂是两亲性分子，是一种天然的农业化合物，可以由微生物合成分泌，表现出具有表面活性剂的特征。生物表面活性剂具有良好的生物降解性、低毒性，在极端温度和pH条件下的生态可接受性和有效性均优于化学表面活性剂。生物表面活性剂的生产成本包括发酵设施和矿盐材料。海水中通常富含各种矿物盐。利用海水和废弃植物油生产生物表面活性剂，具有生产设备简单且生产成本低的特点。在本研究中，通过微生物菌种筛选、摇瓶发酵和5 L液体发酵罐发酵优化后能够获得10 g/L的发酵产量，通过试验证实了这种低成本生产方式的可行性和实用性。本研究还对鼠李糖脂在保护盐渍土农作物中的应用进行了探索。将发酵液分别稀释至1 g/L和0.5 g/L后直接用于防治草莓霜霉病，结果表明，这种发酵液是一种有效的杀菌剂。盆栽试验表明，与空白对照组相比，使用不同稀释浓度发酵液对草莓霜霉病的防治效果分别达到90.8%和87.6%以上，防治结果比使用化学杀菌剂的效果更好。此外，鼠李糖脂发酵液还能够促进草莓根和茎的发育。

[关键词]

海水发酵；耐盐生物表面活性剂产生菌；鼠李糖脂；杀菌剂；白粉病；凤梨

8.1 研究背景

生物表面活性剂是由具有表面活性的微生物分泌的两亲性化合物（Zhang等，2012a，2002b）。研究发现，它们主要是由在水介质中生存的有氧微生物从碳源原料（例如碳水化合物、碳氢化合物、油和其他脂质或其混合物）中产生

X. S. Zhang, B. P. Tang (✉)，中国江苏省盐渍土生物资源重点实验室，盐城师范大学滨海湿地生物资源与环境保护江苏省重点实验室，江苏省盐城师范学院江苏沿海生物农业综合创新中心

的（Zhang 等，2012b）。能够产生表面活性剂的微生物包括芽孢杆菌（*Bacillus*）、假单胞菌（*Pseudomonas*）、不动杆菌（*Acinetobacter*）、无色杆菌（*Achromobacter*）、节杆菌（*Arthrobacter*）、短杆菌（*Brevibacterium*）、棒状杆菌（*Corynebacterium*）、假丝酵母（*Candida*）和红酵母（*Rhodotorula*）（Pirog 等，2012）。目前，已知的生物表面活性剂主要有 5 类，包括糖脂、磷脂、脂肽、聚合表面活性剂和颗粒表面活性剂（Shen 等，2011；Zhang 和 Xiang，2010a）。

与纯化学表面活性剂相比，生物表面活性剂具有很多优点，比如生物可降解、无毒、环保、表面活性高、耐极端环境、临界胶束浓度低等（Silva，2017；Zhang 和 Xiang，2010a）。因此，生物表面活性剂已被用于微生物采油（Bordoloi 和 Konwar，2008；Zhang 和 Xiang，2010b；Zhao 和 Jiang，2004）、石油化工、环境生物修复（Li 等，2000；Zhang，2013）、农业、化妆品工业和食品工业等（Banat 等，2000）。

糖脂是由细菌产生的最常见的生物表面活性剂（Hatha 等，2007；Zhang 等，2012a）。众所周知，鼠李糖脂是目前研究最为广泛的糖脂，具有许多潜在的应用价值（Henkel 等，2012；Huang 等，2009；Sodagari 等，2017；Zhang 和 Lu，2013）。本章主要介绍鼠李糖脂在生物表面活性剂相关领域的筛选、发酵和应用等。

此外，生物表面活性剂（在本文中指鼠李糖脂）也可用作杀虫剂、杀菌剂或杀真菌剂。Ji Ye Ahn 等曾报道称，分离出的一种命名为 SG3 的细菌能够产生鼠李糖脂，其发酵液能抑制所有供试真菌病原菌的菌丝生长，并能有效抑制稻瘟病、番茄灰霉病、番茄晚疫病、小麦叶锈病、大麦白粉病以及红辣椒炭疽病等多种植物病害的发生。但目前其他类似的研究报道并不多。

在温室中草莓白粉病经常导致巨大的产量和品质损失。本研究在实验室条件下探索了鼠李糖脂发酵液对白粉病的抑制效果，以及在田间试验中对草莓白粉病的抑制效果，旨在阐明鼠李糖脂发酵液对草莓白粉病的防治效果。

8.2 生物表面活性剂的低成本生产

理论上，生物表面活性剂的生产可以通过化学合成和微生物发酵两种方式进行。但由于生物表面活性剂结构的复杂性，通过化学合成的手段进行生产很难实现。目前，发酵法是生物表面活性剂生产中的主要方法，包括活细胞法、代谢控制细胞法、静态细胞法和前体添加法等（Sarachat 等，2010；Wu 等，2007）。目前，游离细胞发酵法是最简单、最常用的生产方法。

虽然微生物表面活性剂具有许多优点，但由于生产成本高，限制了其在上述很多应用领域,尤其是农业领域上的应用。因此,降低生产成本是微生物学家的研究热点之一。产品成本问题将极大地影响生物表面活性剂的推广应用。降低成本的生产研究主要是利用低成本的材料作为碳源(Jain 等,2013;Mukherjee 等,2006),例如废油(Nitschke 等,2005)、屠宰场废弃物(Ramani 等,2012)、酿酒厂和凝乳乳清废料(Dubey 和 Juwarkar,2001)、甘蔗渣(Neto 等,2009),以及马铃薯加工废水(Noah 等,2005)等。另据报道,生产成本居高不下也因为好氧发酵过程中起泡倾向严重导致生产效率和产量相对较低。理想条件下,鼠李糖脂的发酵产物能够达到 42 g/L(Sodagari 等,2017),这可能是目前报道的最高产率。

废弃油料(如多次使用的煎炸油,以及公用污水处理厂的积累油相)是最常用的能够降低生产成本的碳源(Nitschke 等,2005;Raza 等,2006)。实际上,它们基本上没有作为二次资源加以回收。如果处置不当会造成环境污染,如果作为食用油非法销售则会对公众健康造成损害。然而,废弃油料可用于生产生物表面活性剂,例如鼠李糖脂可用作乳化剂或破乳剂(Nitschke 等,2005;Yao 和Min,2010)以及生物柴油(Xiong 等,2007)。

虽然国内外对耐盐或耐盐生物表面活性剂的生产进行了大量的研究,但在天然海水中添加或不添加矿物盐生产生物表面活性剂的报道却很少。本章提出了用海水发酵生产鼠李糖脂的新思路并付诸实践,取得了满意的效果。

8.3 生物表面活性剂菌株的筛选

发酵液和矿物盐的成本是总生产成本之一。海水中富含各种矿物盐,然而目前海水淡化的成本较高。利用海水和废弃植物油通过简单的设备生产生物表面活性剂能够大幅度降低成本。利用高盐度海水(不低于30%)发酵在一定程度与极端环境下的发酵有一定的相似性,因此筛选耐海水生物表面活性剂产生菌,利用海水进行发酵在沿海地区具有重要意义。这些菌株在海洋疏水性污染物的生物修复中也有潜在的应用。通常海水发酵菌株的筛选方式如下(Miao 等,2013;Zhang,2014)。

8.3.1 培养基

高盐营养肉汤(nutrient broth,NB)或种子培养基由去离子水、蛋白胨 10 g/L、牛肉提取物 5 g/L 和 NaCl 35 g/L 组成。制备固体营养琼脂(nutrient agar,NA)平板或斜面培养基,可加入 15.0 g/L 琼脂。含 3.5% NaCl 的 NA 培养基作为对照。海水发酵液的组成成分为酵母提取物 1 g/L;pH 为自然状态下数值,个别摇瓶加入 3%(v/v)的煎炸油作为唯一碳源。所有培养基在 121℃高压灭菌 20 min。

8.3.2 嗜盐生物表面活性剂产生菌的分离与筛选

将约 5 g 沿海土壤样品或 5 mL 海水样品在无菌环境下添加到 45 mL 灭菌海水中,并在 NA 培养基上划线。选择溶血圈大且清晰的菌落在斜面培养基上 30 ℃温度条件下孵育,然后在 4 ℃条件下保存备用。将一环培养物接种在 100 mL 烧瓶中的 20 mL 种子培养基中,并在 30℃温度条件和 180 r/min 的旋转摇瓶上孵育 10 h。然后将 2.5 mL(5%)的培养物转移至 250 mL 摇瓶中的 50 mL 海水发酵液中培养 3 d。培养物在 9 000 r/min 的条件下进行离心,测定菌株的生物表面活性剂的产生能力,挑选表现最好的菌株。

8.3.3 生物表面活性剂生产能力的测定

在本节中,下述采用的所有试验步骤均为原作者的实验室条件(Zhang 等,2012b)。

8.3.3.1 铺油法

将约 30 mL 蒸馏水放入培养皿(90 mm)中,然后将 50 μL 柴油滴在水面上形成油膜,在油膜的中央滴入 5 μL 的发酵液,然后用去离子水稀释 10 倍。排油环的直径用滑管测量(精度为 0.02 mm)。选择能够产生较大直径的菌株进行表面张力的测量。该方法通常用于发酵液表面活性剂生产的初步评价。

8.3.3.2 表面张力测量

发酵液上清的表面张力(100 倍稀释液)使用 JYW-200 型表面张力仪测定。

8.3.3.3 硫酸-苯酚反应法

首先将发酵液稀释 100 倍,然后将 2 mL 的稀释液移入 15 mL 的玻璃试管中,再加入 1 mL 苯酚和 5 mL 硫酸后搅拌。然后将试管在水浴中加热 15 min,冷却到室温后,测量混合液在 480 nm 处的光密度。

图 8.1 海水发酵
(左侧,无接种物;右侧,接种菌株后在 30℃下培养 3 d)

8.4 发酵工艺

在实验室水平上,游离细胞发酵有 3 种方法:摇瓶发酵、小发酵罐发酵(液体或固体)和简化发酵。在摇瓶中发酵 3 d 后,就能够看到废油有明显乳化的迹象(图 8.1)。采用海水培养基,摇瓶发酵条件下生产的鼠李糖脂能够达到 10.5 g/L。

发酵液在液体发酵罐中的外观如图 8.2 所示，显示出很好的海水发酵性能。采用这种方法发酵目标物质的最高产量能够达到 33.0 g/L。随着发酵过程的进行，排油圈由 6.61 mm 增加到 14.47 mm，而表面张力则由 78.9 N/m 降低到 36.2 N/m。采用 5 L 发酵罐的发酵效果比摇瓶发酵效果好，这为工业化应用奠定了坚实的基础。

图 8.2　生产鼠李糖脂的 5 L 液体发酵罐

8.5　在盐渍土草莓上的应用

8.5.1　草莓白粉病的处理

如图 8.3 所示，鼠李糖脂发酵液各处理的防治效果均优于对照（A 为代森锰锌；B 为三唑酮），1 g/L 和 0.5 g/L 处理的防治效果分别为 90.8% 和 87.6%，较对照最高提高了 28.6%，防治效果差异非常显著。

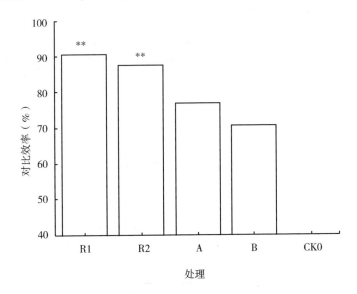

图 8.3　不同处理的比较效率

注：R1—鼠李糖脂发酵液 1.0g/L；R2—鼠李糖脂发酵液 0.5g/L；A—代森锰锌稀释 800 倍；B—三唑酮稀释 1 000 倍；CK0—去离子水；显著水平，$P<0.01$。

8.5.2 对草莓根系发育的影响

如表8.1所示,4个处理都促进了根系发育。尤其是鼠李糖脂发酵液处理的样品根鲜重、根干重、根长、根径和根系呼吸强度等增加了2倍以上。

表8.1 对草莓根系发育的影响

处理	根鲜重 (g)	根干重 (g)	根长 (mm)	根部直径 (mm)	根呼吸强度 [$\mu mol\ O_2/(min \cdot g$ 鲜重$)$]
R1	10.52**	2.95**	352**	0.65**	0.83**
R2	9.37**	2.54**	314**	0.54*	0.70**
A	6.86*	1.33*	229**	0.46	0.60**
B	6.27*	1.29*	201*	0.44	0.52**
对照	4.84	1.18	162	0.42	0.23

注:R1—鼠李糖脂发酵液1.0 g/L;R2—鼠李糖脂发酵液0.5g/L;A—代森锰锌稀释800倍;B—三唑酮稀释1 000倍;对照—去离子水。*表示显著性$P<0.05$;**表示显著性$P<0.01$。

8.5.3 对地上部发育的影响

如表8.2所示,经过处理后地上部分的发育都得到了增强。尤其是R1和R2处理的鲜梢重、干梢重、平均叶面积和叶绿素等含量均有所增加,增加幅度为10.8%~48.2%。

表8.2 对地上部发育的影响

处理	鲜梢重 (g)	鲜梢增长率 (%)	干梢干重 (g)	干梢增长率 (%)	叶绿素含量 (mg/g 鲜重)	叶绿素增长率 (%)
R1	32.78**	48.1	10.67**	48.2	38.63*	16.4
R2	29.40**	32.9	9.57**	32.9	36.78*	10.8
A	25.54*	15.4	8.31*	15.4	35.32	6.4
B	24.47*	10.6	7.96	10.6	34.96	5.3
对照	22.13	0.0	7.2	0.0	33.2	0.0

注:R1—鼠李糖脂发酵液1.0 g/L;R2—鼠李糖脂发酵液0.5g/L;A—代森锰锌稀释800倍;B—三唑酮稀释1 000倍;对照—去离子水。*表示显著性$P<0.05$;**表示显著性$P<0.01$。

8.6 成本评估

成本评估见表8.3。

可以预见，发酵液原料的总成本很低。

表 8.3 发酵原料成本初步评价

项目	原料	所需量	价格	总额
1t 发酵液	海水	1t	0	0
	废弃油料	30L	1 美元/L	30 美元
	酵母提取物	1kg	5 美元/L	5 美元
总和				35 美元

8.6.1 海水发酵的原理

海水富含矿物盐（Orban 等，2007）。虽然一些研究人员发现，借助微生物清理石油泄漏应该考虑到海水营养不足的情况（Atlas 和 Bartha，1972），即海水缺乏可用来处理石油泄漏的微生物的营养元素。海水中最丰富的元素有11种，包括5种阳离子：Na^+、Mg^{2+}、Ca^{2+}、K^+和Sr^{2+}（Li 和 Wu，1993）。通过对海水发酵培养基和合成发酵培养基的比较发现，NO_3^-、PO_4^{3-}和K^+分别作为N、P和K元素的供体，对发酵目标物质的产量影响较大。而海水中过高浓度的Ca^{2+}和Mg^{2+}则可能会抑制生物表面活性剂的产量（Liang 等，2009）。但在本研究中，采用海水发酵则能够显示出良好的效果，其原因如下：①海水中的其他元素可以补偿过高浓度的Ca^{2+}和Mg^{2+}离子造成的负面影响；②细菌对高浓度的阳离子具有抵抗能力。因此，对海水发酵的工业化生产来说，筛选合适的菌种非常重要。海水中的硝酸盐氮含量非常有限（We 等，2010）。本研究发现海水中硝酸盐氮的浓度约为5 mg/L，远低于合成培养基中的浓度，因此成为海水发酵的限制因素之一。为进一步提高生物表面活性剂的产量，应开展添加硝酸盐或尿素对产量影响的研究。

8.6.2 海水发酵鼠李糖脂在农业中的潜在应用

生物表面活性剂具有很强的商业应用潜力（Jamal 等，2014），但其在不同领域的应用中受到生产成本的限制。本文主要从两个方面探讨与生物表面活性剂应用成本相关的问题。首先，降低发酵的成本，采用海水代替纯水，大大降低了生产成本。其次，发酵液直接应用于植物保护，也降低了应用成本。在本研究中利用发酵液对草莓白粉病进行了防治试验，结果表明处理组的防治效果比空白对照组高 87.57%以上，比常用药剂的效果好。此外，鼠李糖脂发酵液还可以促进草莓的根和地上部分的发育。鼠李糖脂是一种极具潜力的天然杀真菌剂。对于鼠李糖脂在草莓栽培中的应用，建议以 0.5~1 g/L 的浓度每 7 d 施用 1 次，共施用 2~3 次以达到更稳定的防治效果。

8.7 结论

作者首次提出了在海水中添加一定量的酵母提取物（生长因子）和足量的废弃植物油（碳源）来进行耐盐菌株的生物表面活性剂发酵。利用海水中的天然矿物盐和从沿海土壤中分离出的一株细菌菌株，大大降低了生产成本，并且鼠李糖脂的产量要高于其他文献报道的采用含油废水的发酵方法（Liang 等，2009）。综上所述，海水可用于发酵鼠李糖脂等生物表面活性剂，并且发酵液对草莓白粉病等多种植物病原菌具有良好的防治效果。本研究为废水发酵液在园艺作物保护中的进一步应用奠定了基础。

[致谢]

感谢江苏省"333"项目、国家星火计划（2015GA690261）、江苏省滨海湿地生物资源与环境保护重点实验室和江苏省盐渍土生物资源重点实验室开放项目的资助。

参考文献

Atlas RM, Bartha R, 1972. Degradation and mineralization of petroleum in sea water: limitation by nitrogen and phosphorous. Biotechnol Bioeng 14 (3): 309-318.

Banat IM, Makkar RS, Cameotra SS, 2000. Potential commercial applications of microbial surfactants. Appl Microbiol Biotechnol 53 (5): 495-508.

Bordoloi NK, Konwar BK, 2008. Microbial surfactant-enhanced mineral oil recovery under laboratory conditions. Colloids Surf B Biointerfaces 63 (1): 73-82. https://doi.org/10.1016/j.colsurfb.2007.11.006.

Dubey K, Juwarkar A, 2001. Distillery and curd whey wastes as viable alternative sources for biosurfactant production. World J Microbiol Biotechnol 17 (1): 61-69.

Hatha A, Edward G, Rahman K, 2007. Microbial biosurfactants- review. J Atmos Sci 3 (2): 1-17.

Henkel M, Müller MM, Kügler JH, Lovaglio RB, Contiero J, Syldatk C, Hausmann R, 2012. Rhamnolipids as biosurfactants from renewable resources: concepts for next-generation rhamnolipid production. Process Biochem 47 (8): 1207-1219.

Huang X, Chen X, Liu J, Lu L, 2009. Recent progress on rhamnolipid produced from fermentation of waste edible oils. Microbiology 36 (11): 1738-1743 (in Chinese).

Jain RM, Mody K, Joshi N, Mishra A, Jha B, 2013. Effect of unconventional carbon sources on biosurfactant production and its application in bioremediation. Int J Biol Macromol 62 (11): 52-58.

Jamal P, Mir S, Alam MZ, Wan NW, 2014. Isolation and selection of new biosurfactant producing bacteria from degraded palm kernel cake under liquid state fermentation [J]. J Oleo Sci 63 (8): 795-804.

Li Y, Wu D, 1993. A study of major seawater chemical constituents in part of Bohai sea. J Qingdao Univ 6 (4): 41-48.

Li G, Huang W, Lerner DN, Zhang X, 2000. Enrichment of degrading microbes and bioremediation of petrochemical contaminants in polluted soil. Water Res 34 (15): 3845-3853.

Liang X, Yao B, Sha R, Zhang H, Meng Q, 2009. Production of rhamnolipids by *Pseudomonas aeruginosa* under high salt concentration conditions. Chin Sci Pap Online 4 (6): 418-422 (in Chinese).

Miao L, Zhang X, Zhou C, Zhang F, Chai X, 2013. Study on production of biosurfactants by seawater fermentation. J Qilu Univ Technol (Nat Sci Ed) (4): 26-30 (in Chinese).

Mukherjee S, Das P, Sen R, 2006. Towards commercial production of microbial surfactants. Trends Biotechnol 24 (11): 509-515. https://doi.org/10.1016/j.tibtech.2006.09.005.

Neto DC, Meira JA, Tiburtius E, Zamora PP, Bugay C, Mitchell DA, Krieger N, 2009. Production of rhamnolipids in solid-state cultivation: characterization, downstream processing and application in the cleaning of contaminated soils. Biotechnol J 4 (4): 748-755.

Nitschke M, Costa SG, Haddad R, G Gonçalves LA, Eberlin MN, Contiero J, 2005. Oil wastes as unconventional substrates for rhamnolipid biosurfactant production by *Pseudomonas aeruginosa* LBI. Biotechnol Prog 21 (5): 1562-1566.

Noah KS, Bruhn DF, Bala GA, 2005. Surfactin production from potato process effluent by *Bacillus subtilis* in a chemostat. Appl Biochem Biotechnol 121 (124): 465-473. doi: ABAB: 122: 1-3: 0465 [pii].

Orban E, Di Lena G, Nevigato T, Casini I, Caproni R, Santaroni G, Giulini G, 2007. Nutritional and commercial quality of the striped venus clam, *Chamelea gallina*, from the Adriatic sea. Food Chem 101 (3): 1063-1070.

Pirog TP, Shevchuk TA, Shuliakova MA, 2012. Glycerol metabolism in surfactants producers *Acinetobacter calcaaceticus* IMV B-7241 and *Rhodococcus erythropolis* IMV Ac-5017. Mikrobiol Z 74 (4): 29-36.

Ramani K, Jain SC, Mandal AB, Sekaran G, 2012. Microbial induced lipoprotein biosurfactant from slaughterhouse lipid waste and its application to the removal of metal ions from aqueous solution. Colloids Surf B Biointerfaces 97: 254-263. https://doi.org/10.1016/j.colsurfb.2012.03.022.

Raza ZA, Khan MS, Khalid ZM, Rehman A, 2006. Production kinetics and tensioactive characteristics of biosurfactant from a *Pseudomonas aeruginosa* mutant grown on waste frying oils.

Biotechnol Lett 28 (20): 1623-1631. https://doi.org/10.1007/s10529-006-9134-3.

Sarachat T, Pornsunthorntawee O, Chavadej S, Rujiravanit R, 2010. Purification and concentration of a rhamnolipid biosurfactant produced by *Pseudomonas aeruginosa* SP4 using foam fractionation. Bioresour Technol 101 (1): 324-330.

Shen Z, Yang HY, Yan XT, Nan FY, Guo YP, Xie HZ, 2011. The application and development trends of biosurfactants in petroleum industry. Appl Chem Ind 40 (10): 1842-1846.

Silva MAM, Silva AF, Rufino RD, Luna JM, Santos VA, Sarubbo LA, 2017. Production of biosurfactants by *Pseudomonas* species for application in the petroleum industry. Water Environ Res 89 (2): 117-126.

Sodagari M, Invally K, Ju LK, 2017. Maximize rhamnolipid production with low foaming and high yield. Enzym Microb Technol 110, 2018. 79-86.

We Q, Liu L, Zhan R, Wei X, Zang J, 2010. Distribution features of the chemical parameters in the southern Yellow sea in summer. Period Ocean Univ China 40 (1): 82-88 (in Chinese).

Wu H, Wang W, Han S, 2007. Research advance on rhamnolipid biosurfactants. Microbiology 34 (1): 148-152 (in Chinese).

Xiong J, Song W, Ye J, 2007. Research advance on decrease the cost of biodiesel production. Chem Ind Eng Prog 26 (6): 774-776 (in Chinese).

Yao J, Min E, 2010. Hazardous effects and resource utilization of waste oil. New Energy 30 (5): 1-6 (in Chinese).

Zhang X, 2013. Enhancement of hydrocarbon degrading and biosurfactant production of *Pseudomonas aeruginosa* strain Z41 by joint employment of low energy ion beam and ultraviolet irradiations. Res J Biotechnol 8 (12): 31-36.

Zhang X, 2014. A pilot study on screening and simplified fermentation of a biosurfactant producing strain with seawater. Res J Biotechnol 9 (3): 74-79.

Zhang X, Lu D, 2013. Response surface analysis of rhamnolipid production by *Pseudomonas aeruginosa* strain Z41 with two response values. Afr J Microbiol Res 7 (22): 2757-2763.

Zhang X, Xiang T, 2010a. Application of physical factors in genetic improvement of biosurfactant producing strains. J Huzhou Vocat Technol Coll 8 (1): 1-5 (in Chinese).

Zhang X, Xiang T, 2010b. Genetic modification of MEOR bacterium *Bacillus licheniformis* H strain by low energy ion beam irradiation. Open Biotechnol J 4: 14-17.

Zhang X, Xu D, Yang G, Zhang H, Li J, Shim H, 2012a. Isolation and characterization of rhamnolipid producing *Pseudomonas aeruginosa* strains from waste edible oils. Afr J Microbiol Res 6 (7): 1466-1471.

Zhang X, Xu D, Zhu C, Lundaa T, Scherr KE, 2012b. Isolation and identification of biosurfactant producing and crude oil degrading *Pseudomonas aeruginosa* strains. Chem Eng J 209: 138-146.

Zhao X, Jiang B, 2004. Brief review on MEOR technology. Pet Sci 1 (4): 17-23.

9 盐渍土有益耐盐微生物商品化的瓶颈及发展前景

Bushra Tabassum, Adeyinka Olawale Samuel, Muhammad Umar Bhatti, Neelam Fatima, Naila Shahid, Idrees Ahmad Nasir

[摘要]

盐胁迫是一种主要的限制农作物产量的非生物胁迫因素。利用耐盐微生物对农作物生长的促进作用是在盐胁迫环境下缓解农作物盐害的一种有效方法。这些耐盐微生物通过各种调节机制不仅能使农作物在盐胁迫环境中良好生长，还能通过产生各种植物激素、溶解磷酸盐和增强固氮作用来促进农作物的生长。但是，利用耐盐微生物等生产的生物产品在农业生产中也存在着一些问题，比如耐盐微生物制剂的田间药效不一致、在作物种类之间的功能性差异以及按行业需求开发此类产品的经济效益成本等，这是制约这些生物产品商业化利用的瓶颈。截至目前，还没有耐盐微生物在农业中应用的商业配方出现。为了保障农业的可持续发展，必须开发和推广以耐盐微生物为主体的农业制剂。本章重点介绍了在盐渍土中利用耐盐微生物提高农作物产量的微生物产品商业化应用现状和前景。

[关键词]

耐盐微生物；盐渍土农业；生物制品；植物生长促生菌；生物肥料

9.1 前言

耐盐微生物具有盐依赖性特征，其细胞构成具有相应的结构特点适应在盐胁迫环境下生存，并且耐盐微生物具有很强的代谢调节和适应能力，能够在各种盐浓度胁迫下茁壮生长，这些微生物包括古生菌（Archaea）、细菌（Bacteria）和真核生物（Eukarya）。根据生长环境盐浓度范围，耐盐微生物可分为极端嗜盐

B. Tabassum (✉), A. O. Samuel, M. U. Bhatti, N. Fatima, N. Shahid, I. A. Nasir, 巴基斯坦拉合尔旁遮普大学分子生物学国家卓越中心

菌、中度嗜盐菌和耐盐菌。能够在盐浓度处于 2.5~5.2 mol/L 范围内生境中生长的被称为极端嗜盐菌；生境盐浓度在 0.5~2.5 mol/L 范围内生存的被称为中度嗜盐菌。这两种嗜盐菌除在某些结构功能上有微小的差异外，两者对盐胁迫的适应程度相似。例如，极端嗜盐盐杆菌（*Halobacterium halobium*）和中度嗜盐沃氏盐杆菌（*Halobacterium volcanii*）在胞外多糖 N-糖苷键结构中添加的糖类存在差异（Mengele 和 Sumper，1992）。极端嗜盐盐杆菌（*Halobacterium halobium*）和盐沼盐杆菌（*Halobacterium salinarum*）在高盐环境中生存的稳定性归因于双酸性磷脂。古生菌细胞膜中甘油甲基磷酸酯的存在，可以防止细胞内容物在高盐浓度下的渗漏（Tenchov 等，2006）。目前有很多研究集中在嗜盐微生物对盐环境的适应和所采用的机制上（Paul 等，2008；Argandona 等，2010；Becker 等，2014）。嗜盐微生物和耐盐微生物能够通过利用两种基本的渗透调节机制在高盐浓度环境中生长。采用这些机制的微生物可以控制细胞内的离子强度和胞外高渗透势环境造成的胞内水分胁迫。在盐的胞质渗透调节机制中，细胞质中的盐浓度被提高到与盐环境中相似的浓度；而在有机渗透调节机制中，不带电的、高度水溶性的有机溶质（如糖、多元醇和氨基酸）等在细胞壁部位积累，通过提高和维持微生物细胞的渗透势，进而应对盐胁迫环境的渗透胁迫。这些非离子态高水溶性的化合物即使在高盐浓度下也不会干扰新陈代谢。因此，不同嗜盐微生物和耐盐微生物的分离和鉴定为提高作物耐盐性提供了有益的途径。

9.2 盐分干扰作物的营养和生殖发育

植物根区土壤饱和提取物（saturation extract，EC_e）的电导率（electrical conductivity，EC）高于 4 dS/m（约 40 mmol/L NaCl），并且可交换性 Na^+ 含量约为 15% 的土壤就可以称为盐渍土。土壤盐分降低了土壤质量，从而降低了其农业生产潜力。世界每天都会因为盐分引起的土壤退化而损失数公顷的农田。盐分能够干扰作物的营养和生殖发育。它通过渗透胁迫或离子毒性抑制植物生长（Läuchli 和 Epstein，1990）。渗透胁迫阶段限制了植物对水分的吸收，因为在盐渍土溶液中，高渗透压降低了水势。当特定离子在植物体内积累一段时间后导致离子毒性或离子不平衡时，就会产生离子毒害效应（Munns 和 Tester，2008）。高盐胁迫降低了叶片膨胀、导致气孔关闭、光合作用降低和由于渗透不平衡引起的水分亏缺造成的生物量损失（James 等，2011；Rahnama 等，2010）。大多数耐盐植物已经进化出渗透调节机制，通过净溶质积累降低细胞渗透势，使它们能够在遭受盐渍化的土壤中生存。其他的适应性机制还包括细胞壁修饰、活性氧解毒、转运蛋白、K^+ 和 NO_3^- 稳态、胞质空泡间隔、相容溶质积累等。Na^+ 的胞质空泡隔离在 NHXs 蛋白的帮助下释放了 Na^+ 的毒性，而 SOS 信号途径能够通过将 Na^+ 输出到细胞外，从而降低 Na^+ 对植物细胞的影响。研究表明，NHX 型蛋白对 K^+ 进入液泡的区域化和细胞 pH 稳态也有重要作用。在拟南芥中 *AtNHX*1 基因的过表达有助于液泡中 Na^+ 的分隔，并已证明其过表达能够提高拟南芥（Apse 等，1999）、番茄和油菜（Zhang 和 Blumwald，2001；Zhang 等，2001）的耐盐性。同样地，过量表达 ROS 清除基因能够降低转基因植物的细胞损伤，并促进盐渍土壤中植物地上部分和根部的生长（Roy 等，2014）。

9.3 农田中的盐分管理

导致土壤中盐分增加的因素包括原生矿物风化作用、较高的水分蒸发速率、过度灌溉、过度施肥和荒漠化过程（Ramadoss 等，2013）。一般来说，根据盐分积累的种类和程度，盐渍土基本上可以分为盐渍土、碱性土和盐碱土。受盐渍化影响的土壤限制了作物的产量，从而造成巨大的经济损失。巴基斯坦印度河流域盐碱地对农作物的影响导致小麦和水稻的总体平均损失分别为 32% 和 48%（Murtaza，2013）。Khodarahmpour 等（2012）报道了在 240 mmol/L NaCl 胁迫环境下，玉米的发芽率和种子活力分别降低 32.4% 和 95%。

为了实现土壤的可持续利用和提高粮食产量，人们已经采用了一些能够减轻盐渍土影响的方法。然而，近年来微生物如细菌、真菌、放线菌、原生动物和藻类等在植物根际定殖的研究受到了广泛关注。细菌是根际最丰富的微生物。一些特别的耐盐微生物能够定殖在植物根系，提高了作物在盐渍化土壤中的生长性能。为了鉴定适合植物生长的嗜盐菌株，我们进行了严格的研究。基本上以下这些特性比如氮固定能力、溶磷能力、1-氨基环丙烷-1-羧酸盐脱氨酶活性以及铁载体和植物激素的产生，通常认为是嗜盐菌能够促进植物生长的原因。

9.3.1 生物固氮

生物固氮（biological nitrogen fixation，BNF）是指在固氮酶复合物存在下，将大气中的氮元素还原为氨。固氮酶复合物包含钼铁（Mo-Fe）蛋白和铁（Fe）蛋白两种金属蛋白成分（Hu 和 Ribbe，2015）。虽然 BNF 是微生物中一种普遍存在的活性蛋白，但在作用机制上存在差异。非异囊蓝藻的 N_2 固定系统与异囊蓝藻的 N_2 固定系统相比有显著差异（Bergman 等，2013）。在有氧条件下，束毛藻属植物（*Cyanobacterium trichodesmium*）固氮酶的活性受到抑制（Staal 等，2007）。与其他生物不同，嗜盐生物或耐盐生物的生物固氮机制需要 Na^+ 存在。已经有人试图确定 Na^+ 缺乏是否直接或间接地影响固氮酶变成非活性构象，以及是否可以通过乙炔预培养来逆转（Apte 和 Thomas，1984）。在没有 Na^+ 的情况下，腺苷三磷酸（Adenosine triphosphate，ATP）供应的丧失表明 Na^+ 在维持腺苷三磷酸供应以支持各种能量转换方面起着重要作用，这在施氏假单胞菌（*Pseudomonas stutzeri*）中得到了证实（Kodama 和 Taniguchi，1976、1977）。有报道称，已经分离并鉴定到了几种能够在 5% 以上 NaCl 浓度下存活的嗜盐固氮细菌（Zahran 等，1995；Kang 等，2015）。

9.3.2 溶磷

磷素（P）是植物生长的必需元素之一，由于磷通常以不溶性的形态存在，因此土壤中可直接用于植物生长的磷很少。最近的研究表明，能够促进磷溶解的盐植物生长促生菌提高了植物对磷的有效利用，包括螯合、离子交换、酶解和通过分泌小分子量有机酸的酸化作用等在内的多种机制，都用来解释微生物是如何溶解不溶性的磷酸盐并增加植物最佳生长所需的有效磷的（Sharma 等，2013；Etesami，2018）。Srinivasan 等（2012）从遭受盐渍化胁迫的土壤中分离出了一些溶磷细菌（Phosphate-solubilizing bacteria，PSB）和溶磷真菌（Phosphate-solubilizing fungi，PSF）。报道称，气球菌属菌株 PSBCRG1-1（*Aerococcus* PSBCRG1-1）无论在何种 NaCl 浓度胁迫下都表现出约 12.12% 的最大的磷增溶作用，明显优于所有其他菌株；曲霉属真菌菌株 PSFNRH-2（*Aspergillus* PSFNRH-2）则表现出最大的磷释放能力（20.81%），在各种 NaCl 浓度胁迫下均明显优于所有其他溶磷真菌菌株。

9.3.3 1-氨基环丙烷-1-羧酸脱氨酶活性

在植物根际发现了几种微生物具有产生 1-氨基环丙烷-1-羧酸脱氨酶的能力，1-氨基环丙烷-1-羧酸脱氨酶能够将乙烯前体 1-氨基环丙烷-1-羧酸分解成 α-酮丁酸盐和铵，因此能够降低寄主植物体内乙烯的含量，并使植物有更多的氮可以利用。阴沟肠杆菌（*Enterobacter cloacae*）和假单胞菌（*Pseudomonas* sp.）在约 10% 浓度的 NaCl 水平下能够产生并分泌大量的 1-氨基环丙烷-1-羧酸脱氨酶（Trung 等，2016）。同样地，荧光假单胞菌 TDK1（*Pseudomonas fluorescens* TDK1）菌株通过产生 1-氨基环丙烷-1-羧酸脱氨酶缓解植物盐胁迫的能力也在花生中得到了评估，与其他没有 1-氨基环丙烷-1-羧酸脱氨酶活性的假单胞菌不同，接种荧光假单胞菌 TDK1（*Pseudomonas fluorescens* TDK1）的花生生长量得到了显著提高（Saravanakumar 和 Samiyappan，2007）。最近，产 1-氨基环丙烷-1-羧酸脱氨酶内生链霉菌 GMKU336（*Streptomyces* sp. GMKU 336）与其 1-氨基环丙烷-1-羧酸脱氨酶缺失突变体在分子水平的相互作用得到了研究。

将这些菌株接种到了用 150 mmol/L NaCl 处理的水稻品种 KDML105 上。结果表明，与未接种和接种 1-氨基环丙烷-1-羧酸脱氨酶缺失突变体的水稻植株相比，接种链霉菌 GMKU336（*Streptomyces* sp. GMKU 336）显著提高了水稻植株的生长，其叶绿素、脯氨酸、K^+、Ca^{2+} 以及水分含量都明显增加；但乙烯、ROS、Na^+、Na^+/K^+ 比值等都降低。此外，参与乙烯途径的基因 *ACO*1 和 *EREBP*1 的表达也显著下调（Jaemsaeng 等，2018）。

9.3.4 铁载体的生产

铁载体是由微生物（细菌和真菌）产生的低分子量铁螯合化合物，用于抵御低铁胁迫并调节其在不同浓度下的吸收。根据其官能团的不同，可将其分为儿茶酚酸酯、羟肟酸酯和羟基羧酸盐。低铁能够激活一系列的基因表达，通过参与铁载体的合成和螯合物的运输等合成很多重要的蛋白质，进而保证细胞存活。铁载体能够在植物体内动员铁的运输，并有效地溶解铁以促进植物吸收，并不像合成螯合物那样容易析出或分解（Siebner-Freibach 等，2004）。

9.3.5 植物激素

植物生长素或植物激素是天然的有机物质，在低浓度下影响植物的生理过程。它们通过调节不同的过程影响植物生长的分化和发育过程（Tabassum 等，2017）。植物激素包括生长素、赤霉素、细胞分裂素、脱落酸和乙烯（Davies，2004；Jha 等，2015）。生长素的增加导致植物对有毒离子的吸收减少，从而改善高盐胁迫条件下的植物生长（Chakraborty 等，2011）。在高盐胁迫条件下接种菜豆（*Phaseolus vulgaris*）的两个耐盐菌株极端东方化假单胞菌（*Pseudomonas extremorientalis*）和绿假单胞菌（*Pseudomonas chlororaphis*）的试验结果证实植物激素（如生长素、脱落酸和赤霉素）对改善作物在遭受盐胁迫下的生长有促进作用（Egamberdieva，2011）。同样地，Ul Hassan 和 Bano（2014）报道了接种假单胞菌（*Pseudomonas* sp.）和蜡状芽孢杆菌（*Bacillus cereus*）后，在盐渍土中生长的小麦叶片中生长素和脱落酸的含量增加。接种 48 h 后，大豆叶片中生长素含量降低，而在盐胁迫下未接种的植株叶片中生长素含量增加（Asim 等，2013）。

9.4 耐盐植物生长促生菌在农业中的应用，提高植物在盐胁迫环境中的存活率

众所周知，耐盐植物生长促生菌涉及到一系列农业相关应用，见表 9.1

(Etesami 和 Beattie，2018)，其中一些应用概述如下。

①上调关键酶的活性，从而激活植物抗氧化防御机制（Jha 和 Subramanian，2013；Islam 等，2016；Qin 等，2016）；②通过固定大气中的氮元素来改善植物营养（Dodd 和 PerezAlfocea，2012；Etesami 和 Beattie，2017；Etesami，2018）；③保持较高的 K^+/Na^+ 比值，从而增强对接种植物的影响（Giri 等，2007；Zuccarini 和 Okurowska，2008；Shukla 等，2012；Islam 等，2016；Etesami，2018）；④通过产胞外多糖的耐盐植物生长促生菌来促进土壤团聚体的形成，改善土壤结构（Watanabe 等，2003；Nunkaew 等，2015）；⑤有助于改变根系结构和形态、水分传导和激素状态（Arora 等，2006，2012）；⑥通过分泌与胁迫有关的挥发性化合物来提高植物生物量和抵御干旱胁迫（Timmosk 等，2014）。

表 9.1 耐盐菌株及其在促进植物生长中的作用概述

耐盐/嗜盐细菌	作用方式	NaCl 浓度	实验植物	对植物的影响	参考文献
沙雷氏菌 SL-12 (Serratia sp. SL-12)	—	150~200 mmol/L	小麦	通过茎/根长度、鲜重/干重和光合色素积累等参数测定，显著促进植物生长	Singh 和 Jha (2016)
荧光假单胞菌 002 (Pseudomonas fluorescens 002)	吲哚-3-乙酸分泌和 1-氨基环丙烷-1-羧酸脱氨酶活性	150 mmol/L	玉米	初生根、侧根和种子根长度和根数以及根干重的增加	Zerrouk 等 (2016)
克雷伯氏菌 SBP-8 (Klebsiella sp. SBP-8)	—	150 mmol/L、175 mmol/L 和 200 mmol/L	玉米	植物生物量和叶绿素含量增加。该菌株引起宿主植物排出 Na^+ (65%) 和吸收 K^+ (84.21%)	Singh 和 Jha (2015)
重氮营养哈特曼杆菌 E19T (Hartmannibacter diazotrophicus E19T)	产生 1-氨基环丙烷-1-羧酸脱氨酶	1%、2% 和 3%	大麦	根 (308%) 和茎 (189%) 干重显著增加	Suarez 等 (2015)
解淀粉芽孢杆菌 NBRISN13 SN13 (Bacillus amyloliquefaciens NBRISN13 SN13)	吲哚-3-乙酸和 1-氨基环丙烷-1-羧酸脱氨酶的产生、磷的增溶和脯氨酸的积累	200 mmol/L	水稻	NADPMe2、EREBP、SOS1、BADH 和 SERK1 基因表达上调以及渗透保护物质的增加	Nautiyal 等 (2013)
解淀粉芽孢杆菌 SQR9 (Bacillus amyloliquefaciens SQR9)	将 Na^+ 隔离在液泡中，从根中排出 Na^+，积累可溶性总糖，提高抗氧化物含量	100 mmol/L	玉米	RBCS、RBCL、H^+-PPase、HKT1、NHX1、NHX2、NHX3 表达上调，显著促进玉米幼苗生长，提高叶绿素含量，上调植物耐盐相关基因表达，下调脱落酸相关基因表达	Chen 等 (2016)
皮氏无色杆菌 (Achromobacter piechaudii)	吲哚-3-乙酸和 1-氨基环丙烷-1-羧酸脱氨酶活性	172 mmol/L	番茄	番茄幼苗鲜重和干重的增加	Mayak 等 (2004)

(续表)

耐盐/嗜盐细菌	作用方式	NaCl 浓度	实验植物	对植物的影响	参考文献
荧光假单胞菌 YsS6（Pseudomonas fluorescens YsS6）和米氏假单胞菌 8R6（P. migulae 8R6）	吲哚-3-乙酸和1-氨基环丙烷-1-羧酸脱氨酶活性	165 mmol/L 和 185 mmol/L	番茄	干鲜生物量、叶绿素含量、花芽数均增加	Ali 等（2014）
巴西固氮螺菌（Azospirillum brasilense）	固氮	40 mmol/L、80 mmol/L 和 120 mmol/L	白三叶草	非盐渍化和盐渍化环境下促进植物生长	Khalid 等（2017）
嗜碱芽孢杆菌 EA1（Bacillus alcalophilus EA1）、苏云金芽孢杆菌 EA3（Bacillus thuringiensis EA3）和纤细芽孢杆菌属（Gracilibacillus saliphilus）	磷的增溶和铁载体的产生	0 mmol/L、510 mmol/L 和 1 030 mmol/L	蝎节木属（Arthrocnemum macrostachyum）	减轻高盐胁迫对植物生长和生理性能的影响	Navarro-Torre 等（2017）
云南微球菌（Micrococcus yunnanensis）、莱比托游动球菌（Planococcus rifietoensis）和争论贪噬菌（Variovorax paradoxus）	1-氨基环丙烷-1-羧酸脱氨酶	50~125 mmol/L	甜菜	甜菜耐盐性显著增强，种子萌发和植株生物量增加，光合能力增强	Zhou 等（2017）
碱湖迪茨氏菌 STR1（Dietzia natronolimnaea STR1）	脱落酸信号、SOS 途径、离子转运体和抗氧化机制的调节	50~150 mmol/L	小麦	盐胁迫诱导基因 TaST 的表达增强	Bharti 等（2016）
克雷伯氏菌（Klebsiella sp.）	RBCL 和 WRKY1 基因表达谱的调控	100 mmol/L	燕麦	接种植株的茎长、根长、茎干重、根干重、相对含水量等生理参数均增加	Sapre 等（2018）
荧光假单胞菌（Pseudomonas fluorescens）	1-氨基环丙烷-1-羧酸脱氨酶活性、铁载体和吲哚-3-乙酸生成	5%	黄瓜	减轻盐分对植物生长的负面影响	Nadeem 等（2016）
肠杆菌 UPMR18（Enterobacter sp. UPMR18）	抗氧化酶活性、活性氧途径基因上调及1-氨基环丙烷-1-羧酸脱氨酶活性	1%~20%	秋葵	发芽率、生长参数和叶绿素含量均增加	Habib 等（2016）
克雷伯氏菌（Klebsiella）、假单胞菌（Pseudomonas）、土壤杆菌（Agrobacterium）和苍白杆菌（Ochrobactrum）	吲哚-3-乙酸与磷酸盐增溶	100 mmol/L	花生	全氮（N）含量显著增加（高达76%）	Sharma 等（2016）

（续表）

耐盐/嗜盐细菌	作用方式	NaCl 浓度	实验植物	对植物的影响	参考文献
苏拉氏短杆菌 JG-06（*Brachybacterium saurashtrense* JG-06）、乳酪短杆菌 JG-08（*Brevibacterium casei* JG-08）和血杆菌 JG-11（*Haererohalobacter* JG-11）	较高的 K^+/Na^+ 比值和较高的吲哚-3-乙酸浓度、Ca^{2+}、磷和氮含量	50 mmol/L	花生	株长、茎长、根长、茎干重、根干重和总生物量均增加	Shukla 等（2012）
地衣芽孢杆菌 A2（*Bacillus licheniformis* A2）	吲哚-3-乙酸与磷酸盐增溶	50~500 mmol/L	花生	鲜生物量、总长度和根长分别增加 28%、24% 和 17%	Goswami 等（2014）
植物内芽孢杆菌（*Bacillus endophyticus*）、特基拉芽孢杆菌（*Bacillus tequilensis*）、莱比托游动球菌（*Planococcus rifietoensis*）、争论贪噬菌（*Variovorax paradoxus*）和运动节杆菌（*Arthrobacter agilis*）	吲哚-3-乙酸与磷酸盐增溶		厚岸草	盐胁迫下促进植物生长	Zhao 等（2016）
芽孢杆菌（*Bacillus* sp.）和滋养节杆菌（*Arthrobacter pascens*）	磷酸盐增溶与铁载体生成	100 mmol/L	玉米	增加渗透压，包括糖和脯氨酸的积累和抗氧化酶的活性	Ullah 和 Bano（2015）
阿氏肠杆菌（*Enterobacter asburiae*）、莫拉菌（*Moraxella pluranimalium*）和施氏假单胞菌（*Pseudomonas stutzeri*）	1-氨基环丙烷-1-羧酸脱氨酶活性	0.25 mol/L	番茄（*Solanum lycopersicum*）	增加幼苗鲜干生物量	Raheem 和 Ali（2015）
芽孢杆菌（*Bacillus* sp.）、耐盐刘志恒氏菌（*Zhihengliuella halotolerans*）、葡萄球菌（*Staphylococcus succinus*）、吉氏芽孢杆菌（*Bacillus gibsonii*）、小鳟鱼大洋芽孢杆菌（*Oceanobacillus oncorhynchi*）、盐单胞菌属（*Halomonas* sp.）和食有机物深海芽孢杆菌（*Thalassobacillus* sp.）	吲哚-3-乙酸生成、固氮和 1-氨基环丙烷-1-羧酸脱氨酶活性	200 mmol/L	小麦（*Triticum aestivum*）	增加植物的根和茎长及总鲜重	Orhan（2016）

(续表)

耐盐/嗜盐细菌	作用方式	NaCl 浓度	实验植物	对植物的影响	参考文献
肠杆菌（*Enterobacter sp.*）和芽孢杆菌（*Bacillus sp.*）	1-氨基环丙烷-1-羧酸脱氨酶活性、胞外多糖分泌和吲哚-3-乙酸生成	0 mmol/L 或 400 mmol/L	藜麦（*Chenopodium quinoa*）	改善植物水分关系和降低 Na^+ 吸收，从而降低渗透势和离子胁迫	Yang 等（2016）

9.5 商业化阶段

目前，还没有可商业化应用的耐盐配方。为了使耐盐微生物产品商业化，还需要完成几个关键步骤，这些步骤包括实地调查、微生物的筛选、微生物菌株之间的相容性、特定区域的地理条件、制备的制剂在实验室条件下的性能研究，以及在大田中现场测试的结构。简而言之，耐盐微生物产品商业化的过程包括如下几个方面。

9.5.1 微生物筛选

合适的微生物筛选和有效性对成功的生物控制/生物刺激剂产品的商业化来说烦琐且重要。这种选择过程包括特定环境和早期试验，以区分微生物的生物控制/生物刺激能力。虽然筛选方法不能单列出来，但根据感兴趣的病理体系可以遵循合理地选择策略进行筛选（Fravel，2005）。这可能包括定殖在叶际/根际（Enya 等，2007；Yoshida 等，2012）或能够促进/保护作物收获的微生物（Janisewicz 和 Korsten，2002）。

评价合适的微生物群对选择潜在的微生物至关重要。值得注意的是，即使存在高效和稳健的筛选方案，也只有不到 1% 的候选细菌能生产出有效的产品（Bailey 和 Falk，2011）。因此，这就需要时间来建立强大的微生物联合体，以确定特定的生物控制/生物刺激剂候选微生物。这可以通过目标定向和宽采样技术来实现。

一个简单的微生物控制/刺激模型可能包括微生物数据库的构建。该数据库将包含有关分离微生物的信息，这些信息可与所需的生物控制/生物刺激剂相关的目标病原体、寄主植物、作用方式和环境生态位等兼容（Glare 等，2012）。然后，每个结果的获得都可以通过对一系列的标准生物进行测定获得，以建立生物控制/生物刺激剂的效能和田间效果。田间效果作为批准和申请的标志的重要性不容忽视。使用体外筛选的方法建立生物防治/生物刺激剂是常见的做法，但是在将体外抑制/促进的结果与田间表现进行关联分析时，应采取谨慎的方法

(Burr等，1996；Milus和Rothrock，1997；Fravel，2005）。可能还存在各种各样的筛选策略和方法，但它们都应该产出一种高效、生态友好和低成本的微生物制剂（Kóhl等，2011；Ravensberg，2011）。

9.5.2 批量生产

要商业化生产这种微生物制剂，必须考虑到一些因素。首先，生产场所需要一个足够大（约200 km²）且相对集中的位置，道路通畅且电力供应顺畅。其次，必须确保所有原材料的不间断供应，这些原材料可能包括母菌、载体材料（后有详解）、玻璃瓶、纸板和生长材料。第三，要建立一个25万L/年的生产设施，这几乎需要1英亩（约4 047 m²）的土地。这片土地用来建立实验室、安装管井、办公室、停车场和其他设施。整个场地必须有带刺铁丝网和合适的入口，能够保证可以覆盖2 000 m²的区域用于常规作业。

必须建造一个单独的封闭式生物肥料装置。该生产单元应设置微生物实验室、生产（发酵）区、包装区、储存区、销售方式等具体渠道。应安装最新的、有效的实验室设备，包括发酵罐、培养基罐、高压灭菌器、锅炉、培养基分配器、除盐装置、空气压缩机等。所有这些设备都应满足总的生产能力。大规模生产耐盐微生物悬浮液的步骤如图9.1所示。

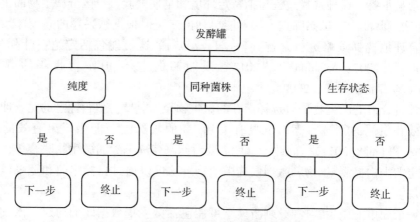

图9.1 批量生产耐盐微生物悬浮液的步骤

9.5.3 母液或发酵剂的制备

通过在温室和田间水平测定代谢物的功能后选定选择菌株的发酵产物。有效的植物生长促生菌菌株的纯化培养在相应的琼脂培养基上进行，并在实验室中保存。母液培养物在更大的烧瓶中进一步繁殖。灭菌后，每个装有肉汤培养基的烧瓶以 1∶5 的比例在无菌条件下接种母液，将烧瓶置于合适的条件下促进微生物的生长。这种每毫升含有 $10^9 \sim 10^{10}$ 个细胞的培养液储存时间不要超过 24 h 或短时间储存在 4℃ 条件下。发酵罐用于大规模生产微生物产品。

对于液体生物制品的生产，发酵罐中的液体培养基直接进入自动灌装机，并根据需要装入贴有适当标签的玻璃瓶（不同体积）中。在生产时必须检查接种剂中微生物的数量。为了达到适当的质量标准，接种剂中的微生物活细胞数目应保持不变。接种剂应由制造商储存在远离热源的阴凉处。有一些微生物种类有一定的质量控制标准，但没有系统的认证。因此，其质量检查监测系统都由制造商自己决定，因此强烈建议各单位为此制定适当的安排和措施。

9.5.4 载体材料

在种子或土壤中接种生物防治剂/生物刺激剂需要载体材料。这些材料通常被加工成粉末状，粒径为 10~40 μm。合适的载体材料具有如下特点：①对接种细菌菌株表现出惰性和无毒性；②具有良好的吸湿能力；③易于加工且不聚集成块；④可以通过高压灭菌或射线辐射毫不费力地进行灭菌；⑤具有相当可观的数量；⑥成本低；⑦具有较高的种子黏附能力以及；⑧可以缓冲总体 pH。最重要的是，它不能对植物有毒。泥炭、褐煤、膨润土和木炭都是用作种子接种剂的常用载体材料。许多国家使用基于泥炭的根瘤菌接种剂，有关其特性和效果的信息很容易获得。

对于土壤接种，通常使用粉末状或网状载体材料（0.5~1.5 mm）。这种泥炭、珍珠岩、木炭或土壤团聚体的颗粒形态适合用于土壤接种。接种菌株能够保持菌株状态持久性的载体选择取决于许多不同的因素，其中包括：①种子一旦被细菌包裹后能够很快晾干，但要保证细菌能够在干燥的条件下生存。干燥后，包衣种子用来播种；②在储存阶段接种细菌也必须保持存活状态；③接种细菌菌剂应装备良好，能够与自然土壤微生物群竞争养分和栖息地。这些微生物菌剂还必须保证自身免受原生动物的捕食。要满足这种生存需求就要求土壤团聚体和木炭中存在微孔结构。因此，这类材料很自然的成为土壤接菌剂的载体。载体在使用之前要选好尺寸，然后经过严格的灭菌处理。这能够确保更多的接菌剂在较长的时间内都能依附在载体上。γ 射线被认为是最适合载体灭菌的方式。这种技术不会改变材料的物理或化学性质。另一种最容易使用、最常用的技术就是高压灭菌。然而，有时载体的特性会发生一些变化，产生一些对某些细菌菌株有毒的物质，给后续使用带来风险。

此外，有研究表明，与没有任何特定载体材料的接种剂与土壤混合后，那些与具有增强土壤团聚体结构的载体相混合的接种剂在土壤中具有极高的存活率。此外，接种以泥炭为载体材料的接种剂的植物表现出更加显著的生长模式。因此，强烈建议能够形成土壤团聚体的材料可能是制备廉价且有效的载体材料，特别是用来固定根瘤菌作为接种剂（El Fattah 等，2013）。选择合适载体材料的步骤如图 9.2 所示。

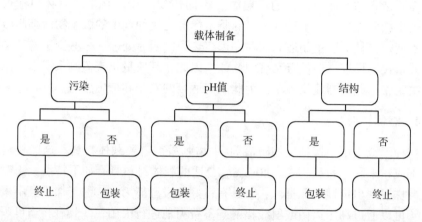

图 9.2 选择合适载体材料的步骤及其大规模生产耐盐微生物悬浮液所需的特性

对于固体配方，要求褐煤、膨润土、木炭、泥炭粉（70~100 目）的规格符合要求。然而，由于产品质量和稳定性的问题，固体配方技术正逐渐被淘汰。固态载体配方对温度敏感，接种剂微生物的数量低于临界值。但与液体载体配方相比，它是一种非常稳定的工艺，产品的自身寿命可以维持在 12 个月以内。

9.5.5 实验室测试

所筛选到的细菌首先在环境稳定的条件下进行盆栽试验评估，随后在大田中进行应用。产品在上市之前已经做了大量的试验。

虽然植物生长促生菌具有潜在的商业化应用前景，但也存在着一些与环境和人类相关的安全问题。因此，商业公司应选择对人类和环境健康都无影响的微生物联合体。政府还应规范筛选商业用途耐盐菌株的程序和法律。在大规模授权、注册和采用植物生长促生菌进行病虫害管理之前，需要澄清安全问题。

9.6 商业化的瓶颈

近年来，随着人们对生物肥料潜力的关注，农民还没有意识到耐盐植物生长促生菌在改善土壤盐分方面的潜力和机遇。即使他们知道，其接受程度仍然很低。因为这项技术还处于初级阶段，还有很多挑战需要克服。

9.6.1 表征挑战

在鉴定不同于其他植物生长促生菌生物体的嗜盐菌方面取得了一些合理的进展。然而，与其他嗜盐菌不同的是，大多数研究仍处于基础研究阶段。对嗜盐菌进行潜在可用性和菌种间的相容性评估是非常重要的，因为能够促进植物生长的各种机制可能是宿主植物特异或菌株特异的。同样地，环境可能会影响嗜盐菌促进植物生长的效率，但在不同的环境条件下这种影响可能不同。由于植物根际含有多种微生物，因此确定的嗜盐菌必须在营养和生存空间上都是良好的竞争者。在不同的盐渍化地区，具有能够有效促进植物生长的嗜盐菌或耐盐菌的复合体，对商业化的成功应用至关重要。

9.6.2 生产挑战

意识缺乏很可能导致有益嗜盐菌产量低下。虽然植物生长促生菌在一般的生产过程中所用到的生产工艺基本相同，但大规模生产仍然具有很大的挑战性。尽管与化肥不同，接种剂生产工厂的建设成本较低，但最近很少有这样的工厂进行生产。为了更好地利用嗜盐微生物，政府应该创造推广生物肥料的有利环境。此外，制定一项可持续的政策，促进对已查明的嗜盐菌进行大规模生产，将增加盐渍化土地的作物产量。还应分配更多的资源以提高对有益嗜盐菌的认识水平。

必须注意的是，一个成功的微生物产品是多种因素共同作用的结果（Ravensberg，2011）。Gelernter 和 Lomer（2000）进行的一项研究提出了一个评估成功的生物控制/生物刺激剂产品的大纲，有效产品的关键特性列举如下。

（1）产品在实验室、温室和田间的功效是实现可重复生物制剂的关键步骤（Nicot 等，2011；Whipps，2001）。除了有效性，在大规模工业生产过程中还应考虑到突变、病毒和相位变化对产品的影响（Takers，2012）。微生物在实验室和温室中发挥的效应和在田间应用时的效果不一定相同。微生物在实验室中数据结果的有效性缺乏与复杂的植物和土壤环境的相互作用，因此，产品在大田中的作用效果也需要确定。除了田间效果问题外，还涉及其他的效率问题，包括生物制剂/生物刺激剂的大规模生产及其在田间的广泛分布。此外，某些耐盐细菌菌株的遗传稳定性也有待研究，因为实验室水平和工业生产之间的差异很大。

（2）微生物菌株也应该有效地定殖在植物上，它们与根的定殖结合也是一个重要步骤（Barea，2015；Companyt 等，2010）。生产耐盐细菌作为盐渍土植物生长的接种物，从实验室到大田都需要高度的关注，还需要对其将冲击的市场进行适当研究和评估。农民使用的每种生物制品的商业化过程需要经过多个阶段才能获得批准，包括选择可用于商业化的微生物菌株。此外，该菌株应具有特定作物所需的活性。同样地，某些地区的天气条件也会影响生物制品的活性。气候条件的变化导致温度升高，不仅对植物有害，而且还能影响微生物的活性，导致生物制剂的生物活性不足。

（3）植物根据其在生长发育过程中的需要，对生物和非生物胁迫的反应方式和土壤中养分的有效性等情况选择使用微生物制剂。因此，任何一种微生物菌株在植物体内的有效性都会受到这些因素的影响。所有这些因素都是选择合适的微生物菌株所需要考虑的，这有利于它们在环境中进行定殖以充分满足植物的需求。在盐渍土中，作为提高植物抗盐能力的耐盐微生物菌株必须具有适应不同盐浓度的联合体。

（4）实用性是微生物接种剂成功的另一个重要因素。产品与农民生产实践中需求的相容性也应得到改善。微生物菌剂在功能上对相关作物的特异性不依赖于其在根中的定殖。因此，对于某一种作物而言，某些菌株的特异性可能较低；而对于另一种作物，它们在相同浓度下的特异性可能要高得多。一旦细菌在植物根部定殖，它们就会产生某些代谢物通过与植物的相互作用，帮助植物抵御逆境环境的胁迫等问题。因此，为某种作物（如甘蔗）准备的微生物接种剂表现出很强的活性，当它应用于另一种作物（如棉花）时，就不一定会有效。其原因可能是菌株不能与作物一起协同生长；或者某种作物的生长模式也会阻碍生物制剂的活性和存活。

（5）保质期长、抗恶劣环境且保护良好的产品能够受到更好的追捧。这样的产品往往能提高微生物在根和叶上的寿命（Xavier 等，2004；Warrior 等，2002；Leggett 等，2015；Ravensberg，2011）。为了使生物产品更快地商业化，它们的配方必须是广谱的、一致的，并且具有更长时间的保质期。开发一种更好的生物制剂，还需要解决一些非常严重的制约因素，包括接种剂配方、批量繁殖、选择合适的菌株、选择载体以及营销和包装。良好的微生物菌剂配方对微生物菌剂的稳定性、作用靶点、菌剂有效性和促进微生物的释放等方面起着至关重要的作用。在接种剂推广过程中，这些生物制品的开发和配方是重要因素之一。

（6）微生物产品在大田中效用的持久性是一个主要问题（Chutia 等，2007）。因为这些产品有更具体的效果目标，但其保质期和有效期通常比化肥/农药短（van Lenteren，2012）。因此，产品的功效需要在短期直接影响和环境持久性之间达成妥协（Barea，2015）。

（7）开发这种生物制剂/生物刺激剂往往需要较高的成本，这妨碍了这些产品的商业化应用（Dalpé 和 Monreal，2004）。因此，生产成本应与此类微生物产品在经济方面的投资回报相一致（Nicot 等，2011）。

因此，有效利用耐盐微生物修复盐渍土需要合理选择微生物，并在改良目标和配方技术等方面进行技术改进。利用遗传工程将耐盐微生物性状在植物中表达，有助于提高作物在盐渍化环境中的耐受和生存。因此，一株具有多种特性的细菌对植物在盐渍化环境中的生长有益，能够促进植物的生长。耐盐菌株能够利用多种机制促进植物生长，这一点已经研究得很清楚。尽管研究重点应侧重于每种机制对促进植物生长有效性的相对贡献，然而，对接种耐盐菌剂的作物进行精心照顾的田间试验，对这些菌株进行最大程度的商业化开发是必要的。总之，微生物菌剂生产行业的成功将取决于先进的企业管理、产品营销、广泛的教育和研究等措施。

参考文献

Ali S, Charles TC, Glick BR, 2014. Amelioration of high salinity stress damage by plant growthpromoting bacterial endophytes that contain ACC deaminase. Plant Physiol Biochem 80: 160-167.

Apse MP, Aharon GS, Snedden WA, Blumwald E, 1999. Salt tolerance conferred by overexpression of a vacuolar Na^+/H^+ antiport in *Arabidopsis*. Science 285: 1256-1258.

Apte SK, Thomas J, 1984. Effect of sodium on nitrogenfixation in Anabaena torulosaand *Plectonema boryanum*. J Gen Microbiol 130: 1161-1168.

Argandona M, Nieto JJ, Iglesias-Guerra F, Calderon MI, Garcia-Estepa R, Vargas C, 2010. Interplay between iron homeostasis and the osmotic stress response in the halophilic bacterium *Chromohalobacter salexigens*. Appl Environ Microbiol 76: 3575-3589. https://doi.org/10.1128/aem.03136-09.

Arora N, Singhal V, Maheshwari D, 2006. Salinity-induced accumulation of poly-β-hydroxybutyrate in rhizobia indicating its role in cell protection. World J Microbiol Biotechnol 22: 603-606.

Arora NK, Tewari S, Singh S, Lal N, Maheshwari DK, 2012. PGPR for protection of plant health under saline conditions. In: Bacteria in agrobiology: stress management. Springer, Heidelberg, pp. 239-258.

Asim M, Aslam M, Bano A, Munir M, Majeed A, Abbas SH, 2013. Role of phytohormones in root nodulation and yield of soybean under salt stress. Am J Res Commun 1: 191-208.

Bailey KL, Falk S, 2011. Turning research on microbial bioherbicides into commercial products-a Phoma story. Pest Technol 5: 73-79.

Barea J, 2015. Future challenges and perspectives for applying microbial biotechnology in sustainable agriculture based on a better understanding of plant-microbiome interactions. J Soil Sci Plant Nutr 15: 261-282.

Becker EA et al., 2014. Phylogenetically driven sequencing of extremely halophilic archaea reveals strategies for static and dynamic osmo-response. PLoS Genet 10: e1004784. https://doi.org/10.1371/journal.pgen.1004784.

Bergman B, Sandh G, Lin S, Larsson J, Carpenter EJ, 2013. Trichodesmium-a widespread marine cyanobacterium with unusual nitrogen fixation properties. FEMS Microbiol Rev 37: 286-302.

Bharti N, Pandey SS, Barnawal D, Patel VK, Kalra A, 2016. Plant growth promoting rhizobacteria *Dietzia natronolimnaea* modulates the expression of stress responsive genes providing protection of wheat from salinity stress. Sci Rep 6: 34768. https://doi.org/10.1038/srep34768.

Burr T, Matteson M, Smith C, Corral-Garcia M, Huang TC, 1996. Effectiveness of bacteria and yeasts from apple orchards as biological control agents of apple scab. Biol Control 6: 151-157.

Chakraborty U, Roy S, Chakraborty AP, Dey P, Chakraborty B, 2011. Plant growth promotion and amelioration of salinity stress in crop plants by a salt-tolerant bacterium. Recent Res Sci Technol 3: 11.

Chen L, Liu Y, Wu G, Veronican Njeri K, Shen Q, Zhang N, Zhang R, 2016. Induced maize salt tolerance by rhizosphere inoculation of *Bacillus amyloliquefaciens* SQR9. Physiol Plant 158 (1): 34-44.

Chutia M, Mahanta J, Bhattacheryya N, Bhuyan M, Boruah P, Sarma T, 2007. Microbial herbicides for weed management: prospects, progress and constraints. Plant Pathol J 6: 210-218.

Compant S, Clément C, Sessitsch A, 2010. Plant growth-promoting bacteria in the rhizo-and endosphere of plants: their role, colonization, mechanisms involved and prospects for utilization. Soil Biol Biochem 42: 669-678.

Dalpé Y, Monreal M, 2004. *Arbuscular mycorrhiza* inoculum to support sustainable cropping systems. Online. In, 2004.

Davies PJ, 2004. The plant hormones: their nature, occurrence, and functions. In: Davis PJ (ed) Plant hormones biosynthesis, signal transduction, action, vol 1. Kluwer Academic/Springer, Dordrecht, pp. 1-15.

Dodd IC, Pérez-Alfocea F, 2012. Microbial amelioration of crop salinity stress. J Exp Bot 63 (9): 3415-3428.

Egamberdieva D, 2011. Survival of *Pseudomonas extremorientalis* TSAU20 and *P. chlororaphis* TSAU13 in the rhizosphere of common bean (*Phaseolus vulgaris*) under saline conditions. Plant Soil Environ 57: 122-127.

El-Fattah DAA, Eweda WE, Zayed MS, Hassanein MK, 2013. Effect of carrier materials, sterilization method, and storage temperature on survival and biological activities of *Azotobacter chroococcum* inoculant. Ann Agric Sci 58: 111-118.

Enya J, Shinohara H, Yoshida S, Tsukiboshi T, Negishi H, Suyama K, Tsushima S, 2007. Culturable leaf-associated bacteria on tomato plants and their potential as biological control agents. Microb Ecol 53: 524-536.

Etesami H, 2018. Can interaction between silicon and plant growth promoting rhizobacteria benefit in alleviating abiotic and biotic stresses in crop plants? Agric Ecosyst Environ 253: 98-112. https://doi.org/10.1016/j.agee.2017.11.007.

Etesami H, Beattie GA, 2017. Plant-microbe interactions in adaptation of agricultural crops to abiotic stress conditions. In: Probiotics and plant health. Springer, pp. 163-200.

Etesami H, Beattie GA, 2018. Mining halophytes for plant growth-promoting halotolerant bacteria to enhance the salinity tolerance of non-halophytic crops. Front Microbiol 9: 148.

Fravel D, 2005. Commercialization and implementation of biocontrol. Annu Rev Phytopathol 43: 337-359.

Gelernter WD, Lomer CJ, 2000. Success in biological control of above-ground insects by pathogens. In: Gurr G, Wratten S (eds) Biological control: measures of success. Springer, Dordrecht, pp. 297-322.

Giri B, Kapoor R, Mukerji KG, 2007. Improved tolerance of *Acacia nilotica* to salt stress by *Arbuscular mycorrhiza*, *Glomus fasciculatum* may be partly related to elevated K/Na ratios in root and shoot tissues. Microb Ecol 54 (4): 753-760.

Glare T et al., 2012. Have biopesticides come of age? Trends Biotechnol 30: 250-258.

Goswami D, Dhandhukia P, Patel P, Thakker JN, 2014. Screening of PGPR from saline desert of Kutch: growth promotion in *Arachis hypogea* by *Bacillus licheniformis* A2. Microbiol Res 169: 66-75.

Habib SH, Kausar H, Saud HM, 2016. Plant growth-promoting rhizobacteria enhance salinity stress tolerance in okra through ROS-scavenging enzymes. Biomed Res Int 2016: 6284547.

Hu Y, Ribbe MW, 2015. Nitrogenase and homologs. J Biol Inorg Chem 20 (2): 435-445.

Islam F, Yasmeen T, Arif MS, Ali S, Ali B, Hameed S, Zhou W, 2016. Plant growth promoting bacteria confer salt tolerance in *Vigna radiata* by up-regulating antioxidant defense and biological soil fertility. Plant Growth Regul 80: 23-36.

Jaemsaeng R, Jantasuriyarat C, Thamchaipenet A, 2018. Molecular interaction of 1-aminocyclopropane-1-carboxylate deaminase (ACCD) -producing endophytic *Streptomyces* sp. GMKU 336 towards salt-stress resistance of *Oryza sativa* L. cv. KDML105. Sci Rep 8: 1950.

James RA, Blake C, Byrt CS, Munns R, 2011. Major genes for Na^+ exclusion, *Nax*1 and *Nax*2 (wheat *HKT*1; 4 and *HKT*1; 5), decrease Na^+ accumulation in bread wheat leaves under saline and waterlogged conditions. J Exp Bot 62: 2939-2947. https://doi.org/10.1093/jxb/err003.

Janisiewicz WJ, Korsten L, 2002. Biological control of postharvest diseases of fruits. Annu Rev Phytopathol 40: 411-441.

Jha Y, Subramanian R, 2013. Paddy plants inoculated with PGPR show better growth physiology and nutrient content under saline condition. Chilean J Agric Res 73: 213-219.

Jha B, Singh VK, Weiss A, Hartmann A, Schmid M, 2015. *Zhihengliuella somnathii* sp. nov., a halotolerant actinobacterium from the rhizosphere of a halophyte *Salicornia brachiata*. Int J Syst Evol Microbiol 65: 3137-3142.

Kang SR, Srinivasan S, Lee SS, 2015. *Vibrio oceanisediminis* sp. nov., a nitrogen-fixing bacterium isolated from an artificial oil-spill marine sediment. Int J Syst Evol Microbiol 65: 3552-3557.

Khalid M, Bilal M, Hassani D, Iqbal HMN, Wang H, Huang D, 2017. Mitigation of salt stress in white clover (*Trifolium repens*) by *Azospirillum brasilense* and its inoculation effect. Bot Stud 58 (1): 5.

Khodarahmpour Z, Ifar M, Motamedi M, 2012. Effects of NaCl salinity on maize (*Zea mays* L.) at germination and early seedling stage. Afr J Biotechnol 11: 298-304.

Kodama T, Taniguchi S (1976) Sodium-dependent growth and respiration of a nonhalophilic bacterium, *Pseudomonas stutzeri*. J Gen Microbiol 96 (1): 17-24.

Kodama T, Taniguchi S (1977) Sodium-controlled coupling of respiration to energy-linked functions in *Pseudomonas stutzeri*. Microbiology 98 (2): 503-510.

Köhl J, Postma J, Nicot P, Ruocco M, Blum B, 2011. Stepwise screening of microorganisms for commercial use in biological control of plant-pathogenic fungi and bacteria. Biol Control 57: 1-12.

Läuchli A, Epstein E, 1990. Plant responses to saline and sodic conditions. Agric Salinity Assess Manag 71: 113-137.

Leggett M, Newlands N, Greenshields D, West L, Inman S, Koivunen M, 2015. Maize yield response to a phosphorus-solubilizing microbial inoculant infield trials. J Agric Sci 153: 1464-1478.

Mayak S, Tirosh T, Glick BR, 2004. Plant growth-promoting bacteria confer resistance in tomato plants to salt stress. Plant Physiol Biochem: PPB 42 (6): 565-572. https://doi.org/10.1016/j.plaphy.2004.05.009.

Mengele R, Sumper M, 1992. Drastic differences in glycosylation of related S-layer glycoproteins from moderate and extreme halophiles. J Biol Chem 267: 8182-8185.

Milus E, Rothrock C, 1997. Efficacy of bacterial seed treatments for controlling Pythium root rot of winter wheat. Plant Dis 81: 180-184.

Munns R, Tester M, 2008. Mechanisms of salinity tolerance. Annu Rev Plant Biol 59: 651-681. https://doi.org/10.1146/annurev.arplant.59.032607.092911.

Murtaza G, 2013. Economic aspects of growing rice and wheat crops on salt-affected soils in the Indus Basin of Pakistan (unpublished data). Institute of Soil and Environmental Sciences, University of Agriculture, Faisalabad, Pakistan.

Nadeem SM, Ahmad M, Naveed M, Imran M, Zahir ZA, Crowley DE, 2016. Relationship between in vitro characterization and comparative efficacy of plant growth-promoting rhizobacteria for improving cucumber salt tolerance. Arch Microbiol 198: 379-387. https://doi.org/10.1007/s00203-016-1197-5.

Nautiyal CS, Srivastava S, Chauhan PS, Seem K, Mishra A, Sopory SK, 2013. Plant growth promoting bacteria *Bacillus amyloliquefaciens* NBRISN13 modulates gene expression profile of leaf and rhizosphere community in rice during salt stress. Plant Physiol Biochem 66: 1-9.

Navarro-Torre S et al., 2017. Assessing the role of endophytic bacteria in the halophyte *Arthrocnemum macrostachyum* salt tolerance. Plant Biol (Stuttg) 19: 249-256. https://doi.org/10.1111/plb.12521.

Nicot P, Blum B, Köhl J, Ruocco M, 2011. Perspectives for future research-and-development projects on biological control of plant pests and diseases. Class Augmentative Biol Control Against Dis Pests: Crit Status Anal Rev Factors Influencing Success 2011: 68-70.

Nunkaew T, Kantachote D, Nitoda T, Kanzaki H, Ritchie RJ, 2015. Characterization of exopolymeric substances from selected *Rhodopseudomonas palustris* strains and their ability to adsorb sodium ions. Carbohydr Polym 115: 334-341.

Orhan F, 2016. Alleviation of salt stress by halotolerant and halophilic plant growth-promoting bacteria in wheat (*Triticum aestivum*). Braz J Microbiol 47: 621-627.

Paul S, Bag SK, Das S, Harvill ET, Dutta C, 2008. Molecular signature of hypersaline adaptation: insights from genome and proteome composition of halophilic prokaryotes. Genome Biol 9: R70. https://doi.org/10.1186/gb-2008-9-4-r70.

Qin Y, Druzhinina IS, Pan X, Yuan Z, 2016. Microbially mediated plant salt tolerance and microbiome-based solutions for saline agriculture. Biotechnol Adv 34: 1245-1259.

Raheem A, Ali B, 2015. Halotolerant rhizobacteria: beneficial plant metabolites and growth enhancement of *Triticum aestivum* L. in salt-amended soils. Arch Agron Soil Sci 61: 1691-1705.

Rahnama A, James RA, Poustini K, Munns R, 2010. Stomatal conductance as a screen for osmotic stress tolerance in durum wheat growing in saline soil. Funct Plant Biol 37: 255-263.

Ramadoss D, Lakkineni VK, Bose P, Ali S, Annapurna K, 2013. Mitigation of salt stress in wheat seedlings by halotolerant bacteria isolated from saline habitats. Springerplus 2: 6. https://doi.org/10.1186/2193-1801-2-6.

Ravensberg WJ, 2011. A roadmap to the successful development and commercialization of microbial pest control products for control of arthropods, vol 10. Springer Science & Business Media, Dordrecht.

Roy SJ, Negrao S, Tester M, 2014. Salt resistant crop plants. Curr Opin Biotechnol 26: 115-124.

Sapre S, Gontia-Mishra I, Tiwari S, 2018. *Klebsiella* sp. confers enhanced tolerance to salinity and plant growth promotion in oat seedlings (*Avena sativa*). Microbiol Res 206: 25-32. https://doi.org/10.1016/j.micres.2017.09.009.

Saravanakumar D, Samiyappan R, 2007. ACC deaminase from *Pseudomonas fluorescens* mediated saline resistance in groundnut (*Arachis hypogea*) plants. J Appl Microbiol 102: 1283-1292.

Sharma SB, Sayyed RZ, Trivedi MH, Gobi TA, 2013. Phosphate solubilizing microbes: sustainable approach for managing phosphorus deficiency in agricultural soils. Springerplus 2: 587. https://doi.org/10.1186/2193-1801-2-587.

Sharma S, Kulkarni J, Jha B, 2016. Halotolerant rhizobacteria promote growth and enhance salinity tolerance in peanut. Front Microbiol 7: 1600. https://doi.org/10.3389/fmicb.2016.01600.

Shukla PS, Agarwal PK, Jha B, 2012. Improved salinity tolerance of *Arachis hypogaea* (L.) by the interaction of halotolerant plant-growth-promoting rhizobacteria. J Plant Growth Regul 31: 195-206.

Siebner-Freibach H, Hadar Y, Chen Y, 2004. Interaction of iron chelating agents with clay minerals. Soil Sci Soc Am J 68: 470-480.

Singh RP, Jha PN, 2015. Plant growth promoting potential of ACC deaminase rhizospheric bacteria isolated from *Aerva javanica*: a plant adapted to saline environments. Int J Curr Microbiol App Sci 4 (7): 142-152.

Singh RP, Jha PN, 2016. A halotolerant bacterium *Bacillus licheniformis* HSW-16 augments induced systemic tolerance to salt stress in wheat plant (*Triticum aestivum*). Front Plant Sci 7: 1890.

Srinivasan R, Yandigeri MS, Kashyap S, Alagawadi AR, 2012. Effect of salt on survival and P-solubilization potential of phosphate solubilizing microorganisms from salt affected soils. Saudi J Biol Sci 19: 427-434.

Staal M, Rabouille S, Stal LJ, 2007. On the role of oxygen for nitrogenfixation in the marine cyanobacterium *Trichodesmium* sp. Environ Microbiol 9: 727-736.

Suarez C, Cardinale M, Ratering S, Steffens D, Jung S, Montoya AMZ, Geissler-Plaum R,

Schnell S, 2015. Plant growth-promoting effects of *Hartmannibacter diazotrophicus* on summer barley (*Hordeum vulgare* L.) under salt stress. Appl Soil Ecol 95: 23-30. https://doi.org/10.1016/j.apsoil.2015.04.017.

Tabassum T, Farooq M, Ahmad R, Zohaib A, Wahid A, 2017. Seed priming and transgenerational drought memory improves tolerance against salt stress in bread wheat. Plant Physiol Biochem 118: 362-369.

Takors R, 2012. Scale-up of microbial processes: impacts, tools and open questions. J Biotechnol 160: 3-9.

Tenchov B, Vescio EM, Sprott GD, ZeidelJo ML, Mathai JC, 2006. Salt tolerance of Archaeal extremely Halophilic lipid membranes. J Biol Chem 281: 10016-10023. https://doi.org/10.1074/jbc.M600369200.

Timmusk S et al., 2014. Drought-tolerance of wheat improved by rhizosphere bacteria from harsh environments: enhanced biomass production and reduced emissions of stress volatiles. PLoS One 9: e96086. https://doi.org/10.1371/journal.pone.0096086.

Trung N, Hieu H, Thuan N, 2016. Screening of strong 1-aminocyclopropane-1-carboxylate deaminase producing bacteria for improving the salinity tolerance of cowpea. Appli Micro Open Access 2: 2.

Ul-Hassan T, Bano A, 2014. Role of plant growth promoting rhizobacteria and L-tryptophan on improvement of growth, nutrient availability and yield of wheat (*Triticum aestivum*) under salt stress. Int J Appl Agric Res 4: 30-39.

Ullah S, Bano A, 2015. Isolation of plant-growth-promoting rhizobacteria from rhizospheric soil of halophytes and their impact on maize (*Zea mays* L.) under induced soil salinity. Can J Microbiol 61: 307-313.

Van Lenteren JC, 2012. The state of commercial augmentative biological control: plenty of natural enemies, but a frustrating lack of uptake. BioControl 57: 1-20. https://doi.org/10.1007/s10526-011-9395-1.

Warrior P, Konduru K, Vasudevan P, 2002. Formulation of biological control agents for pest and disease management. In: Biological control of crop diseases. Marcel Dekker, New York, pp. 421-442.

Watanabe M, Kawahara K, Sasaki K, Noparatnaraporn N, 2003. Biosorption of cadmium ions using a photosynthetic bacterium, *Rhodobacter sphaeroides* S and a marine photosynthetic bacterium, *Rhodovulum* sp. and their biosorption kinetics. J Biosci Bioeng 95: 374-378. https://doi.org/10.1016/S1389-1723 (03) 80070-1.

Whipps JM, 2001. Microbial interactions and biocontrol in the rhizosphere. J Exp Bot 52: 487-511.

Xavier IJ, Holloway G, Leggett M, 2004. Development of rhizobial inoculant formulations. Crop Manag 3. https://doi.org/10.1094/cm-2004-0301-06-rv.

Yang H, Hu J, Long X, Liu Z, Rengel Z, 2016. Salinity altered root distribution and increased diversity of bacterial communities in the rhizosphere soil of Jerusalem artichoke. Sci Rep 6: 20687.

Yoshida S, Ohba A, Liang YM, Koitabashi M, Tsushima S, 2012. Specificity of *Pseudomonas* isolates on healthy and fusarium head blight-infected spikelets of wheat heads. Microb Ecol 64: 214-225. https://doi.org/10.1007/s00248-012-0009-y.

Zahran H, Ahmad M, Afkar E, 1995. Isolation and characterization of nitrogen-fixing moderate halophilic bacteria from saline soils of Egypt. J Basic Microbiol 35: 269-275.

Zerrouk IZ, Benchabane M, Khelifi L, Yokawa K, Ludwig-Müller J, Baluska F, 2016. A *Pseudomonas* strain isolated from date-palm rhizospheres improves root growth and promotes root formation in maize exposed to salt and aluminum stress. J Plant Physiol 191: 111-119.

Zhang HX, Blumwald E, 2001. Transgenic salt-tolerant tomato plants accumulate salt in foliage but not in fruit. Nat Biotechnol 19: 765-768. https://doi.org/10.1038/90824.

Zhang HX, Hodson JN, Williams JP, Blumwald E, 2001. Engineering salt-tolerant *Brassica* plants: characterization of yield and seed oil quality in transgenic plants with increased vacuolar sodium accumulation. Proc Natl Acad Sci U S A 98: 12832-12836. https://doi.org/10.1073/pnas.231476498.

Zhao S, Zhou N, Zhao ZY, Zhang K, Wu GH, Tian CY, 2016. Isolation of endophytic plant growth-promoting bacteria associated with the halophyte *Salicornia europaea* and evaluation of their promoting activity under salt stress. Curr Microbiol 73: 574-581.

Zhou N, Zhao S, Tian CY, 2017. Effect of halotolerant rhizobacteria isolated from halophytes on the growth of sugar beet (*Beta vulgaris* L.) under salt stress. FEMS Microbiol Lett 364. https://doi.org/10.1093/femsle/fnx091.

Zuccarini P, Okurowska P, 2008. Effects of mycorrhizal colonization and fertilization on growth and photosynthesis of sweet basil under salt stress. J Plant Nutr 31 (3): 497-513.

10 耐盐微生物在盐胁迫条件下促进植物生长的作用

Zahir Ahmad Zahir, Sajid Mahmood Nadeem,
Muhammad Yahya Khan, Rana Binyamin, Muhammad Rashid Waqas

[摘要]

包括细菌和真菌在内的耐盐微生物具有在盐胁迫环境中促进植物生长的能力，主要通过多种植物促生机制实现，包括能够合成胞外多糖，合成和分泌 1-氨基环丙烷-1-羧酸脱氨酶、渗透调节物质，增强农作物对营养成分的获取，增加盐胁迫下植株细胞内抗氧化酶的活性，以及维持较高的 K^+/Na^+ 比值等。具有这些功能的耐盐微生物才能成为促进植物在盐胁迫环境下生长的有益微生物的候选者。此外，这些微生物具备的一些其他能力，比如固氮、合成和分泌铁载体、溶解不溶性养分（比如磷素和钾素），以及能够有针对性对各种农作物病原体进行防御，进一步巩固了这类有益微生物在整个农业生产系统中的重要性。为了能够降低在不利环境中农作物生长受到的危害，如今在农业生产系统中人们正在加快对这些环境友好且具有耐逆境能力的有益微生物的研究和利用，比如将这些有益微生物开发为生物肥料使用。因此，耐盐细菌和真菌就很有可能是克服农作物生产中所面临的盐胁迫问题，是一种经济且有效的方法。耐盐微生物用作生物农药和生物肥料在农业生产中应用，可以在很大程度上减少我们目前对化学合成肥料和农药的依赖，这些耐盐微生物也能够对受到污染的环境进行生物修复。本章节着重介绍了在盐胁迫条件下和对污染环境进行生物修复的过程中，耐盐微生物在改善农作物生长中的重要作用。通过选定其他研究人员的研究实例讨论了耐盐微生物在盐胁迫下耐受盐分、促进农作物生长的具体机制。同时，本章节内容还回顾了这些微生物在环境科学中的作用，以及详细讨论了目前科学研究中存在的一些问题和需要进一步研究探索的方向，并对该技术在未来的应用前景进行了

Z. A. Zahir (✉)，巴基斯坦费萨拉巴德农业大学土壤与环境科学研究所，E-mail：zazahir@uaf.edu.pk

S. M. Nadeem, M. Y. Khan, R. Binyamin, M. R. Waqas, 巴基斯坦费萨拉巴德农业大学布雷瓦拉-韦哈里校区

展望。

[关键词]

耐盐；细菌；菌根；胁迫；农作物；生长；环境

10.1 前言

农作物在生长过程中，面临各种生物和非生物胁迫，其中非生物逆境胁迫严重威胁着农作物的正常生长，这些非生物逆境胁迫包括盐碱胁迫、干旱胁迫、冷害、养分缺乏、重金属胁迫等。在逆境胁迫环境下，由于农作物应激反应而产生的细胞内激素水平变化以及营养吸收的不平衡，导致农作物细胞内产生特定的离子毒性（如 Na^+）以及土壤溶液中渗透势升高导致农作物根系吸收水分困难，因此，农作物的正常生长往往受到很严重地抑制。尤其是在干旱和半干旱地区，由于降水量稀少且环境温度较高，加剧了土壤水分的蒸发，导致这些地区的土壤环境变得尤为恶劣，极易形成盐渍化土壤，导致农作物生长不良。另外，由于缺乏优质的农业灌溉用水，农民在农作物种植的过程中不得不使用劣质的地面蓄水满足灌溉的需要。然而，大部分地区的地面蓄水经常掺杂着各种杂质和污染，其水质经常达不到农业用水的标准。正是因为劣质用水被循环用于农作物灌溉，因此土壤中的盐分物质浓度才不断积累增加（Shakirova 等，2003）。

为了适应这样的盐胁迫环境，并且能够完成正常的生长过程，农作物在漫长的进化过程中形成了许多生理生化层面的机制以抵御不良环境的胁迫影响。这些机制包括合成渗透调节物质、细胞内毒性离子的区域化分隔、抗氧化酶系统生物合成和激活、激素调节、离子排斥效用以及 Na^+ 进入细胞的阻断等（Tuteja，2007；Gupta 和 Huang，2014）。除了农作物本身具有的各种耐盐机制以外，土壤中的各种微生物群落在诱导和增强农作物耐盐性方面也起着非常关键的作用。这些微生物除了自身能够耐受一定的盐胁迫，可以在盐胁迫环境中正常生长以外，还具有许多促进农作物生长的特性，这些促生特性是这类微生物长期与农作物共生的过程中形成的，通过多种特殊的作用机制实现，能够对在恶劣环境下农作物的生长产生十分显著的积极影响。在这些微生物种群中，有一些细菌和菌根真菌菌株具有在盐水环境中生存的能力，具有这种特性的微生物一般被称为耐盐和/或嗜盐微生物。耐盐性是指生物体的正常生长并不需要大量的盐分，在非盐胁迫条件下能够更好地生长，但是能够耐受一定浓度的盐胁迫，并具有在盐胁迫条件下保持生长的能力；而嗜盐性则是指生物体的正常生长需要大量的盐分，在一定程度盐分条件下的生长比在非盐分胁迫条件下表现得更好，盐分是这类生物体正常生长所必需的。这些耐盐/嗜盐微生物已经进化出很多特殊的机制，使它们能够在盐胁迫环境中生存。因此，这些微生物才具有商业开发价值，能够通过培养、改造和驯化等生物技术手段进行培育，使其在盐渍化土壤中进行利用。这些微生物还具有生物修复的能力，能够在清洁污染环境中发挥重要作用。

包括细菌和真菌在内的多种耐盐微生物都具有在不同盐浓度条件下生存的能力。这些微生物包括表皮短杆菌（Brevibacterium epidermidis）、蜡状芽孢杆菌（Bacillus cereus）和盐盐单胞菌（Halomonas salina）等，都是具有抗盐能力的细菌，这些微生物能够在高达20%的盐浓度下生存（Behera等，2012，2014）。一些常见的耐盐细菌菌株还包括海水芽孢杆菌VITP4（Bacillus aquimaris VITP4）、克劳氏芽孢杆菌I-52（Bacillus clausii I-52）、枯草芽孢杆菌RSKK96（Bacillus subtilis RSKK96）以及地衣芽孢杆菌Shahed-07（Bacillus licheniformis Shahed-07）（Joo和Chang，2005；Rasooli等，2008；Shivan和Jayaraman等，2009；Akcan和Uyar，2011）。Mandal（2014）研究报道指出，他们在27株三叶草根瘤菌（Rhizobium trifolii）中，筛选出了5株菌株能够在3%的NaCl浓度下正常生存。Akhter等（2012）在研究固氮菌（Azotobacter）的耐盐机理时，报道称他们在15株菌株中筛到5株菌株能够耐受6%的NaCl溶液，其中还有2株菌株甚至能够在10%的NaCl浓度下保持生长。Tippannavar等（1989）也报道了固氮菌（Azotobacter）具有很强的耐盐能力。一般而言，耐盐细菌对盐胁迫的耐受范围通常为3%~30%的NaCl浓度，与微生物发挥耐盐作用相关的蛋白酶活性都有良好的表现，因此，这一类耐盐微生物具有在盐胁迫农业中应用的潜力（Ventosa等，1998）。Banik等（2018）从盐胁迫环境中分离得到很多耐盐芽孢杆菌（Bacillus）和嗜盐芽孢杆菌（Halobacillus sp.），研究指出，这些细菌能够促进花生在盐渍土和重金属污染土壤条件下的生长。

与耐盐细菌类似，真菌菌株如正茁芽短梗霉（Aureobasidum pullulans）和嗜鱼壁菌（Wallemia ichthyophaga）等也被认为是进行耐盐研究的合适生物（Gunde Cimerman和Zalar，2014）。Sengupta和Chaudhuri（1990）以及Hildebrandt等（2001）通过研究还报道了盐沼植物中存在菌根真菌。Landwehr等（2002）也报道了在碱性土壤中存在大量的菌根真菌孢子等情况。因此，耐盐菌根真菌也有很大的潜力能够用来改善盐渍化土壤条件下农作物的生长。Manga等（2017）通过研究8种菌根真菌对盐胁迫下金合欢属植物（Acacia seyal）生长的影响发现，根内嗜根菌（Rhizophagus intraradices）能够通过提高对矿物质养分的吸收量，促进金合欢属植物（Acacia seyal）在高达680 mmol/L NaCl下的生长。进一步的研究指出，真菌孢子的存在与土壤类型和盐胁迫程度无关；不同菌根真菌在农作物根系上的定殖程度不同。

在各种不同的耐盐真菌中，球囊霉属（*Glomus* spp.）真菌是一种比较常见的菌根真菌（Allen 和 Cunningham，1983；Wang 等，2004）。Aliasgharzadeh 等（2001）研究报道指出，在伊朗大不里士平原受盐渍化影响的土壤中存在大量的球囊霉属（*Glomus* spp.）真菌。在这些受到高浓度盐胁迫危害的耕作土壤中，富含丰富的丛枝菌根菌（*Glomus etunicatum*）、变形球囊霉属（*Glomus versiform*）和丛枝菌根真菌（*Glomus intraradices*）。Wang 等（2004）报道了球囊霉属（*Glomus* spp.）能够在盐水条件下生存。这些微生物能够通过各种调节机制减缓农作物受到的盐胁迫危害，甚至促进农作物在盐渍化胁迫土壤中生长的潜力。Porras Soriano 等（2009）的研究工作也支持这一观点。他们组织测试了包括球囊霉属（*Glomus* spp.）的丛枝菌根真菌（*Glomus intraradices* 和 *Glomus mosseae*）和近明球囊霉（*Glomus claroideum*）在内的 3 种真菌菌株对农作物植株生长促进的潜力。通过研究发现，在这 3 种微生物真菌中，丛枝菌根真菌（*Glomus mosseae*）是提高橄榄树耐盐性最有效的菌种。

同一真菌菌株的耐盐能力可能因其所处的生长阶段不同而不同。从球囊霉属（*Glomus* spp.）对农作物促生的例子中可以看出，这些菌株在 300 mmol/L NaCl 水平下能够在农作物根系中定殖，然而这种真菌的孢子并没有在该盐度水平下生长（Juniper 和 Abbott，2006）。研究者认为，这可能是由于微生物在不同的生长阶段存在能力差异，以及各个生命阶段对盐胁迫的反应和耐受程度不同或在盐胁迫条件下启动各生命发育阶段具有差异导致的。

通过上述论述可以说明，在自然环境中存在大量的细菌和菌根真菌具有能够在盐水环境中生存的能力。这些微生物种类不仅能够在盐渍化条件下维持自身的正常生长，而且还能够有助于减轻盐胁迫条件对农作物生长发育的影响。以下章节将讨论这些微生物耐受盐胁迫和促进农作物在盐胁迫下生长发育的机理机制。

10.2 耐盐微生物耐盐机理

为了能够发挥促进农作物在逆境胁迫环境下生长的作用，微生物自身必须有能够在恶劣的环境中维持生长的能力。耐盐微生物（细菌和菌根真菌）在进化过程中形成了许多机制来耐受盐胁迫环境，并在这种环境下促进农作物的生长。这些微生物对盐渍化土壤环境的适应性是其促进农作物生长的前提条件，与其特有的一些耐盐机制有关，这些耐盐机制能够使该类微生物较其他土壤微生物种群在盐渍化土壤条件中具有更强的生长竞争优势。

Kunte（2012）在研究微生物通过渗透调节作用机制来抵御盐胁迫的过程中发现，嗜盐微生物应对盐胁迫的机制包括两种：一是将盐分在细胞质中特定的位置分区储存；二是利用有机渗透调节物质进行调节。根据他们的研究结果，这些机制保证了嗜盐微生物能够在盐胁迫环境中生长，甚至能够直接在不同浓度的盐溶液中生长。在通过细胞质调节 Na^+ 积累的机制中，嗜盐微生物能够在细胞质中开辟专门的区域储存 Na^+，而使胞质内其他的区域几乎不含有 Na^+，并且这些微生物还在细胞中大量积聚 K^+。K^+ 的积累使细胞质中 Na^+ 的相对含量降低，减轻了 Na^+ 毒害作用对微生物存活的负面影响。而在有机渗透调节物质进行渗透调节的作用机制下，微生物细胞积累大量的有机溶质分子，比如糖、多元醇和氨基酸等，提高胞内的渗透势以抵御外界环境中因盐分积累而造成的高渗透势。此外，这些非离子溶质分子即使在很高的浓度下也不会干扰细胞正常的新陈代谢，因此才会在长期的进化发展中被保留下来。研究者认为，与嗜盐微生物（即必须有足够的盐分才能维持细胞正常生长的微生物）相比，通过利用有机渗透调节剂作用机制的微生物在抵御盐胁迫等逆境胁迫具有更大的灵活性，这是因为它们既可以在高盐浓度下生长，又能够在低盐浓度下生长。然而，与嗜盐微生物分区储存盐分离子相比，通过有机渗透调节作用机制来抵御盐胁迫则需要更多的腺苷三磷酸功能。

为了在盐胁迫环境中生存，耐盐细菌在细胞中积累了甘氨酸、甜菜碱、脯氨酸、胞外碱和谷氨酸等渗透调节溶质；然而大多数耐盐真菌则选择在细胞内储存大量的甘油作为渗透调节物质（Blomberg 和 Adler，1992；Burg 等，2007）。许多细菌和一些真核生物在细胞中大量积累渗透相容溶质以提高胞内渗透势，目的有两个：一是维持胞内高渗透势，与盐胁迫环境造成的胞外高渗透势相抗衡，避免细胞失水造成死亡；二是减少细胞从外界环境中吸收过量的 Na^+，使胞内 Na^+ 浓度维持在一个低于产生离子毒性的水平。在耐盐细菌抵御盐胁迫危害的过程中，微生物在细胞质中积累各种有机溶质，有效地降低胞质内的 Na^+ 水平，这对于耐盐微生物在盐胁迫环境中的生存具有重要的意义。除了能够提高胞内渗透势降低 Na^+ 含量之外，这些渗透调节物质还能在盐胁迫的环境下起到保护细胞质膜的作用，使细胞处在一定程度的盐胁迫环境中不能发生质壁分离的现象（Kempf 和 Bremer，1998）。而在异养型真菌细胞中，小分子的外源蛋白是效果最好的渗透调节物质（Galinski，1995）。此外，除了甘氨酸和甜菜碱外，海藻糖在生物体耐受逆境胁迫方面也起着非常关键的作用（Turan 等，2012）。海藻糖最开始是在盐绿外硫红螺菌（*Ectothiorhodospira halochloris*）中发现的（Galinski 等，1985）。

Mendpara 等（2013）从农用耕作土壤中分离出了 6 种细菌，根据研究结果，其中有 2 种细菌分别属于微杆菌属（*Exiguobacterium* sp.）和沙雷氏菌属（*Serratia* sp.），这两种细菌能够在高达 10% 的 NaCl 溶液胁迫条件下生存。进一步的研究证明，这些细菌的高耐盐能力可能就像早期的研究人员报道的那样，它们能够在细胞内合成大量的甜菜碱、胞外素和海藻糖等（Oren, 2008; Kondepudi 和 Chandra, 2011）。Patel 等（2018）研究报道了耐盐细菌深海微小杆菌 phm11（*Exiguobacterium profundum* phm11）对盐胁迫环境的耐受机理。根据他们的研究结果表明，这种细菌是通过在胞内积累 L-脯氨酸来抵御盐胁迫危害的。他们通过进一步的试验报道称，该菌株的耐盐能力也与编码参与相关代谢途径物质的基因表达水平的微调有关。在 100 mmol/L 的 NaCl 胁迫处理下，微生物的生长速率、生物量和代谢物产生量达到最大，但是当 NaCl 的浓度达到 150 mmol/L 的水平时，这些生长指标受到了严重地抑制。另外据报道，在温度范围为 12~55℃的冷热环境中已经鉴定到了属于微杆菌属（*Exiguobacterium* sp.）的细菌（Vishnivetskaya 等, 2009）。

Firth 等（2016）研究了维尔氏菌 34H（*Colwellia psychrerythraea* 34H）和嗜冷杆菌（*Psychrobacter* sp. 7E）通过渗透调节抵御盐胁迫危害的机制。他们使用 ^{14}C 胆碱作为渗透调节物质并观察这种相容性溶质在维持细菌抵御盐胁迫过程中的作用，即细菌短期和长期暴露在盐胁迫条件下对其生长的影响。通过试验，他们证明胆碱能够提高微生物耐盐性的原因在于细菌在进行呼吸作用时，可以将胆碱中的氮转化为微生物体生长所必须的铵，而铵的再生尤其是硝化作用可能会增强微生物生命活动中特定的一些生物化学过程，这也许是胆碱类物质提高微生物耐盐能力的原因之一。

细菌在盐胁迫下生长和发育的另外一个非常重要的机制是钾素的积累。耐盐微生物细胞质中积累的 K^+ 能够与谷氨酸相结合，含有谷氨酸的 K^+ 的积累在大肠杆菌和其他细菌中也有报道（Oren, 1999; Ventosa 等, 1998; Dinbier 等, 1988）。类似地，Amir 等（1993）也通过研究报道了一些微生物菌株甚至能够在高达 11% 的 NaCl 溶液中生存，他们具有极强的盐耐受性。研究者们还注意到，在盐胁迫条件下微生物细胞中的 Na^+ 在细胞内积累，为了维持细胞内 Na^+ 的平衡，细菌增加了对 K^+ 和脯氨酸等物质的积累。然而，K^+ 的大量积累可能会带来另外一个问题，即干扰细胞中多种酶的正常功能（细胞内酶功能的正常发挥需要 K^+ 的存在，但是需要 K^+ 维持在一个相对的范围内，过高或者过低对酶功能的发挥都有影响）。为了解决这个问题，耐盐微生物又采取了另外一种策略，即在细胞内大量积累渗透调节物质（Lucht 和 Bremer, 1994; Galinski, 1995; Ventosa 等, 1998; Saum 和 Muller, 2007）。

Sandhya 和 Ali（2015）通过研究指出，耐盐细菌可以通过合成特定的代谢物质以抵御恶劣的生长环境对自身生长带来的不利影响。一个编码豌豆 DNA 解旋酶 45 的基因 *PDH*45 在多种农作物耐盐性中的作用已经有很多研究进行了报道

(Sanan Mishra 等，2005；Sahoo 等，2012；Augustine 等，2015）。然而，Tajrishi 等（2011）通过研究证实它在细菌的耐盐性中也起作用。他们证明了大肠杆菌 BL21（*Escherichia* BL21）的耐盐能力是由于 *PDH45* 基因的活跃表达所决定的。另外，他们还观察到微生物对盐胁迫的应激反应具有特异性，因为当用 LiCl 取代 NaCl 进行盐胁迫时，细菌则无法生长，说明该基因对 Na^+ 和 Li^+ 引起的盐胁迫反应具有特异性。

Tripathi 等（2002）在具有耐盐能力的巴西偶氮螺菌（*Azospirillum brasilense*）中也观察到了脯氨酸、甘氨酸、谷氨酸和海藻糖等渗透调节物质在胞质中的积累。他们通过研究证明脯氨酸在细菌耐盐胁迫过程中起着重要的作用。同时，研究还发现，随着渗透胁迫的增加，渗透调节物质由谷氨酸占主导慢慢转变为脯氨酸占据主导地位。Reina-Bueno 等（2012）还研究了海藻糖在提高微生物耐盐能力方面的作用，他们通过对需盐色盐杆菌（*Chromohalobacter salexigens*）的研究来揭示海藻糖对提高微生物的盐胁迫耐受能力。通过研究他们发现，当编码海藻糖-6-磷酸合成酶的基因缺失时，突变体菌株不受盐胁迫的影响；但缺乏胞外素和海藻糖的双突变体菌株对盐胁迫敏感。因此，他们报道了海藻糖在细菌耐盐性方面起作用。除了提高微生物适应由于盐胁迫引起的渗透胁迫之外，渗透调节物质在保护生物大分子乃至生物体组织和器官等方面也起着非常重要的作用（Da Costa 等，1998；Welsh，2000a、2000b）。

Tiquia 等（2007）通过试验研究分离出 127 株嗜盐细菌，这些嗜盐细菌分别隶属于葡萄球菌属（*Staphylococcus*）、芽孢杆菌属（*Bacillus*）、类芽孢杆菌（*Paenabacillus*）、嗜盐芽孢杆菌属（*Halobacillus*）、梭菌属（*Clostridium*）和盐单胞菌属（*Halomonas*）。研究者发现，在这些细菌菌株中的大多数菌株中均能检测到明胶酶、β-半乳糖苷酶和色氨酸脱氨酶的存在，并且他们还指出，精氨酸二氢酶是从这些嗜盐细菌分离物中鉴定到的主要酶。根据他们的研究观点，这些存在于耐盐细菌中功能各异的各种酶是这些耐盐细菌能够在盐胁迫环境中生存的重要原因。Upadhyay 等（2012）研究了分离到的一些细菌菌株的耐盐能力，试验结果表明，大约有33%的菌株能够在8%的 NaCl 浓度下存活；19%的菌株在较高的 NaCl 浓度下能够表现出促进农作物生长的特性。此外，他们进一步研究了这些耐盐细菌的耐渗机理，指出这些耐盐菌株中存在的高浓度脯氨酸和蛋白质类物质是它们能够耐受高浓度盐胁迫的主要原因。

与细菌相似，某些真菌菌株也能够在盐胁迫环境下生存，表现出一定的耐盐特性。这些菌株能够在盐渍化环境中生存也是通过各种耐盐机制来实现的，使它们能够在含盐的环境中生存。Petrovic 等（2002）通过研究指出，甘油是真菌菌株中最重要的渗透相容性溶质之一，在威尼克外瓶霉（*Hortaea werneckii*）的耐盐性中发挥了重要作用。根据 Plemenitas 等（2014）的研究指出，高渗甘油（High osmolarity glycerol，HOG）途径是真菌菌株响应盐胁迫而做出的主要的应激途径。研究指出，嗜鱼壁菌（*Wallemia ichthyophaga*）和威尼克外瓶霉（*Hortaea werneckii*）通过产生与甘油相容的其他种类的有机溶质稀释胞内的 Na^+ 浓度，从而使胞内一直维持着低 Na^+ 浓度。在盐胁迫等逆境条件下，真菌菌株也会主动引起细胞膜性质的改变，这也是真菌在面对恶劣的生存环境所进化出的一种重要机制。真菌这种细胞膜性质的改变主要是指在盐渍化等逆境环境中膜流动性的变化。已经有很多研究指出，在处于盐胁迫的情况下，耐盐威尼克外瓶霉（*Hortaea werneckii*）的质膜维持着更多的流动性（Turk 等，2004，2007）。他们通过对盐胁迫下质膜成分的进一步研究指出，在盐胁迫下威尼克外瓶霉（*Hortaea werneckii*）细胞质膜的总固醇含量保持不变，其膜流动性变化主要归因于构成质膜的组分磷脂结构的改变。此外，与细菌通过在胞内合成和储存大量的渗透调节物质以改变胞内渗透势抵御盐胁迫危害类似，真菌也会在细胞内积聚多元醇和游离氨基酸及其衍生物等多种渗透调节物质，以耐受胁迫的条件，这也是真菌耐受盐胁迫危害的一种重要机制。

上述讨论表明，耐盐微生物能够利用多种机制耐受盐胁迫环境。而在这些机制中，渗透相容溶质的积累是微生物最经常采用的一种耐性机制。此外，一些其他的微生物菌株还能够在细胞内积累 K^+ 和 Ca^{2+} 等离子物质，以维持胞内离子平衡，减少 Na^+ 积累对细胞造成的毒害作用。此外，微生物细胞中还能够通过积累一些特殊的酶来保护微生物免受盐胁迫的有害影响。

10.3 逆境农业促进植物生长和产量有效性的机制

在农业生产中，农作物的根际环境与粒块状土壤环境的理化性质不同。正是由于这种差异，农作物根际和微生物之间发生了许多相互作用包括拮抗作用和协同作用。农作物根际与微生物发生何种效用取决于所涉及的微生物菌株类型和农作物物种与该菌株的相互作用。这些相互作用可能发生在农作物和细菌之间，或者农作物和真菌之间，甚至也能够发生在农作物根际的真菌和细菌之间。相互作用中的协同作用是一种相互促进的关系。在农作物与微生物的协同作用中，双方通过为对方提供生长发育所必需的物质来实现最终的互相帮助，从而促进微生物和农作物的生长（Finlay，2007；Beattie，2007；Franche 等，2009；Nadeem 等，2014；Shin 等，2016）。

土壤微生物与农作物通过协同作用相互促进生长和生存。在盐胁迫等逆境环境下，农作物能够与其根际不同的微生物之间产生有益的联系，并最终通过协同效应降低农作物受到的盐胁迫危害（Badri，2009）。耐盐微生物包括促进农作物生长的植物生长促生细菌和真菌，这些促生微生物能够通过保护植物免受因盐分诱导而产生的对生长发育的负面影响，在农作物抵御盐胁迫等逆境环境中发挥重要的作用（Lugtenberg 和 Kamilova，2009）。就微生物促进农作物生长的机理而言，耐盐微生物主要是通过提高胞内渗透压、合成和分泌胞外多糖、产生铁载体以及相关的植物激素等，并增强细胞内抗氧化酶的活性，溶解农作物根际必需的营养物质（如氮和磷）等，通过提高农作物对养分的获取能力、水分利用效率，以及促进根系导水率等作用，提高农作物的耐逆能力，减少盐胁迫对植物生长的有害影响。耐盐细菌和真菌对盐渍化环境下促进农作物生长的一些精选实例分别见表 10.1 和表 10.2。

表 10.1 接种植物生长促生菌诱导植物耐盐抗性

作物种类	细菌株系	耐性机理	参考文献
水稻 (*Oryza sativa*)	泛养副球菌 SRV4 菌株 (*Curtobacterium albidum* SRV4 strain)	提高脯氨酸含量、膜系统稳定性以及光合速率	Vimal 等（2018）
		增强水稻相关抗氧化酶活力、K^+ 吸收能力	
		通过 EPS 的产生减少乙烯应激和 ACC 脱氨酶的活力和 Na^+ 的可获取性	
	肠杆菌属 P23 菌株 (*Enterobacter* sp. P23)	抗氧化酶活性和胁迫应激乙烯产生的降低	Sarkar 等（2018）
		溶磷作用增强	
		生长素和铁蛋白含量增加	
	解淀粉芽孢杆菌 NBRISN-13 菌株 (*Bacillus amyloliquefaciens* NBRISN13)	膜系统完整性提高	Tiwari 等（2017）
		渗透保护剂积累提高	
		标记基因表达量提高	
	纺锤形赖氨酸芽孢杆菌 BPC2 菌株 (*Lysinibacillus* sp. BPC2) 和铜绿假单胞菌 PRR1 和 PHL3 (*Pseudomonas aeruginosa* PRR1/PHL3)	提高植物干重、鲜重和次生根数目	Kumar 等（2017）
		提高叶绿素的含量	
		溶磷作用增强	
		生长素和铁蛋白含量增加	
芸薹 (*Brassica napus* L.)	阴沟肠杆菌菌株 HSNJ4 (*Enterobacter cloacae* HSNJ4)	增加叶绿素含量、侧根数目、茎长以及根长	Li 等（2017）
		减少丙二醛含量	
		增强抗氧化酶活力并提高脯氨酸含量	
		产生更多的内源生长素	
		通过调节 ACC 脱氨酶活性减少乙烯释放	

（续表）

作物种类	细菌株系	耐性机理	参考文献
西瓜（*Citrullus lanatus*）、拟南芥（*Arabidopsis thaliana*）和黄瓜（*Cucumis sativus*）	芽孢杆菌菌株 LBEndo1（*Bacillus sp.* LBEndo1）和假单胞菌菌株 liniKBEcto4（*Pseudomonas liniKBEcto4*）	在正常环境和盐胁迫环境下均能提高生长能力 诱导合成生长素的组成型基因的表达，并提高根中生长素的含量 溶磷作用增强 生长素和铁蛋白含量增加	Palacio Rodríguez 等（2017）
辣椒（*Capsicum annuum*）	解磷假单胞菌菌株 OS261（*Pseudomonas frederiksbergensis* OS261）	通过调节 ACC 脱氨酶活性减少乙烯释放 抗氧化酶活性提高 减少 H^+ 含量	Chatterjee 等（2017）
拟南芥（*Arabidopsis thaliana*）	伯克霍尔德氏菌菌株 PsJN（*Burkholderia phytofirmans* PsJN）	增加脯氨酸含量 上调活性氧清除相关基因的表达 上调响应脱落酸信号途径相关基因的表达 上调解毒相关基因的表达 下调茉莉酸信号途径相关基因的表达 改变离子稳态调整相关基因的表达模式（如高亲和 K^+ 转运蛋白、拟南芥 K^+ 转运蛋白等）	Pinedo 等（2015）
落花生（*Arachis hypogaea*）	假单胞菌（*Pseudomonas*）、克雷伯氏菌（*Klebsiella*）、苍白杆菌（*Ochrobactrum*）和土壤杆菌（*Agrobacterium*）	提高总氮含量并增强生长 调节离子稳态平衡 减少活性氧积累 生长素含量增加并增强溶磷作用 ACC 脱氨酶活性改变	Sharma 等（2016）
小麦（*Triticum aestivum* L.）	嗜麦芽糖寡养单胞菌菌株 SBP-9（*Stenotrophomonas maltophilia* SBP-9）	增加叶绿素含量 减少丙二醛和脯氨酸的含量 增强抗氧化酶如 CAT、SOD 以及 POD 等的活性 减少根和茎中 Na^+ 浓度 增强对 K^+ 离子的吸收	Singh 和 Jha（2017a, b）

(续表)

作物种类	细菌株系	耐性机理	参考文献
小麦（*Triticum aestivum* L.）	三叶草根瘤菌菌株 CIAT899（*Rhizobium tropici* strain CIAT899）和巴西固氮螺菌菌株 Ab-V5 和 Ab-V6（*Azospirillum brasilense* strains Ab-V5/Ab-V6）	增强了抗氧化酶的活性，这些酶比如 CAT、SOD 和 APX 等对活性氧具有解毒作用 减少丙二醛和脯氨酸的含量 通过调节基因表达（通常是进行上调），这些基因一般编码叶中 SOD2、APX1、SOD4 和 CAT1 以及根中的 APX2 等抗氧化酶基因 下调基因表达，如与病原相关的 prp2、PRI 和 prp4 基因 下调叶中和根中热激蛋白 hsp70 基因的表达	Fukami 等（2018）
	克雷伯氏菌菌株 SBP-8（*Klebsiella* sp. SBP-8）	增加总蛋白、可溶性糖和脯氨酸的含量 减少盐胁迫诱导丙二醛的含量 增加抗氧化酶如 POX、CAT 和 SOD 等的活性 增强寄主植物中 Na^+ 外排和 K^+ 吸收 增强植物生物量和叶绿素含量	Singh 等（2015），Singh 和 Jha（2017a，b）
甜菜（*Beta vulgaris* L.）	争论贪噬菌（*Variovorax paradoxus*）、莱比托游动球菌（*Planococcus rifietoensis*）和云南微球菌（*Micrococcus yunnanensis*）	植物生物量增加和种子萌发率提高 光合作用增强 减少胁迫诱导乙烯的产生 通过 ACC 脱氨酶活性降低宿主植物 ACC 的含量	Zhou 等（2017）
番茄（*Solanum lycopersicum*）	短小芽孢杆菌菌株 AM11（*Bacillus pumilus* AM11）和微杆菌属菌株 AM25（*Exiguobacterium* sp. AM25）	更高的生物量 增强光合作用效率 增强叶绿素的积累 降低脂质过氧化反应 增加谷胱甘肽过氧化氢酶和过氧化物酶的活性	Ali 等（2017）

（续表）

作物种类	细菌株系	耐性机理	参考文献
豌豆（Pisum sativum L.）	动物性杆菌菌株 MSSA-10（Planomicrobium sp. MSSA-10）	降低活性氧的含量 增强抗氧化酶活性 增加叶绿素和蛋白含量 溶磷作用增强 生长素产生 ACC 脱氨酶活性改变	Shahid 等（2018）

表 10.2　接种丛枝菌根真菌诱导植物耐盐抗性

作物	真菌菌株	作用机理	参考文献
印度田菁（Sesbania sesban）	摩西管柄囊霉（Funneliformis mosseae）、根内嗜根菌（Rhizophagus intraradices）和幼套近明囊霉（Claroideoglomus etunicatum）	增加结节的数量和重量 增强固氮酶的活性 通过清除活性氧降低抗氧化伤害 增加植物激素（如生长素、赤霉素）和非酶抗氧化物质的含量水平	AbdAllah 等（2015）
水稻（Oryza sativa）	幼套近明囊霉菌株 EEZ163（Claroideoglomus etunicatum EEZ163）	从胞质中排出 Na^+ 从木质部中卸载 Na^+ 并隔离到液泡中 降低根到茎中的 Na^+ 分布	Porcel 等（2016）
	幼套近明囊霉菌株 EEZ163（Claroideoglomus etunicatum EEZ163）	光合、气孔导度和蒸腾作用增强 提高 PSII 光化学量子产率并降低非光化学猝灭量子产率	Porcel 等（2015）
小麦（Triticum aestivum L.）	丛枝菌根真菌（Glomus mosseae）、根内球囊霉菌（Glomus intraradices）和丛枝菌根菌（Glomus etunicatum）	调节 Na^+ 和 Cl^- 的吸收 增加对必需大量和微量元素的吸收	Mardukhi 等（2015）
烟草（Nicotiana tabacum）	不规则嗜根菌（Rhizophagus irregularis）	通过细胞色素氧化酶途径增加呼吸作用 提高合成 ATP 的能力 提高植物的生物量 提高磷素的同化为 ATP 的合成和植物生长提供基础	Del-Saz 等（2017）
鹰嘴豆（Cicer arietinum L.）	摩西管柄囊霉菌（Funneliformis mosseae）	提高植物生长、产量和营养吸收 保护光系统 II 色素避免受到伤害 增加核酮糖 1,5-二磷酸羧化酶活性	Garg 和 Bhandari（2016）

（续表）

作物	真菌菌株	作用机理	参考文献
罗勒（Ocimum basilicum）	沙漠菌根真菌（Glomus deserticola）	提高 K/Na 和 Ca/Na 稳态平衡 提高叶绿素含量和光合能力 提高气体交换和水利用效率	Elhindi 等（2017）
	近明球囊霉菌（Glomus clarum Nicol. & Schenck）	产生更多的生物量 增加了营养元素、还原性糖、可溶性碳水化合物、光和色素、脯氨酸和蛋白的积累 减少了 Na^+ 的积累	Elhindi 等（2016）
木豆 [Cajanus cajan（L.）Millsp.]	摩西管柄囊霉菌（Funneliformis mossseae）和不规则嗜根菌（Rhizophagus irregularis）	提高对营养元素的吸收 显著提高生物量和产量 提高膜稳定性	Garg 和 Pandey（2015）
	摩西管柄囊霉菌（Funneliformis mosseae）	强化抗氧化系统 调节非酶组分物质含量和比率	Garg 和 Chandel（2015）
田菁（Sesbania cannabina）	摩西管柄囊霉菌株 BGC-NM03D（Funneliformis mosseae BGCNM03D）	降低 H_2O_2 含量 增加铁蛋白积累 调控累积铁蛋白增强耐盐性	Kong 等（2017）
碱茅（Puccinellia tenuiflora）	摩西管柄囊霉菌（Funneliformis mosseae）和幼套近明囊霉（Claroideoglomus etunicatum）	增加植物生物量 增加植物茎磷和钾的浓度 减少茎 Na^+ 浓度	Liu 等（2018）
芦竹（Arundo donax）	摩西管柄囊霉菌（Funneliformis mosseae）和不规则嗜根菌（Rhizophagus irregularis）	增加脯氨酸积累 提高异戊二烯和过氧化氢释放	Pollastri 等（2017）
大冀橙（Citrus macrophylla）	摩西管柄囊霉菌（Funneliformis mosseae）和根内球囊霉菌（Glomus intraradices）	降低胞内二氧化碳和氧化胁迫的水平 增加总叶绿素含量 增加叶子中水分含量	Navarro 等（2015）
羊草（Leymus chinensis）	丛枝菌根真菌菌株 BGCHEB02（Glomus mosseae BGCHEB02）	增加植物生物量 提高光合作用相关参数 提高光合色素含量	Lin 等（2017）

农作物根际细菌可以定殖在农作物根系，通过直接和间接两种机制来促进共生农作物的生长（Muthukumarasamy 等，2007）。农作物根际细菌可以通过多种形式促进农作物的生长，比如增加对农作物根系环境中磷矿和钾矿的溶解能力，

以释放更多的磷素和钾素，合成 1-氨基环丙烷-1-羧酸脱氨酶减少乙烯合成的前体，合成并向周围环境中释放吲哚-3-乙酸、胞外多糖、脯氨酸和铁载体等物质，促进农作物在盐渍化环境中的生长（Mohamed 和 Gomaa，2012；Nunkaew 等，2014；Palaniyandi 等，2014；Munoz 等，2014；Kang 等，2014a，2014b）。在众多的植物生长促生细菌中，芽孢杆菌（*Bacillus*）和假单胞菌（*Pseudomonas*）在提高农作物耐盐性方面应用最广泛。有研究指出，在 150 mmol/L 浓度的 NaCl 胁迫下，与未接种这类生长促生菌的农作物相比，接种恶臭假单胞菌 UW4（*Pseudomonas putida* UW4）的试验处理油菜的茎鲜重显著增加（Cheng 等，2007）。从这个实例中就可以看出，耐盐菌株的应用是减缓盐胁迫对植物生长影响的一种较好选择。因为耐盐微生物中也有耐盐基因的存在，因此这种能力也会世代传递。在生产中应用的耐盐细菌都是能够与农作物成功发生相互作用的植物生长促生菌。另外有研究报道，在使用绿色荧光蛋白标记的生脂固氮螺菌（*Azospirllum lipoferum*）接种小麦后，在盐渍化土壤上生长的小麦生长状况发生了明显地改善，表明促生微生物降低了盐胁迫对小麦生长的影响，促进了小麦植株的生长（Bacili 等，2004）。

同时，为了减少盐胁迫对植株生长的不利影响，农作物也会采取一些措施降低受到的盐胁迫危害。比如，有农作物会在细胞液泡中积累大量的离子，通过积累的大量离子提高胞内渗透势并维持离子平衡。有研究报道，为了中和液泡中积累的大量 Na^+ 离子对植株生长的影响，农作物将有机溶质和 K^+ 大量积累到细胞质中，维持了细胞中的离子稳态，提高了农作物在盐胁迫环境下的生存能力（Hasegawa 等，2000）。在盐胁迫条件下，接种植物生长促生细菌能够有效降低农作物中的 Na^+/K^+ 比值。因为植物生长促生细菌能够提高农作物组织对 K^+ 的吸收，从而间接地降低了细胞和组织中 Na^+ 的浓度，使细胞和组织中维持合适的 Na^+/K^+ 比值，避免离子平衡被打破。施用植物生长促生菌之后，农作物细胞和组织中的 Na^+/K^+ 比值降低表明这类细菌有提高农作物对 K^+ 吸收的能力（Govindarajan 等，2006）。

在盐胁迫条件下促进农作物体内有机溶质的积累也是植物生长促生菌促进农作物生长的一种有效机制。在胁迫条件下，植物生长促生菌能够诱导农作物产生渗透相容溶质。渗透相容物质一半都是一些低分子量的化合物，这些化合物在细胞内大量积累，能够提高细胞内的渗透势，从而保护细胞免受胞外高渗环境的胁迫。有研究指出，在盐胁迫环境中分离到的固氮螺菌（*Azospirillum*）中有大量的甘氨酸、脯氨酸、甜菜碱、谷氨酸和海藻糖的积累，说明这些物质能够有效地减少盐胁迫对微生物生长的影响（Tripathi 等，2002）。此外，农作物接种植物生长促生菌，能显著提高盐胁迫下农作物叶片和根系中可溶性糖和脯氨酸的含量

（Azarmi 等，2016）。这说明这些渗透相容溶质的积累能够缓解农作物遭受的盐胁迫危害，在一定程度上保证农作物的正常生长。同样地，细胞中的可溶性蛋白通过维持渗透调节、细胞内外水势和调节细胞膨胀性等改善农作物遭受的盐胁迫危害也有大量的报道（Tester 和 Davenport，2003）。通过给农作物接种耐盐细菌菌株，利用微生物与农作物协同互作效应，增加农作物细胞内水势、减少电解质渗漏和降低 Na^+ 浓度，改善了盐胁迫下农作物的生长状况（Kang 等，2014a，2014b）。

根际细菌合成的胞外多糖能够促进土壤团聚体的形成，改善土壤状况也有利于盐胁迫条件下农作物的生长。因此，高密度的根际胞外多糖产生菌在农作物根际周围的集聚有助于减少盐胁迫对农作物生长的影响。也有研究指出，土壤细菌通过产生胞外多糖促进了土壤结构的稳定和聚集的形成（Gouzou 等，1993）。此外，植物生长促生菌产生的胞外多糖对降低农作物吸收 Na^+ 的效果十分明显。接种能够产生胞外多糖的门多萨假单胞菌（*Pseudomonas mendocina*）可以降低农作物对 Na^+ 的吸收，从而改善盐胁迫环境下农作物的生长（Kohler 等，2006）。Ashraf 等（2004）也观察到接种了能够产生胞外多糖的菌株的农作物其体内的 Na^+ 含量显著减少。

另外，植物生长促生菌对农作物的促生作用还表现在能够通过产生各种不同的化合物刺激农作物的生长。这些化合物包括植物激素类物质，如生长素、赤霉素、细胞分裂素、铁载体以及各种抗菌化合物等，它们能够在减轻农作物遭受到的生物和非生物胁迫等方面发挥重要作用（De-Salamon 等，2001；Mitter 等，2013；Cassan 等，2014）。有研究报道指出，用耐盐细菌拌种小麦种子后，在盐渍化土壤上种植小麦可以观察到小麦幼苗生长良好，并且含有较高的生长素含量（Afrasayab 等，2010）。Mazhar 等（2013）还报道了接种蓝藻细菌（Cyanobacteria）能够产生生长素促进农作物的生长，提高农作物幼苗的生物量并促进根系和茎的生长，减轻农作物受到的盐渍化危害。

在盐胁迫条件下，农作物因为应激响应会在细胞内合成大量的乙烯，而过高浓度的乙烯会对农作物根系的生长产生非常严重的负面影响，主要表现在抑制农作物根系的伸长和根数量的生长（Abeles 等，1992）。而在农作物根际定殖的根际细菌产生的多种促生长物质中，1-氨基环丙烷-1-羧酸脱氨酶的存在能够降低受到盐胁迫危害的农作物细胞中乙烯的含量，解除高浓度乙烯对农作物根系生长的抑制，从而使农作物在盐胁迫环境中保持正常的生长（Nadeem 等，2010a，2010b；Glick，2010）。能够产生 1-氨基环丙烷-1-羧酸脱氨酶的促生菌能够降低农作物细胞内的乙烯水平，从而促进农作物生长的原因在于 1-氨基环丙烷-1-羧酸脱氨酶能够水解乙烯合成的直接前体，即 1-氨基环丙烷-1-羧酸。乙烯的合成前体减少了，因此农作物细胞内能够合成的乙烯也就相应减少。农作物细胞内乙烯浓度降低之后，解除了对根系生长的抑制，从而刺激根系伸长，进而促进农作物生长（Glick 等，1995，1999）。有研究报道指出，在 75 mmol/L 的 NaCl 浓度胁迫下，给黄瓜幼苗接种能够产生 1-氨基环丙烷-1-羧酸脱氨酶的细菌后，黄瓜的生长得到了明显改善（Gamalero 等，2010）。同样地，除了能够增加番茄的根系和幼苗地上部生物量之外，接种可产生 1-氨基环丙烷-1-羧酸脱氨酶的植物生长促生菌还能够提高盐胁迫下番茄种子的发芽率（Chookietwattana 和 Maneewan，2012）。有报道指出，耐盐细菌莱比托游动球菌（*Planococcus rifietoensis*）在盐胁迫条件下促进了小麦的生长和产量（Rajput 等，2013）。该菌株能在浓度为 65 g/L 的 NaCl 溶液下生长，并能利用 1-氨基环丙烷-1-羧酸作为唯一的氮源，在这种情况下，农作物的生物量增加 37%。此外，该细菌还具有溶磷特性，能够分泌生化物质，将不溶于水的磷酸钙溶解来获取磷素营养，促进农作物生长，通过该途径莱比托游动球菌（*Planococcus rifietoensis*）对农作物生长的增强效率达到 63%。当通过培养基培养细菌时，在培养基中添加 1-氨基环丙烷-1-羧酸后，该细菌的生长提高率超过 60%，这说明菌株的促生长能力还取决于其他因素。此外，也有很多研究人员都提到过，在盐胁迫条件下产 1-氨基环丙烷-1-羧酸脱氨酶细菌具有降低农作物因应激反应而过量产生乙烯的潜力（Mayak 等，2004；Glick 等，2007；Nadeem 等，2010a，2010b；Shin 等，2016）。

盐胁迫对农作物的生长造成不利影响还表现在能够通过产生活性氧对农作物造成氧化损伤（Zhu 等，2007）。而为受到盐胁迫危害的农作物接种耐盐细菌，则可以利用耐盐细菌与农作物的协同作用提高农作物细胞内抗氧化酶的活性，进而减轻活性氧对农作物细胞的氧化损伤。这样的研究已经有过报道。有研究指出，利用能够促进植物生长的耐盐细菌接种秋葵后，秋葵细胞内的抗氧化酶活性得到提高，通过清除活性氧物质增强了秋葵对盐胁迫的耐受性。此外还观察到了秋葵对水分的利用效率也相应提高，这可能也是秋葵增强对盐胁迫耐受性的另外一个原因（Habib 等，2016）。农作物在遭受盐胁迫时，接种植物生长促生菌的农作物细胞内的抗氧化酶蛋白浓度明显提高，这些抗氧化酶包括过氧化物酶、过氧化氢酶以及超氧化物歧化酶等（Sen 和 Chandrasekhar，2015）。另外，Jha 和 Subramanian（2013）也通过试验发现，在盐胁迫下给农作物接种植物生长促生菌，可以改善农作物的生长状况。这种生长状况的改善也是由于接种的耐盐细菌调节了农作物细胞内的离子浓度，并提高了抗氧化酶的活性。

与耐盐细菌相似，丛枝菌根真菌与农作物也能够发生非常密切的互作效用，也是能够提高农作物抗盐能力的一种有益微生物。丛枝菌根真菌能够通过直接和间接的机制保护农作物免受生物和非生物等逆境胁迫。将丛枝菌根真菌（*Glomus mosseae*）在柑橘（*Citrus reticulata*）上接种以后，发现柑橘的根系形态、光合活性以及离子稳态都得到了提高，并且其耐盐能力也得到了改善（Wu 等，2010）。类似地，也有很多研究人员将不同的耐盐细菌接种到各种农作物上进行试验，接种后的农作物基本上都表现出根系生长提高，地上部鲜重和干重生物量增加，以及对水分的吸收和利用率的提高（Ghoulam 等，2002；Cho 等，2006；Al-Karaki，2006）。

目前，对丛枝菌根真菌提高宿主农作物抗盐能力机制的研究已经有了很多深入的报道（Liu 等，2015a，2015b；Wu 等，2014；Meng 等，2015；De Almeida 等，2016）。丛枝菌根真菌的菌丝从侵染农作物根系开始，与农作物建立共生关系。真菌菌丝渗透到农作物根系皮层细胞后大量增殖，并形成专门的结构，即丛枝。丛枝菌根真菌在宿主农作物根系定殖后，真菌的菌丝体开始向土壤溶液中渗透，在外部环境中寻找水分和养分（Breuninger 和 Requena，2004；He 和 Nara，2007）。有研究指出，在盐胁迫条件下，丛枝菌根真菌的外生菌根和内生菌根都能显著提高接种农作物植株的幼苗生长和生物量（Diouf 等，2005）。丛枝菌根真菌还能够增强农作物的抗氧化系统，以减轻因盐胁迫诱导而在农作物细胞内产生的各类活性氧物质对细胞产生的有害影响，并减少因氧化损伤而带来的盐胁迫对农作物生长的负面影响（Rabie 和 Almadini，2005；Manchanda 和 Garg，2011；Wu 等，2014；Ahmad 等，2015a，2015b）。丛枝菌根真菌还能够通过其他途径来提高农作物耐受盐胁迫的能力，这些途径包括改善农作物养分获取（尤其是磷）的方式（Meng 等，2015；Evelin 等，2009）、诱导获得性系统耐受性（Hashem 等，2016a）、提高农作物的水分吸收率（Ruiz Lozano 和 Azcon，2000）、提高农作物根系渗透压（Ibrahim 等，2011；Evelin 等，2013）、减少农作物根系对有害离子的吸收（如 Cl^- 和 Na^+）（Al-Karaki，2006；Daei 等，2009）、通过提高对 K^+ 的吸收来维持农作物细胞内的离子平衡（Evelin 等，2012）、提高农作物对水分的使用效率（Hajiboland 等，2010），以及改变农作物的根系形态以增加水分和养分的吸收（Aroca 等，2013；Ahanger 等，2014）等。

有研究指出，丛枝菌根真菌与黄瓜共生后能够降低盐胁迫对黄瓜生长的负面影响（Hashem 等，2018）。研究人员通过试验证实，丛枝菌根真菌是通过诱导提高农作物抗氧化系统的活性来增强农作物清除细胞内各种活性氧物质，从而减少活性氧对细胞造成的各种氧化损伤。许多研究者的早期研究也表明，丛枝菌根真菌与农作物共生后改善农作物的抗氧化系统，有助于提高农作物对盐胁迫的耐受能力，从而缓解盐胁迫对农作物生长造成的危害，维持盐胁迫环境中农作物的生长（Alguacil 等，2003；He 等，2007）。Sarwat 等（2016）证明了芥菜幼苗的耐盐性正是由于抗氧化系统得到改善、生理生化参数重组以及次生代谢物和植物激素的产生而获得的。Yang 等（2014）也通过研究指出，在盐胁迫条件下，丛枝菌根真菌接种的农作物中抗坏血酸过氧化物酶和过氧化氢酶的活性都比未接种的农作物中的活性更强。Abd Allah 等（2015）在研究盐胁迫条件下丛枝菌根真菌对田菁生长的影响时发现，田菁的生长状况和生物量（产量）的改善是由于细胞内抗氧化系统和非酶抗氧化剂的激活所导致的。

超氧化物歧化酶能够清除细胞内的超氧化物活性氧，从而保证农作物细胞膜免受氧化损伤的危害。另外，与不规则嗜根菌（*Rhizophagus irregularis*）共生的农作物具有更高的超氧化物歧化酶活性，可以降低细胞内脂质过氧化水平，从而保护农作物细胞膜免受损伤。与丛枝菌根真菌共生的农作物细胞中活性氧水平较低是由于活性氧清除酶（包括超氧化物歧化酶、过氧化氢酶和愈创木酚过氧化物酶等）的快速产生和激活所导致的（Pandey 和 Garg，2017）。而在与不规则嗜根菌（*Rhizophagus irregularis*）共生的农作物中，研究人员发现异戊二烯的排放量越高，农作物细胞中各类活性氧物质和过氧化氢的积累就越多（Pollastri 等，2017）。而在番茄中，由于丛枝菌根真菌与根系的结合会诱导农作物产生大量的抗氧化剂，如超氧化物歧化酶和过氧化氢酶等，能够迅速降解细胞内的各种活性氧物质，从而缓解农作物受到的盐胁迫危害（He 等，2007）。

此外，盐胁迫还影响农作物的多种生理过程，比如气体交换、光合作用、水分和养分的吸收、物质的跨膜运输等（Aroca 等，2006；Porcel 等，2006）。丛枝菌根真菌对提高盐胁迫下农作物根系对水分和养分的吸收和运输具有积极作用，农作物通过提高对水分和养分的吸收和分配来改善身处盐胁迫环境带来的生长压力，并积极促进自身的生长和发育（Ruiz Lozano 和 Azcon，2000；Ruiz Lozano，2003）。另外，丛枝菌根真菌（*Glomus intraradices*）在农作物根系的定殖也能够起到稀释农作物根际环境盐分的作用，降低根系周围的渗透势，保护农作物根系细胞免于脱水。此外，与丛枝菌根真菌共生的农作物具有较低的水分饱和临界点，并且丛枝菌根真菌对水分的吸收也可以被农作物利用，因此丛枝菌根真菌与农作物的共生改善了农作物对水分的吸收和利用状况（Sheng 等，2008；Al Garni，2006）。有研究指出，在接种了丛枝菌根真菌的农作物中检测到了相对较高的含水量，这是由于丛枝菌根真菌的存在提高农作物根系对水分的吸收能力（Sheng 等，2008；Jahromi 等，2008；Kapoor 等，2008）。在盐胁迫条件下，正如以番茄为研究对象进行的试验所得到的结果那样，接种丛枝菌根真菌后会对农作物的各种生理过程产生影响，比如减少农作物对毒性离子的大量吸收，通过这些生理过程的变化来保护农作物免受盐胁迫的危害，并改善盐胁迫条件下农作物的生长，丛枝菌根真菌减轻了盐胁迫对农作物生长的负面影响（Balliu 等，

2015；Hashem 等，2016）。在盐胁迫条件下，接种丛枝菌根真菌的瓜类植物与未接种的处理相比，能够明显提高瓜类植物对水分和养分的吸收和利用状况，可以更好地提高瓜类植物的生长、产量以及质量（Colla，2008）。另外有研究指出，与丛枝菌根真菌共生的莲藕的耐盐能力也得到了很大程度的提升（Sannazzaro 等，2007）。在接种丛枝菌根真菌农作物的根细胞中，其电导率值也较未接种的处理要高，这说明接种后农作物根细胞中离子含量很丰富（Garg 和 Manchanda，2008）。这种情况在玉米中也有相关报道。有研究指出，接种丛枝菌根真菌的玉米根细胞质膜的电解质通透性较高，产生这种现象的原因可能是丛枝菌根真菌激活了玉米细胞内的抗氧化酶活性，并且提高了其对磷素的吸收能力，从而促进了根细胞对水分和营养物质等的吸收，因此根细胞质膜的通透性发生了变化（Feng 等，2002）。

植物在光合作用中二氧化碳的固定速率是植物获取生长动力的一个重要的原因（Querejeta 等，2007）。而接种丛枝菌根真菌能够缓解盐胁迫对农作物造成的生长危害的重要原因之一是丛枝菌根真菌能够恢复或提高盐胁迫条件下农作物降低的光合速率。接种丛枝菌根真菌的农作物在盐胁迫条件下具有更强的叶黄素活性，这是由于在盐胁迫条件下丛枝菌根真菌诱导农作物产生了特定的酶所致（Hajbagheri 和 Enteshari，2011）。在盐胁迫条件下，接种丛枝菌根真菌的农作物叶片中的叶绿素含量一般都很高（Sannazzaro 等，2006；Sheng 等，2008）。有研究指出，丛枝菌根真菌与玉米共生能够增加盐胁迫条件下玉米植株中叶绿素的含量和光合能力（Sheng 等，2008）。植物中光合系统效率的提高是一个复杂的过程，依赖多种不同的生理生化途径共同完成。Borde 等（2010）观察到，在盐胁迫条件下接种丛枝菌根真菌的大蒜植株的光合作用速率明显增强。接种丛枝菌根真菌的农作物还表现出对镁素吸收的增加。众所周知，镁素是叶绿素发挥正常功能所必需的元素，农作物对镁素吸收的增加则进一步促进了叶绿素的合成。为了能够更好地促进农作物提高光合作用，与丛枝菌根真菌共生的农作物还能够抑制 Na^+ 的运输，从而减轻在叶片内积累过量的 Na^+ 导致的毒害作用，从而提高光合作用速率（García-Garrido 和 Ocampo，2002）。Sharma 等（2017）通过试验观察到，接种丛枝菌根真菌能够促进农作物对营养元素的吸收，不仅如此，农作物中的光合色素、磷酸酶和过氧化物酶的活性等也都受到积极影响，因此农作物的耐盐能力得到了提高。另外，研究者还发现，丛枝菌根真菌与农作物共生后也会减少细胞质膜的损伤。Elhindi 等（2017）也通过试验证明了丛枝菌根真菌能够通过改变农作物的气体交换、光合效率以及水分利用效率等机制促进农作物在盐胁迫条件下的生长。

此外，丛枝菌根真菌还可以通过改变农作物自身的激素水平来减少盐胁迫对农作物生长产生的负面影响。脱落酸是植物中一种非常重要的生长激素，能够在农作物耐旱和耐盐中发挥重要的作用。有研究指出，丛枝菌根真菌有能力改变接种农作物细胞内脱落酸的含量水平（Bothe 等，1994；Estrada-Luna 和 Davies，2003），这也是丛枝菌根真菌提高农作物耐盐胁迫能力的一种重要机制。类似地，在接种丛枝菌根真菌的农作物的根和芽中也鉴定到了较高浓度的细胞分裂素（Allen 等，1980）。Allen 等（1980）通过研究还指出，盐胁迫条件下由于细胞分裂素含量的升高，从而促进农作物生长。

逆境胁迫下农作物中碳水化合物的积累是维持其正常生长的另一个重要机制。在盐胁迫条件下，用束状球囊霉菌（*Glomus fasciculatum*）接种澳洲芦苇后观察到，澳洲芦苇中碳水化合物的含量大量增加（Al-Garni，2006）。Porcel 和 Ruiz Lozano（2004）的研究也指出，在盐胁迫条件下，接种了丛枝菌根真菌（*Glomus intraradices*）的大豆中碳水化合物的含量也大量增加，得到了相似的结果。Evelin 等（2009）研究指出，在逆境胁迫条件下，接种了丛枝菌根真菌的农作物与未接种丛枝菌根真菌的农作物相比，在盐胁迫条件下植株中可溶性糖的含量更高。

盐胁迫降低了农作物对矿物养分尤其是磷素的吸收，因为磷素在盐渍化土壤中极易与钙和镁等金属离子形成不溶于水的沉淀，降低了土壤中磷素的可用性。由于丛枝菌根真菌的菌丝体直径很小，可以延伸到农作物根际范围之外并且渗透到土壤中，从而间接增加了农作物根系的表面积。这种丛枝菌根真菌细密的菌丝在改良了农作物根系表面积的同时，也有助于农作物从逆境环境中吸收足够的养分和水分。据估计，通过丛枝菌根真菌菌丝从周围土壤溶液中吸收的营养可满足农作物约 80% 的磷素需求（Matamoros 等，1999）。因此，接种了丛枝菌根真菌的农作物比未接种丛枝菌根真菌的农作物在逆境条件下具有更好的竞争优势，能够在盐胁迫等逆境环境中更好地生长。Plenchette 和 Duponnois（2005）以及 Sharifi 等（2007）也通过研究证实了可以通过改善农作物养分获取，特别是磷素的可获得性改善农作物的生长。

丛枝菌根真菌在盐胁迫条件下对维持与其共生的农作物中的离子稳态也很重要（Estrada 等，2013）。丛枝菌根真菌能够通过改善农作物对营养元素的吸收提高农作物组织中的 K^+/Na^+ 比值、提高光合作用和水分利用效率、增加渗透物质的产生、对离子进行区域化储存以及维持各种代谢过程中参与酶的活性等方式，增强农作物对盐胁迫的耐受性（Rabie 和 Almadini，2005；Al-Karaki，2006；Porcel 等，2012）。组织中高 K^+/Na^+ 比值是耐盐性的一个重要指标，接种丛枝菌根真菌有助于在盐胁迫条件下保持农作物组织中的高 K^+/Na^+ 比值（Giri 等，2007；Zhang 等，2011）。有研究报道指出，沙漠菌根真菌（*Glomus deserticola*）在盐胁迫条件下可以提高罗勒（*Ocimum basilicum*）植株中 K^+/Na^+ 和 Ca^{2+}/Na^+ 的比例，并且提高植株对水分的利用效率（Elhindi 等，2017）。Yang 等（2014）研究也发现，接种丛枝菌根真菌的农作物中的 K^+/Na^+ 比例比未接种丛枝菌根真菌的农作物中的要高。这是因为丛枝菌根真菌通过抑制农作物对 Na^+ 和 Cl^- 的吸收，并且降低这些离子向质膜、液泡膜的运输以及向芽等幼嫩组织的移动，降低盐胁迫对农作物的危害（Al-Karaki，2006；Lee 等，2015）。

丛枝菌根真菌能够在最大程度上降低盐分胁迫对农作物生长的影响并提高农作物的生产能力（Evelin 等，2009；Garg 和 Pandey，2015）。有报道指出，丛枝菌根真菌能够与豆科植物田菁形成共生关系，提高了田菁对盐胁迫的耐受能力（Ren 等，2016）。为了能够成功地建立农作物-丛枝菌根真菌联合体，双方需要在分子水平上进行物质交流（Andreo Jimenez 等，2015）。丛枝菌根真菌能够诱导提高农作物植株中粗内酯（strigolactone，SL）的水平，并因此诱导提高了农作物的抗盐性，甚至能够在一定程度上恢复因逆境胁迫而损失的生物量和生理损伤（Kong 等，2017）。独脚金内酯作为植物根际信号分子在物种互作中的作用已经有很多研究报道（Xie 等，2010）。在盐胁迫条件下，接种了不规则嗜根菌（*Rhizophagus irregularis*）的生菜植株中也鉴定到了独脚金内酯水平的升高（Aroca 等，2013）。据报道，农作物不仅会产生独脚金内酯，还会根据农作物的种类不同、遭受到的胁迫因素不同等而产生各种不同的内酯成分（Xie 等，2010；Ruyter-Spira 等，2013）。另外，还有很多研究已经证明，低浓度水平的 H_2O_2 也是丛枝菌根真菌用来促进农作物生长和抵抗盐胁迫的一种机制（Hajiboland 等，2010；Garg 和 Bhandari，2012）。因为 H_2O_2 是一种过氧化物，浓度过高会引起氧化损伤作用，危害农作物的正常生长；而低浓度的 H_2O_2 则可以起到信号分子的作用，在农作物与农作物之间，或者农作物与微生物之间进行信息交流（Xia 等，2009；Torres 和 Dangl，2005）。例如，在大麻幼苗中，丛枝菌根真菌在幼苗根际定殖后，遇到盐胁迫条件时幼苗根中会有 H_2O_2 的积累。独脚金内酯会通过一系列复杂的信号转录途径诱导农作物产生耐盐性。在盐胁迫条件下，H_2O_2 作为一种常见的信号分子，其产生是接种丛枝菌根真菌的农作物 NADPH 氧化酶激活的结果（Kong 等，2017）。农作物根系直接接触土壤溶液，是植株中受盐胁迫危害最为严重的部位，因此农作物根系要比叶片产生更多的活性氧物质。这些活性氧物质比如 O_2^- 和 H_2O_2 有可能会引起植株细胞器和细胞膜中脂质膜系统的过氧化，从而造成细胞损伤，影响细胞正常功能的发挥（Pedranzani 等，2016）。

丛枝菌根真菌菌株摩西管柄囊霉（*Funneliformis mosseae*）和嗜根菌（*Rhizophagus*）能够通过诱导农作物增加脯氨酸的产生来改善农作物受到的盐胁迫危

害。脯氨酸是众所周知的渗透调节物质之一，在各种逆境胁迫中都能起到缓冲的作用，降低农作物受到的胁迫危害（Porcel 等，2012；Pollastri 等，2017）。由于脯氨酸是一种良好的渗透相容性溶质，能够平衡农作物体内积累的盐分，因此其耐盐性作用目前被广泛研究（Szabados 和 Savoure，2010；Deinlein 等，2014）。目前，已有研究报道能够在接种丛枝菌根真菌的农作物中提取到了脯氨酸（Jindal 等，1993；Sharifi 等，2007）。Sharifi 等（2007）研究表明，在盐胁迫条件下，与未接种丛枝菌根真菌的大豆相比，接种的大豆芽中具有更高的脯氨酸含量。脯氨酸的积累可以对农作物根系在水分吸收和保持内外环境渗透平衡等方面起着非常关键的作用。甜菜碱是多种农作物在包括盐胁迫在内的逆境胁迫条件下产生的另一种渗透调节物质。有研究指出，接种了丛枝菌根真菌的农作物中甜菜碱的含量显著增加（Al-Garni，2006）。甜菜碱在逆境条件下能够起到稳定酶活、蛋白质复合物结构完整性以及膜系统完整性的作用，在农作物抵御盐胁迫方面同样也发挥着重要的作用。

有研究指出，梨状孢霉在逆境胁迫下会产生脱落酸和细胞分裂素，产生的这些激素会参与提高宿主农作物的耐盐性中（Crafts 和 Miller，1974；Nishiyama 等，2011）。脱落酸是一种植物激素，在植物的生长、发育，甚至是盐胁迫应激反应中都发挥着非常重要的作用。据报道，接种丛枝菌根真菌的农作中脱落酸的含量发生了变化（Estrada Luna 和 Davies，2003）。在逆境条件下真菌细胞也会产生 1-氨基环丙烷-1-羧酸脱氨酶，降解乙烯前体 1-氨基环丙烷-1-羧酸，保护农作物的正常生长，并通过调节 HKT1 基因的表达维持盐胁迫条件下细胞内的 Na^+ 稳态，从而增强农作物的耐盐能力（Contreras-Cornejo 等，2009；Viterbo 等，2010）。

接种丛枝菌根真菌的麻疯树在盐胁迫条件下其叶片含水量和根系导水性表现正常（Kumar 等，2015）。在温室模拟盐胁迫条件下，对两个椰枣品种（Nakhla hamra 和 Tijib）接种了 5 种不同的丛枝菌根真菌，即（丛枝菌根真菌（*Glomus intraradices*）、聚集球囊霉（*Glomus aggregatum*）、丛枝菌根真菌（*Glomus mosseae*）、维氏球囊霉（*Glomus verriculosum*）和束状球囊霉菌（*Glomus fasciculatum*），对这些真菌提高椰枣的耐盐性进行鉴定和评价（Diatta 等，2014）。结果表明，不同的椰枣品种在接种不同的丛枝菌根真菌后对盐胁迫的反应并不相同。在盐胁迫条件下，接种丛枝菌根真菌（*Glomus intraradices*）和束状球囊霉菌（*Glomus fasciculatum*）的椰枣生长较好。进一步的研究表明，椰枣品种的耐盐能力是由于脯氨酸等渗透相容溶质在植株细胞中的积累所致。而脯氨酸的积累浓度取决于用于接种的丛枝菌根种类和盐胁迫的强度。

Manga 等（2017）还发现，在盐胁迫环境下，丛枝菌根真菌对盐胁迫的反应也有不同的表现。在对 8 种丛枝菌根真菌提高金合欢耐盐性效果的研究中，研究人员观察到接种了丛枝菌根真菌的农作物吸收矿质营养的效率是存在变化的。其效率取决于接种的丛枝菌根类型和受到的盐胁迫水平。结果表明，用来接种的丛枝菌根真菌的类型、多样性和非生物胁迫强度都能够对农作物与丛枝菌根真菌的相互作用产生影响。在盐胁迫条件下接种丛枝菌根真菌不仅能够提高农作物的产量，而且还能够提高果实的品质（Kaya 等，2009）。

然而，我们必须清楚地认识，并不是所有的丛枝菌根真菌都具有减少盐胁迫对农作物生长的危害的能力。在某些特殊的盐胁迫条件下，丛枝菌根真菌也并不能保持正常的生长，其活性也会受到影响。在这种情况下丛枝菌根真菌并不能减轻农作物所遭受到的盐胁迫危害（Juniper 和 Abbott，2006；Sheng 等，2008）。丛枝菌根真菌在逆境环境下的生长也具有一定的差异性。Nazareth 等（2012）在研究从印度果阿红树林和盐场分离到的耐盐真菌对金属的耐受性时指出，与曲霉属真菌（*Aspergillus*）相比，青霉属（*Penicillium*）真菌的耐盐性和对重金属的抗性要好很多；而拟青霉属真菌（*Paecilomyces*）对盐胁迫的耐受性则较低。这说明真菌对逆境环境的耐受能力与种属类型有关。

10.4　耐盐微生物的协同应用

尽管植物生长促生菌在农业领域的应用已经有很长一段时间，但它们通过与本土微生物的区系竞争在农作物根际环境中的生存以及它们在根际环境中的有效定殖能力仍需进一步的研究，以使人们对植物生长促生菌在农业中的应用了解得更多（Bashan，1998）。此外，土壤溶液中较高浓度的盐分也有可能进一步导致微生物正常的生命活动受到抑制，不能与农作物发生有效的互作作用（Bremer 和 Kramer，2000）。在进行植物生长促生菌促进农作物提高耐盐能力的试验过程中，由于不能一次性引入很多的单一微生物菌株的种群数量，以及它们需要适应新的环境，并与土著微生物种群展开激烈的生存位的竞争，因此单一微生物菌株/制剂的性能在进行效果评估时，其鉴定结果经常表现不一致（Felici 等，2008）。此外，进行接种时不同的环境因素（包括生物和非生物环境）和接种后微生物因适应新环境而导致的数量迅速减少以及土壤环境中积累的大量微生物酶，也会导致接种引入的微生物菌株种群数量迅速减少（Johannes 等，1997），这些因素都会导致接种失败。另外，单菌种接种的有效性和稳定性也是接种失败的主要问题之一。因此，特别是在逆境环境条件下，提高农作物根际微生物的存活率是接种工作要面临的首要问题。

利用两种或多种有益微生物的组合进行接种，是提高微生物接种成功率和可靠性的重要保证。因为多样性的根际微生物群落能够建立更好的多种群协同机制，并在各种环境胁迫条件下提高生存能力（Larkin 和 Fravel，1998）。因此，多类型菌株微生物接种比单菌株接种对农作物生长具有更加明显的促进作用。多类型菌株微生物接种对农作物生长的刺激作用是由于两种或多种有益细菌和/或有益真菌/农作物根际之间的协同作用产生的。它们具有多种促进农作物生长的机制，如产生多种类型的植物激素促进农作物的生长、提高农作物对各种养分的利用率（Raja 等，2006）、增强固氮酶的活性（Alam 等，2001）以及提高微生物在农作物根际的定殖能力（Govindarajan 等，2007）等。接种混合微生物不仅可以改善正常条件下农作物的生长，而且还能够在逆境胁迫条件下诱导农作物产生系统性耐受性（Ruiz-Sanchez 等，2011）。表 10.3 列出了在盐胁迫条件下细菌和丛枝菌根真菌共同接种对不同农作物生长的影响。

表 10.3 植物生长促生菌和丛枝菌根真菌共接种诱导植物耐盐抗性

作物	真菌株系	植物生长促生菌株系	机理机制	参考文献
木豆（Cajanus cajan）	摩西管柄囊霉（Funneliformis mosseae）和不规则嗜根菌（Rhizophagus irregularis）	中华根瘤菌株 AR-4（Sinorhizobium fredii strain AR-4）	提高生物量积累，固氮能力，增加氮和磷的吸收 丛枝菌根菌定殖能力增强 通过降低海藻糖酶活性，增加真菌结节中海藻糖的含量	Garg 和 Pandey（2016）
玉米（Zea mays L.）	丛枝菌根菌（Glomus etunicatum）	甲基杆菌株系 CB-MB20（Methylobacterium oryzae CBMB20）	增加生物量干重和营养吸收 减少脯氨酸含量和 Na^+ 吸收	Lee 等（2015）
	珠状巨孢囊霉菌 S-23（Gigaspora margarita S-23）和幼套近明囊霉菌株 S-11（Claroideoglomus etunicatum S-11）	假单胞菌株系 S2CB35（Pseudomonas koreens S2CB35）	增加植物干重、增加根茎营养元素吸收 减少根茎中脯氨酸和 Na^+ 的积累 维持高 K^+/Na^+ 比率 改变 ZmAKT2、ZmSOS1 和 ZmSKOR 基因的表达量	Selvakumar 等（2018）
金合欢属（Acacia gerrardii Benth）	幼套近明囊霉菌（Claroideoglomus etunicatum）、根内嗜根菌（Rhizophagus intraradices）和摩西管柄囊霉（Funneliformis mosseae）	芽孢杆菌株系 BE-RA71（Bacillus subtilis BERA71）	增加丛枝菌根真菌的定殖能力 增加总脂质、酚类和纤维含量 渗透保护剂如甘氨酸、甜菜碱和脯氨酸含量增加 减少脂质过氧化	Hashem 等（2016b）

（续表）

作物	真菌株系	植物生长促生菌株系	机理机制	参考文献
野薄荷（Mentha arvensis L.）	聚集球囊霉（Glomus aggregatum）、丛枝菌根真菌（Glomus mosseae）、束状球囊霉菌（Glomus fasciculatum）和根内球囊霉菌（Glomus intraradices）	嗜盐单胞菌株系STR8（Halomonas desiderata STR8）和微杆菌属株系STR36（Exiguobacterium oxidotolerans STR36）	提高植物鲜重；增加植物油分产量；丛枝菌根菌定殖能力增强；增加氮素积累	Bharti 等（2016a，b）
燕麦（Avena sativa）	根内球囊霉菌（Glomus intraradices）	不动杆菌（Acinetobacter sp.）	降低植物中丙二醛和脯氨酸含量；增强抗氧化酶如 SOD、POD 和 CAT 的活性；增强土壤酶活	Xun 等（2015）
蚕豆（Vicia faba）	光壁无梗囊霉（Acaulospora laevis）、地球囊霉（Glomus geosporum）、丛枝菌根真菌（Glomus mosseae）和黄色盾巨孢囊霉（Scutellospora armeniaca）	豆科三叶草根瘤菌株系 STDF-Egyp19（Rhizobium leguminosarum biovar viciae STDF-Egyp19）	增加真菌结节数目、早呢更加固氮酶活性；增加丛枝菌根真菌的定殖能力；增加植物根茎的干重	Abd-Alla 等（2014）
菜豆（Phaseolus vulgaris L.）	丛枝菌根真菌（Glomus mosseae Nicol. & Gerd.）	荧光假单胞菌（Pseudomonas fluorescence）	提高植物茎中 K^+ 浓度并降低 Na^+ 浓度；提高茎中脯氨酸含量；提高 POX 和 CAT 等抗氧化酶活性	Younesi 和 Moradi（2014）
罗勒（Ocimum basilicum）	聚集球囊霉（Glomus aggregatum）	枯草芽孢杆菌（Bacillus subtilis）	对植物生长、油产量和营养吸收产生积极影响	AbdelRahman 等（2011）
	根内球囊霉菌（Glomus intraradices）	碱湖迪茨氏菌株系 STR1（Dietzia natronolimnaea STR1）	对本地微生物群落结构有积极影响；提高挥发油含量、出油率和中草药产量	Bharti 等（2016a，b）

将微生物直接接种到土壤中，微生物群落产生的多种代谢活性物质主要负责微生物与农作物根际进行相互作用识别，产生协同作用以后改善农作物的生长（Adesemoye 和 Kloeper，2009）。接种的微生物通过生物合成各种植物激素、促进

固氮作用以及促进农作物根系糖和氨基酸的释放增加等作用，多菌株微生物联合体对农作物生长的促进效应就可以得到增强（Raja 等，2006；Ruiz Sanchez 等，2011）。与单独接种和未接种微生物的对照相比，进行丛枝菌根真菌和固氮菌（*Azotobacter*）联合接种的小麦其粒重和穗粒数均有增加（Bahrani 等，2010）。小麦产量的提高可能与多菌株微生物联合接种对小麦根际土壤环境的改善和对小麦植株的刺激作用有关。另外，还有报道指出，中慢生根瘤菌（*Mesohizobium*）和芽孢杆菌（*Bacillus*）的联合接种提高了小麦的分蘖数、穗长和最终的粮食产量。Akhtar 等（2013）研究表明，联合接种根瘤菌（*Rhizobia*）和芽孢杆菌（*Bacillus*）能够提高农作物对磷利用的有效性。此外，还有研究指出，巴西偶氮螺菌（*Azospirillum brasilense*）和亚甲基杆菌（*Methylobacterium* sp.）共同接种番茄后，与单独接种这些菌株的番茄相比，番茄的生长明显得到加强（Joe 等，2014）。在莴苣的研究中，有报道称用能够产生抗坏血酸过氧化物酶和谷胱甘肽还原酶的沙雷氏菌（*Serratia* sp.）和根瘤菌（*Rhizobium* sp.）联合接种，能够提高莴苣的耐盐性。此外，与接种单一类型的微生物相比，联合接种还能提高农作物中的矿物质含量和光合速率（Han 和 Lee，2005）。

Afrasayab 等（2000）通过研究指出，在盐胁迫条件下微生物多菌株联合接种对小麦的生长有较大的促进作用。这可能是由于多种微生物联合接种促进了小麦生长素、可溶性蛋白的合成和对 K^+ 吸收的增加，并且减少吸收过量的 Na^+ 所致。Bulut（2013）研究报道指出，通过接种不同的微生物菌株，对小麦生长的促进作用几乎等同于施用化肥的效果。通过多菌株细菌联合接种比接种单一菌株更能促使小麦植株吸收更多的氮素，从而促进小麦的生长。由荧光假单胞菌（*Pseudomonas flourescens*）、生脂固氮螺菌（*Azospirillum lipoferum*）和巨大芽孢杆菌（*Bacillus megaterium*）组成的细菌联合体能够通过改善农作物中氨基酸、还原糖和植物激素的生物合成以及固氮能力的提高，增加这些细菌菌株在水稻根际中的定殖效率（Raja 等，2006）。

大多数豆科农作物对盐胁迫表现敏感。丛枝菌根真菌与豆科农作物在盐胁迫下的共生关系提高了微生物在接种农作物根际的定殖能力，从而对农作物的生长产生显著的影响（Garg 和 Chandel，2011）。为评估丛枝菌根真菌和根瘤菌联合接种对苜蓿耐盐性的提升效果进行了一项试验研究，结果表明这两种菌株的联合接种显著增加了苜蓿的根瘤数量和重量，并提高了苜蓿的产量（Zhu 等，2016）。采用多菌种联合解菌的宿主农作物表现出较高的脯氨酸含量。丛枝菌根真菌与细菌之间的协同关系促进了土壤溶液中氮素和磷素含量的增加，从而增加了植物生长所必须的营养物质（Giri 等，2002）。

Koc 等（2016）还研究了细菌和丛枝菌根真菌联合应用对草莓耐盐性的影

响。盐浓度的增加能够显著降低农作物生长参数。丛枝菌根真菌与细菌联合应用接种改善了盐胁迫对草莓生长的负面影响。研究人员指出，微生物菌株的共接种通过提高农作物中脯氨酸和花青素的水平来保护农作物免受盐胁迫的伤害。他们还观察到，盐胁迫程度越高，这种影响就越明显。在 Rabie 等（2005）的早期研究中，丛枝菌根真菌在高盐浓度胁迫条件下对莴苣的促生长作用更加明显。在许多情况下，多种微生物菌种联合接种被证明是能够改善农作物生长的。除了多种微生物菌株的共接种之外，丛枝菌根真菌与多胺的联合应用也能够减轻盐胁迫对农作物生长的负面影响。Abdel Fattah 等（2013）通过在盐胁迫条件下进行的盆栽试验，评估了苔藓球藻和精胺联合接种对小麦生长的影响和小麦生理层面的变化。结果表明，苔藓球藻和精胺联合施用对小麦的生长具有积极影响。小麦耐盐性的提高与微生物在其根系的定殖、小麦植株中脯氨酸、蛋白质以及叶绿素含量的提高有关。

在逆境胁迫条件下，营养失衡对农作物的生长发育具有负面影响。盐胁迫环境中的高浓度 Na^+ 抑制农作物对 K^+ 和 Ca^{2+} 等必需营养元素的吸收。而接种丛枝菌根真菌和细菌则有助于提高农作物对养分的吸收效率并维持农作物植株中的养分平衡（Gamalero 等，2010）。Najafi 等（2012）报道了植物生长促生菌和丛枝菌根真菌联合接种增强了农作物对水分和养分的吸收。丛枝菌根真菌在农作物根际的定殖，有助于改善土壤养分，尤其是改善农作物根际磷素的有效性。接种丛枝菌根真菌可以提高豆科植物抗氧化酶的产生，也能够增强豆科植物的结瘤、固氮能力以及植株的生长和最终的产量（Garg 和 Manchanda，2008）。类似地，接种丛枝菌真菌还能够提高农作物对氮吸收和利用的有效性（Govindarajulu 等，2005）。根据研究人员的发现，这可能是由于接种的微生物通过改变与之生命活动相关的酶的活性和其他物质的分泌而引起农作物根际周围土壤中氮素含量的变化。除了氮素和磷素之外，给农作物接种丛枝菌根真菌还提高了农作物对其他必需营养元素（包括钙、镁、钾、铜、锰、锌和铁等）吸收和利用的有效性（Al-Karaki 等，2001；Mardukhi 等，2011）。

有研究指出，丛枝菌根真菌与细菌共接种可以促进三叶草的结瘤，并最终促进其生长（Meyer 和 Linderman，1986）。丛枝菌根真菌可以减轻盐胁迫对农作物固氮过程的有害影响，并促进农作物根系结瘤数的显著增加（Giri 和 Mukerji，2004；Manchanda，2008）。盐胁迫下农作物固氮能力的提高是由于固氮酶活性和血红蛋白含量的提高导致的。Soliman 等（2012）在研究丛枝菌根真菌与根瘤菌（*Rhizobium*）共接种对农作物生长的影响时指出，共接种微生物能够通过诱导宿主农作物减少 Na^+ 的吸收，并在细胞内积累脯氨酸的方式保护农作物减轻/免受盐胁迫对其生长造成的负面影响。

Hashem 等（2016b，2016c）通过研究农作物内生细菌与丛枝菌根真菌的相互作用，评估盐胁迫下共接种对金合欢属植物（*Acacia gerrardii*）生长的影响。结果表明，共接种可增加农作物根冠重、根系根瘤数和血红蛋白的含量。接种联合微生物的农作物和未接种农作物中固氮酶和亚硝酸盐还原酶的活性也有显著差异。接种联合微生物的农作物的磷酸酶活性明显得到提升，并且农作物对氮、磷、钾、镁、钙的吸收能力也有所提高，同时还降低了 Na^+ 和 Cl^- 在植株中的积累浓度，从而促进了盐胁迫下金合欢属植物（*Acacia gerrardii*）的生长。类似地，在盐胁迫下，丛枝菌根真菌和磷溶性真菌的联合接种对土壤中磷素含量、接种农作物植株的磷酸酶活性以及农作物对必需营养素的吸收和利用效率也有显著影响（Xueming 等，2014）。同时，研究者还指出，施行微生物联合接种的农作物根系周围富集了大量的溶磷细菌，溶磷菌能够通过多种途径将土壤中沉淀的磷素溶解出来，从而提高盐胁迫条件下可供农作物利用的磷素含量。

菌根圈是农作物根系和真菌菌丝共同形成的一个区域，是真菌、细菌以及农作物根系相互作用的场所（Johansson 等，2004）。该区域微生物种群间的相互作用不仅对农作物生长产生积极的影响，而且还能够促进彼此之间的生存能力（Artursson 等，2006；Yusran 等，2009）。在这种相互作用的过程中，细菌可以通过产生某些化合物来增强细胞膜的通透性，刺激真菌菌丝的生长，并促进真菌穿透农作物根系细胞完成真菌与农作物的相互作用（Jeffries 等，2003）。在大多数情况下，这些相互作用的发生是双向的。细菌与真菌的相互作用有助于真菌与农作物共生体的发育，并提高真菌在农作物根际的有效定殖能力（Hildebrandt 等，2002；Jaderlund 等，2008）。例如，将巴西类芽孢杆菌（*Paenibacillus brasilensis*）与丛枝菌根真菌（*Glomus mosseae*）共同接种，能够增强真菌在三叶草根际的定殖（Artursson，2005）；而丛枝菌根真菌则可增强细菌对土壤颗粒的解磷和增强农作物根系的固氮能力（Linderman，1992）。

丛枝菌根真菌与农作物之间的相互联系受到包括盐胁迫在内的多种环境逆境条件的强烈影响（Gupta 等，2000；Cavagnaro 等，2015）。在这些恶劣条件的影响下，丛枝菌根真菌与细菌的联合接种能够进一步提高丛枝菌根真菌在农作物根系的定殖效率和对农作物生长的促进能力。在盐胁迫条件下，耐盐费氏中华根瘤菌（*Bradyrhizobium*）与丛枝菌根真菌（*Glomus mosseae*）联合接种绿豆（*Vigna radiata* L.）后能够明显促进绿豆在盐胁迫条件下的生长（Singh 等，2011）。研究还指出，与单独接种丛枝菌根真菌相比，联合接种费氏中华根瘤菌（*Bradyrhizobium*）可以显著增加农作物茎和根的长度、根系上根瘤的数量、农作物总生物量，以及丛枝菌根真菌在农作物根系的定殖效率和数量。通过对这两种菌株的联合接种应用，绿豆（*Vigna radiata* L.）的生长和产量都得到了显著地提高。这说明丛枝菌根真菌与根瘤菌的协同作用对提高农作物固氮效率非常有效，并最终对豆科植物的生长和产量产生显著的增加效用。

通过以上讨论可以总结出，细菌和丛枝菌根真菌共接种有助于促进盐胁迫下农作物的生长。在微生物-农作物这种协同互作作用中，双方相互促进生长。在逆境胁迫条件下，联合接种对农作物生长的促进作用似乎更为有效。这是因为在恶劣的逆境环境胁迫下，单一菌株有时不能完全适应恶劣的环境，因此发挥不出对农作物生长促进的潜力。

10.5 耐盐微生物在环境科学中的作用

由于人类活动、人口增长，以及工业化和城镇化的不断发展，人们生存的自然环境所遭受的污染正在不断增加（Tabak 等，2005；Ahemad 等，2009）。在进行农药、化学品、石油以及天然气等产品的生产和加工过程中产生的废水会含有盐分离子，并且当这些废水释放到环境中会造成土壤或水体中盐分含量的不断波动。传统的微生物处理工艺不能完全适应盐度变化条件的污染处理，因为微生物只在一定程度的盐分环境中才能保持良好的活性。由于嗜盐细菌存在久远，他们参与过地球的生物化学进化过程，因此嗜盐细菌在处理环境中的盐分污染中能够发挥重要的作用（Capone，2002）。Das 等（2011）研究了盐渍化区域红树林中嗜盐微生物随土壤盐分的分布规律。根据他们的研究发现，分布在表层土壤中的微生物比分布在中下层土壤中的微生物对盐分波动的耐受性更强。同时他们还观察到，在土壤底部厌氧微生物的耐盐能力高于土壤表层的好氧微生物。

Lowe 等（1993）的研究表明，厌氧微生物能够更好地适应极端环境条件，比如 pH 值、温度、盐度等。这是因为厌氧菌比好氧菌具有更好的节能机制，使其在极端逆境环境下的适应能力变得更好。根据他们的研究成果，由于全球变暖所引起的水分中盐度的波动对需氧菌的影响要大于对厌氧菌的影响。微生物非常适于清理受到污染的环境，因为一些微生物能够分泌特殊的酶，这些酶可以非常高效地将环境中的一些造成污染的化合物进行分解，并且分解后的物质可以用来供这类微生物生长。因此，耐盐微生物由于具有特殊的酶和具有耐受恶劣环境的能力可以用作生物环境修复剂。Qin 等（2012）研究了盐胁迫条件下利用耐盐微生物对石油烃的生物修复。他们发现，微生物所处环境的盐分程度对生物修复的效果有很大影响。在初始阶段，中等程度的盐胁迫能够促进石油烃的生物降解，但随着盐胁迫程度的逐渐升高石油烃的降解速率逐渐下降。将从盐胁迫条件中分离到了两株真菌毛孢子菌属（*Trichosporon* sp.）和橘青霉（*Penicillium citrinum*）混合加入反应体系中，能够提高脱氢酶的活性，并提高石油烃的生物降解效率，这说明脱氢酶的活性与石油烃的降解效率呈正相关。试验结果也表明盐胁迫环境中添加耐盐菌剂，能有效地促进烃类物质的生物修复。

Abdughafurovich 等（2010）在好氧条件下研究了耐盐芽孢杆菌（*Bacillus*）菌株的生物修复潜力。研究采用氚标记多氯联苯的同源物，添加微生物后在高盐培养物中追踪氚标记。结果表明，大部分芽孢杆菌（*Bacillus*）菌株都能在一定程度上降解多氯联苯。气相色谱分析结果也表明了在添加微生物后多氯联苯发生了降解。这说明该菌株在盐渍化土壤下具有很好的生物修复功能。也有研究评价了在海水中添加或不添加生物表面活性剂和微量营养素对海水微生物降解红树林沼泽中的石油污染物的影响，以寻找海水微生物降解石油烃的最佳条件（Okoro，2010）。该项研究表明，应用生物表面活性剂和微量营养素能够提高微生物进行生物修复的效果。

Arakaki 等（2013）的研究证实盐胁迫能够抑制他们筛选到的 10 株生物修复菌株中的 7 株菌株的生长，另外 3 株菌株猪苓菌（*Polyporus* sp.）、雅致匙盖假花耳菌（*Dacryopinax elegans*）和赤藓（*Datronia stereoides*）在含有盐水的培养基中能够产生很高的生物量。在培养基中加入甘油和甘露醇，可明显提高微生物的生物量和木质素酶的活性。这些微生物在 NaCl 和多元醇存在的条件下仍可降解农药敌草隆。类似地，在另外的 26 株耐盐菌株中，有 2 株需氧芽孢杆菌表现出对四价硒胁迫的抗性和还原四价硒的潜力。这两株菌株均能够在 40 h 之内将四价硒完全还原为亚硒酸盐。在这些生物修复细菌的细胞内和周围发现了纳米级的球形硒颗粒，并通过共聚焦显微镜得到进一步证实。结果说明这些微生物菌株是完成硒生物修复的重要途径。

由于目前纺织工业的快速发展，工业废水中含有大量的合成染料，为了保护土壤免受污染，需要从废水中将这些染料去除，目前来说是一个不小的挑战。一旦这些染料排放到外界环境中，不仅会降低土壤质量，还会对地下水水质造成负面影响。而利用微生降解消除废水中的染料是一项很有前途和应用价值的技术。盐胁迫的程度对生物修复过程能够产生非常显著的影响，并对参与降解过程的微生物种群的生长产生不利影响，从而减缓降解的过程。即当工业废水中存在大量的盐分后，并非所有的细菌菌株都能有效地对这些染料化合物进行生物降解。而出于环境保护和控制技术成本，在这种条件下筛选合适的耐盐细菌是一种行之有效的方法（Parshetti 等，2006）。目前，研究人员已经分离出一批能降解偶氮染料的细菌。这些细菌菌株属于不同的细菌类型，包括微杆菌属（*Exiguobacterium*）、肺炎克雷伯氏菌（*Klebsiella pneumoniae*）、不动杆菌（*Acinetobacter*）和纤细芽孢杆菌属（*Gracilibacillus*）等（Salah 等，2007；Tan 等，2009；Bibi 等，2012；Agarwal 和 Singh，2012）。

从被污染的环境中进行微生物的分离，然后在同一环境条件下利用这些分离到的微生物进行生物降解，是一种环境友好且高效的技术策略。Lalnunhlimi 和 Krishnaswami（2016）分析了从盐胁迫环境中下分离得到的一些细菌菌株对不同浓度（100~300 mg/L）的偶氮染料的降解效果，对这些细菌菌株降解化学染料的能力进行了评估。细菌的生长和繁殖需要碳源和氮源，并且需要比较稳定的 pH 值和温度。有研究指出，以蔗糖和酵母抽提物作为碳源和氮源时，降解菌群对偶氮染料的脱色效果更为明显。Nicholson 和 Fathepure（2004）在有氧条件下评价了耐盐菌的生物降解能力。该研究以嗜盐菌为主体，对海洋杆菌降解苯、乙苯、甲苯和二甲苯的能力进行了研究。结果表明，这些嗜盐细菌在试验条件下完全降解了添加的苯、乙苯、甲苯和二甲苯。另外，研究人员还报道了以酵母提取物作为促进降解菌生长的营养物质，其降解能力显著增强。类似地，Oren 等（1992）和 Bertrand 等（1990）也通过试验证实了嗜盐古菌在盐胁迫条件下降解芳香烃和正构烷烃的能力。

Feng 等（2014）分离到了一株耐盐菌株库德里阿兹威毕赤酵母（*Pichia kudriavzevii*），他们通过试验对该菌株的盐胁迫适应能力、偶氮染料的降解能力，以及菌株生长的最适温度和 pH 值进行了评估。结果表明，该菌株的最适生长温度为33 ℃，最适生长 pH 值为 5.0，最适生长的盐浓度为 10%，在初始浓度为 100 mg/L 时，对污染物的降解率能够达到 94% 以上。Asad 等（2007）通过筛选研究，从 27 株耐盐菌筛选出 3 株菌株在不同浓度的 NaCl、不同 pH 值和不同温度下对偶氮染料有显著的脱色能力。因此，这些菌株有可能用于对化学染料的生物降解。同时，研究人员对微生物降解化学染料的机制进行了深入的研究，他们指出是由于化学染料分子的偶氮键在微生物作用下被还原，因此导致化学染料分子的降解。

多环芳烃具有的致癌性质，因而一旦泄漏到水体中就会对水环境和生物的发育造成严重的影响（Perugini 等，2007）。由于这些多环芳烃具有疏水性，因此这些化合物可以吸附在颗粒物上逐渐形成沉积物，给海洋生态造成严重的破坏（Yu 等，2005；Osuji 和 Ezebuiro，2006）。Arulazhagan 等（2010）研究了苍白杆菌属（*Ochrobactrum* sp.）、阴沟肠杆菌属（*Enterobacter cloacae*）和嗜麦芽糖寡养单胞菌（*Stenotrophomonas maltophilia*）所构成的耐盐菌群对萘、芴、蒽和菲等多种多环芳烃的生物降解能力。在 NaCl 浓度为 30 g/L 的高盐浓度条件下对不同浓度的多环芳烃（即 5 mg/kg、10 mg/kg、20 mg/kg、50 mg/kg 和 100 mg/kg）使用微生物进行生物降解以评估微生物联合体的降解潜力。研究表明，微生物联合体在 4 d 内就能够将多环芳烃完全降解，并且在培养液中田间酵母抽提物之后微生物的降解能力得到了提高，进一步促进了多环芳烃的降解。

Sowmya 等（2014）将分离得到的 35 株嗜盐菌按照对盐胁迫的耐受程度分为中度和极度耐盐菌。并且对分离到的嗜盐细菌进行了铅和镉的生长耐受性试验。结果表明，这些菌株对铅的耐受性比镉强。这种耐受性与培养基中盐的浓度有关。用所选的菌株进行重金属富集去除试验，结果表明采用的菌株对镉的富集效果要远远小于对铅的富集效果。进一步的研究表明，菌株对不同重金属富集效果的差异是由多种原因造成的，比如不同类型的重金属对菌株的毒性不同以及不同菌株之间存在的差异所致。另外，除细菌之外，也有研究发现，在盐胁迫条件下使用真菌也能够进行石油烃的降解修复（Qin 等，2012）。Zhang 等（2016）分离到了一株耐盐菌变幻棒状杆菌（*Corynebacterium variabile*），以生物炭为载体将该菌株固定，并观察菌株对不同类型石油烃的降解潜力。结果表明，该细菌对正构烷烃和多环芳烃的降解十分有效。类似地，Jenab 等（2016）研究了耐盐真菌对石油烃的生物降解能力。在 0、2.5% 和 5% 的盐度下，以原油和芘作为生物降解对象，该菌株在 7 d 内分别能够降解 77%、83.4% 和 77.4% 的原油；而在这些盐度条件下，该菌株对芘的降解效率分别为 82.4%、88.3% 和 95.2%。

上述讨论表明，耐盐微生物除了能改善农作物在盐胁迫条件下的生长外，还可以成功地用于修复受到污染的环境。这些微生物具有广泛的耐逆特性，使其能够在逆境条件下生存的同时，还具有净化各种土壤污染物的能力。耐盐微生物具有环境友好、成本低廉的特点，可以开发成为一种适合于商业化应用的用于环境污染物降解的产品。

10.6 展望

上述研究表明，农作物和微生物的生长都会受到盐胁迫的影响。文献报道的大多数研究也表明，微生物已经在进化过程中形成了很多特殊的机制，使它们能够在包括盐胁迫等逆境环境中生存和生长。尽管农作物本身也具有多种抗盐和耐盐的机制，但微生物与农作物的相互作用进一步协助了农作物在恶劣的生长环境中的生存。这些耐盐微生物由于具有多种促生长机制，因此在提高农作物耐盐性方面发挥着重要的作用。通过诱导 1-氨基环丙烷-1-羧酸脱氨酶的合成降低农作物中的乙烯胁迫，提高外源多糖对 Na^+ 吸附的有效性，增强农作物细胞中抗氧化系统的活性，以及促进农作物产生渗透调节保护剂等，都是减轻盐胁迫对农作物生长造成危害的有效机制，通过这些途径能够促使农作物在盐胁迫条件下良好地生长。农作物与这些微生物菌株通过共生关系可以提高农作物产生渗透调节相容溶质的能力，还能够提高抗氧化酶的活性、提高农作物根系对养分的获取和水分的吸收和利用，维持农作物植株的正常生长。本章综述了在盐胁迫条件下微生物接种对农作物生长的积极影响,并指出接种了微生物的农作物对盐胁迫的耐受性更强。通过总结其他研究还发现,耐盐微生物对于降解造成环境污染的有害化合物同样具有十分重要的作用。与非耐盐微生物相比,这些耐盐微生物菌株在盐胁迫条件下具有非常强的竞争优势,因为它们能够应对恶劣的环境胁迫条件,并且耐盐微生物还具有一些特殊的特性,使它们在生物修复/降解的过程中表现得更加出色。

从报道的文献中我们也能总结到,大多数的研究都是在无菌条件下进行的,且进行试验的环境条件是可控的。然而,关于这些微生物在自然环境中的作用情况目前的研究还很少。耐盐细菌作为生物肥料和生物修复剂在农业和环境生态领域的应用,在其最佳生存的环境条件、与本地微生物种群的竞争能力等方面还有待于进一步探索。同样地,无论是在实验室还是在野外条件下,都需要对多种菌株接种的效果进行评估,而不是单纯的使用某种单一菌株。另外,在使用多菌株混合物试验之前,需要先对它们之间的生存兼容性进行测试,以便获得满足微生物生长的最佳条件。为了能够使从自然环境中筛选到的微生物种群在实际应用中发挥出最大的价值,利用生物技术对目标微生物进行适当改造也是同样需要特别关注的课题。

参考文献

Abd-Alla MH, El-Enany AW, Nafady NA, Khalaf DM, Morsy FM, 2014. Synergistic interaction of *Rhizobium leguminosarum* bv. *viciae* and *Arbuscular mycorrhizal* fungi as a plant growth promoting biofertilizers for faba bean (*Vicia faba* L.) in alkaline soil. Microbiol Res 169: 49-58.

Abd-Allah EF, Hashem A, Alqarawi AA, Bahkali AH, Alwhibi MS, 2015. Enhancing growth performance and systemic acquired resi, stance of medicinal plant *Sesbania sesban* (L.) Merr using *Arbuscular mycorrhizal* fungi under salt stress. Saudi J Biol Sci 22: 274-283.

Abdel-Fattah GM, Ibrahim AH, Al-Amri SM, Shoker AE, 2013. Synergistic effect of *Arbuscular mycorrhizal* fungi and spermine on amelioration of salinity stress of wheat (*Triticum aestivum* L. cv. gimiza 9). Aust J Crop Sci 7: 1525.

Abdel-Rahman SS, Abdel-Kader AA, Khalil SE, 2011. Response of three sweet basil cultivars to inoculation with *Bacillus subtilis* and *Arbuscular mycorrhizal* fungi under salt stress conditions. Nat Sci 9: 93-111.

Abdughafurovich RB, Andreevich KA, Adrian L, Tashpulatovich YK, 2010. Biodegradation of tritium labeled polychlorinated biphenyls (PCBs) by local salt tolerant mesophilic*Bacillus* strains. J Environ Protect 1: 420-425.

Abeles FB, Morgan PW, Saltveit ME Jr, 1992. Ethylene in plant biology. Academic, London.

Adesemoye AO, Kloepper JW, 2009. Plant microbes interactions in enhanced fertilizer-use efficiency. Appl Microbiol Biotechnol 85: 1-12.

Afrasayab S, Faisal M, Hasnain S, 2000. Synergistic growth stimulatory effect of mixed culture bacterial inoculation on the early growth of *Triticum aestivum* var, Inqalab 91 under NaCl salinity. Pak J Biol Sci 3: 1016-1023.

Afrasayab S, Faisal M, Hasnain S, 2010. Comparative study of wild and transformed salt tolerant bacterial strains on *Triticum aestivum* growth under salt stress. Braz J Microbiol 41: 946-955.

Agarwal T, Singh R, 2012. Bioremedial potentials of a moderately halophilic soil bacterium. J Pharm Biomed Sci 19: 1-6.

Ahanger MA, Hashem A, Abd-Allah EF, Ahmad P, 2014. *Arbuscular mycorrhiza* in crop improvement under environmental stress. In: Ahmad P, Rasool S (eds) Emerging technologies and management of crop stress tolerance, vol 2. Academic, Amsterdam, pp. 69-95.

Ahemad M, Khan MS, Zaidi A, Wani PA, 2009. Remediation of herbicides contaminated soil using microbes. In: Khan MS, Zaidi A, Musarrat J (eds) Microbes in sustainable agriculture. Nova Science Publishers, New York, pp. 261-284.

Ahmad P, Hashem A, Abd-Allah EF, Alqarawi AA, John R, Egamberdieva D, Gucel S, 2015a. Role of *Trichoderma harzianum* in mitigating NaCl stress in Indian mustard (*Brassica juncea* L) through antioxidative defense system. Front Plant Sci 6: 1-15.

Ahmad P, Sarwat M, Bhat NA, Wani MR, Kazi AG, Tranl S, 2015b. Alleviation of cadmium toxicity in *Brassica juncea* L. (Czern. &Coss.) by calcium application involves various physiological and biochemical strategies. PLoS One 10 (1): e0114571.

Akcan N, Uyar F, 2011. Production of extracellular alkaline protease from *Bacillus subtilis* RSKK96 with solid state fermentation. Eurasia J Biosci 5 (64): 72.

Akhtar N, ArshadI SMA, QureshiMA SJ, Ali L, 2013. Co-inoculation with *Rhizobium* and *Bacillus* sp. to improve the phosphorus availability and yield of wheat (*Triticum aestivum* L.). J Anim Plant Sci 23: 190-197.

Akhter MS, Hossain SJ, Amir-Hossain SK, Datta RK, 2012. Isolation and characterization of salinity tolerant *Azotobacter* sp. Greener J Biol Sci 2: 42-51.

Al-Garni SMS, 2006. Increased heavy metal tolerance of cowpea plants by dual inoculation of an *Arbuscular mycorrhizal* fungi and nitrogen-fixer *Rhizobium* bacterium. Afr J Biotechnol 5: 133-142.

Al-Karaki GN, 2006. Nursery inoculation of tomato with *Arbuscular mycorrhizal* fungi and subsequent performance under irrigation with saline water. Sci Hortic 109: 1-7.

Al-Karaki GN, Hammad R, Rusan M, 2001. Response of two tomato cultivars differing in salt tolerance to inoculation with mycorrhizal fungi under salt stress. Mycorrhiza 11: 43-47.

Alam MS, Cui Z, Yamagishi T, Ishii R, 2001. Grain yield and related physiological characteristics of rice plants (*Oryza sativa* L.) inoculated with free-living rhizobacteria. Plant Prod Sci 4: 126-130.

Alguacil MM, Hernandez JA, Caravaca F, Portillo B, Roldan A, 2003. Antioxidant enzyme activities in shoots from three mycorrhizal shrub species afforested in a degraded semi-arid soil. Physiol Plant 118: 562-570.

Ali A, Shahzad R, Khan AL, Halo BA, Al-Yahyai R, Al-Harrasi A, Al-Rawahi A, Lee IJ, 2017. Endophytic bacterial diversity of *Avicennia marina* helps to confer resistance against salinity stress in *Solanum lycopersicum*. J Plant Interact 12: 312-322.

Aliasgharzadeh N, Rastin SN, Towfighi H, Alizadeh A, 2001. Occurrence of *Arbuscular mycorrhizal fungi in saline soils of the Tabriz Plain of Iran in relation to some physical and chemical properties of soil*. Mycorrhiza 11: 119-122.

Allen A, Cunningham GL (1983) Effects of vesicular-*Arbuscular mycorrhizae* on *Distichlis spicata* under three salinity levels. New Phytol 93: 227-236.

Allen MF, Moore JTS, Christensen M, 1980. Phytohormone changes in *Bouteloua gracilis* infected by vesicular-*Arbuscular mycorrhizae*: I. Cytokinin increases in the host plant. Can J Bot 58: 371-374.

Amir R, Ahmed N, Talat M, 1993. Salt-tolerance in mangrove soil bacteria. Pak J Mar Sci 2: 129-135.

Andreo-Jimenez B, Ruyter-Spira C, Bouwmeester H, Lopez-Raez JA, 2015. Ecological relevance of strigolactones in nutrient uptake and other abiotic stresses, and in plant-microbe interactions below-ground. Plant Soil 394: 1-19.

Arakaki RL, Monteiro DA, Boscolo M, Dasilva R, Gomes E, 2013. Halotolerance, ligninase production and herbicide degradation ability of basidiomycetes strains. Braz J Microbiol 44: 1207-1214.

Aroca R, Ferrante A, Vernieri P, Chrispeels MJ, 2006. Drought, abscisic acid and transpiration rate effects on the regulation of PIP aquaporin gene expression and abundance in *Phaseolus vulgaris* plants. Ann Bot 98: 1301-1310.

Aroca R, Ruiz-Lozano JM, Zamarreno AM, Paz JA, Garcia-Mina JM, Pozo MJ, Lopez-Raez JA, 2013. *Arbuscular mycorrhizal* symbiosis influences strigolactone production under salinity and alleviates salt stress in lettuce plants. J Plant Physiol 170: 47-55.

Artursson V, 2005. Bacterial-fungal interactions highlighted using microbiomics: potential application for plant growth enhancement (dissertation). Swedish University of Agricultural Sciences.

Artursson V, Finlay RD, Jansson JK, 2006. Interactions between *Arbuscular mycorrhizal* fungi and bacteria and their potential for stimulating plant growth. Environ Microbiol 8: 1-10.

Arulazhagan P, Vasudevan N, Yeom IT, 2010. Biodegradation of polycyclic aromatic hydrocarbon by a halotolerant bacterial consortium isolated from marine environment. Int J Environ Sci Technol 7: 639-652.

Asad S, Amoozegar MA, Pourbabaee A, Sarbolouki MN, Dastgheib SMM, 2007. Decolorization of textile azo dyes by newly isolated halophilic and halotolerant bacteria. Bioresour Technol 98: 2082-2088.

Ashraf M, Hasnain S, Berge O, Mahmood T, 2004. Inoculating wheat seedlings with exopolysaccharide-producing bacteria restricts sodium uptake and stimulates plant growth under salt stress. Biol Fertil Soils 40 (3): 157-162.

Augustine SM, Narayan JA, Syamaladevi DP, Appunu C, Chakravarthi M, Ravichandran V, Tuteja N, Subramonian N, 2015. Introduction of pea DNA helicase 45 into sugarcane (*Saccharum* spp. Hybrid) enhances cell membrane thermostability and upregulation of stressresponsive genes leads to abiotic stress tolerance. Mol Biotechnol 57: 475-488.

Azarmi F, Mozafari V, Abbaszadeh Dahaji P, Hamidpour M, 2016. Biochemical, physiological and antioxidant enzymatic activity responses of pistachio seedlings treated with plant growth promoting rhizobacteria and Zn to salinity stress. Acta Physiol Plant 38 (1):

1-16.

Bacilio M, Rodriguez H, Moreno M, Hernandez JP, Bashan Y, 2004. Mitigation of salt stress in wheat seedlings by agfp-tagged *Azospirillum lipoferum*. Biol Fertil Soils 40: 188-193.

Badri DV, Weir TL, van der Lelie D, Vivanco JM, 2009. Rhizosphere chemical dialogues: plant- microbe interactions. Curr Opin Biotechnol 20: 642-650.

Bahrani A, Pourreza J, Joo MH, 2010. Response of winter wheat to co-inoculation with *Azotobacter* and *Arbescular mycorrhizal* fungi (AMF) under different sources of nitrogen fertilizer. J Agric Environ Sci 8: 95-103.

Balliu A, Glenda S, Boris R, 2015. AMF inoculation enhances growth and improves the nutrient uptake rates of transplanted, salt–stressed tomato seedlings. Sustainability 7: 15967-15981.

Banik A, Pandya P, Patel B, Rathod C, Dangar M, 2018. Characterization of halotolerant, pigmented, plant growth promoting bacteria of groundnut rhizosphere and its in–vitro evaluation of plant-microbe protocooperation to withstand salinity and metal stress. Sci Total Environ 630: 231-242.

Bashan Y, 1998. Inoculants of plant growth promoting bacteria for use in agriculture. Biotechnol Adv 16: 729-770.

Beattie GA, 2007. Plant associated bacteria: survey, molecular phylogeny, genomics and recent advances. In: Ganamanickam SS (ed) Plant associated bacteria. Springer, Dordrecht.

Behera BK, Das P, Maharana J, Paria P, Mandal S, Meena DK, Sharma AP, Jayarajan R, Dixit V, Verma A, Vellarikkal SK, Scaria V, Sivasubbu S, Rao AR, Mohapatra T, 2014. Draft genome sequence of the extremely halophilic bacterium *Halomonas salina* strain CIFRI1, isolated from the East Coast of India. Genome Announc: e01321-14.

Behera BK, Das P, Patra A, Sharma AP, 2012. Transcriptome analysis of *Enterobacter aerogenes* KCTC 2190 in response to elevated salt stress In: Abstracts of 99th the Indian Science Congress, held at Bhubaneswar during 3-7th January.

Bertrand JC, Almallah M, Acquaviva M, Mille G, 1990. Biodegradation of hydrocarbons by an extremely halophilic archaebacterium. Lett Appl Microbiol 11: 260-263.

Bharti N, Barnawal D, Shukla S, Tewari SK, Katiyar RS, Kalra A, 2016a. Integrated application of *Exiguobacterium oxidotolerans*, *Glomus fasciculatum*, and vermicompost improves growth, yield and quality of *Mentha arvensis* in salt–stressed soils. Ind Crop Prod 83: 717-728.

Bharti N, Barnawal D, Wasnik K, Tewari SK, Kalra A, 2016b. Co-inoculation of Dietzia natronolimnaea and *Glomus intraradices* with vermicompost positively influences *Ocimum basilicum* growth and resident microbial community structure in salt affected low fertility soils. Appl Soil Ecol 100: 211-225.

Bibi R, Arshad M, Asghar HN, 2012. Optimization of factors for accelerated biodegradation of reactive Black-5 Azo dye. Int J Agric Biol 14: 353-359.

Blomberg A, Adler L, 1992. Physiology of osmotolerance in fungi. Adv Microb Physiol 33: 145-212.

Borde M, Dudhane M, Jite PK, 2010. AM fungi influences the photosynthetic activity, growth and antioxidant enzymes in *Allium sativum* L. under salinity condition. Not Sci Biol 2: 64-71.

Bothe H, Klingner A, Kaldorf M, Schmitz O, Esch H, Hundeshagen B, Kernebeck H,

1994. Biochemical approaches to the study of plant-fungal interactions in *Arbuscular mycorrhizas*. Experientia 50: 919-925.

Bremer E, Kramer R, 2000. Coping with osmotic challenges: osmoregulation through accumulation and release of compatible solutes in bacteria. In: Storz G, Hengge - Areonis R (eds) Bacterial stress response. ASM Press, Washington, DC, pp. 79-97.

Breuninger M, Requena N, 2004. Recognition events in AM symbiosis: analysis of fungal gene expression at the early appressorium stage. Fungal Genet Biol 41: 794-804.

Bulut S, 2013. Evaluation of efficiency parameters of phosphorous-solubilizing and N-fixing bacteria inoculations in wheat (*Triticum aestivum* L.). Turk J Agric For 37: 734-743.

Burg MB, Ferraris JD, Dmitreva NI, 2007. Cellular response to hyperosmotic stresses. Physiol Rev 87: 1441-1474.

Cai F, Chen FY, Tang YB, 2014. Isolation, identification of a halotolerant acid red B degrading strain and its decolorization performance. Apcbee Procedia 9: 131-139.

Capone DG, 2002. Microbial nitrogen cycle. In: Aurstic CJ, Crawford RL (eds) Manual of environmental microbiology. ASM Press, Washington, DC, p. 439.

Cassán F, Vanderleyden J, Spaepen S, 2014. Physiological and agronomical aspects of phytohormone production by model plant-growth-promoting rhizobacteria (PGPR) belonging to the genus *Azospirillum*. J Plant Growth Regul 33: 440-459.

Cavagnaro TR, Bender SF, Asghari HR, van der Heijden MG, 2015. The role of *Arbuscular mycorrhizas* in reducing soil nutrient loss. Trends Plant Sci 20 (5): 283-290.

Chatterjee P, Samaddar S, Anandham R, Kang Y, Kim K, Selvakumar G, Sa T, 2017. Beneficial soil bacterium *Pseudomonas frederiksbergensis* OS261 augments salt tolerance and promotes red pepper plant growth. Front Plant Sci 8: 1-9.

Cheng Z, Park E, Glick BR, 2007. 1 - Aminocyclopropane - 1 - carboxylate deaminase from *Pseudomonas putida* UW4 facilitates the growth of canola in the presence of salt. Can J Microbiol 53: 912-918.

Cho K, Toler H, Lee J, Ownley B, Stutz JC, Moore JL, Auge RM, 2006. Mycorrhizal symbiosis and response of sorghum plants to combined drought and salinity stresses. J Plant Physiol 163: 517-528.

Chookietwattana K, Maneewan K, 2012. Selection of efficient salt-tolerant bacteria containing ACC deaminase for promotion of tomato growth under salinity stress. Soil Environ 31: 30-36.

Colla G, Rouphael Y, Cardarelli M, Tullio M, Rivera CM, Rea E, 2008. Alleviation of salt stress by *Arbuscular mycorrhizal* in zucchini plants grown at lowand high phosphorus concentration. Biol Fertil Soil 44: 501-509.

Contreras-Cornejo HA, Macias-Rodriguez L, Cortes-Penagos C, Lopez-Bucio J, 2009. *Trichoderma virens*, a plant beneficial fungus, enhances biomass production and promotes lateral root growth through an auxin - dependent mechanism in *Arabidopsis*. Plant Physiol 149: 1579-1592.

Crafts CB, Miller CO, 1974. Detection and identification of cytokinins produced by mycorrhizal fungi. Plant Physiol 54: 586-588.

Da Costa MS, Santos H, Galinski EA, 1998. An overview of the role and diversity of compatible solutes in bacteria andarchaea. In: Biotechnology of extremophiles. Springer, Berlin, pp. 117-153.

Daei G, Ardekani M, Rejali F, Teimuri S, Miransari M, 2009. Alleviation of salinity stress on wheat yield, yield components, and nutrient uptake using *Arbuscular mycorrhizal* fungi under field conditions. J Plant Physiol 166: 217-225.

Das S, De M, Ray R, Ganguly D, Kumar Jana T, De TK, 2011. Salt tolerant culturable microbes accessible in the soil of the Sundarban mangrove forest, India. Open J Ecol 1: 35.

De Almeida AMM, Gomes VFF, Mendes PF, de Lacerda CF, Freitas ED, 2016. Influence of salinity on the development of the banana colonized by *Arbuscular mycorrhizal* fungi. Rev Cienc Agron 47: 421-428.

De-Salamon IEG, Hynes RK, Nelson LM, 2001. Cytokin in production by plant growth promoting rhizobacteria and selected mutants. Can J Microbiol 47: 404-411.

Deinlein U, Stephan AB, Horie T, Luo W, Xu G, Schroeder JI, 2014. Plant salt-tolerance mechanisms. Trends Plant Sci 19: 371-379.

Del-Saz NF, Romero-Munar A, Alonso D, Aroca R, Baraza E, Flexas J, Ribas-Carbo M, 2017. Respiratory ATP cost and benefit of *Arbuscular mycorrhizal* symbiosis with *Nicotiana tabacum* at different growth stages and under salinity. J Plant Physiol 218: 243-248.

Diatta ILD, Kane A, Agbangba CE, Sagna M, Diouf D, Aberlenc-Bertossi F, Duval Y, Borgel A, Sane D, 2014. Inoculation with *Arbuscular mycorrhizal* fungi improves seedlings growth of two sahelian date palm cultivars (*Phoenix dactylifera* L., cv. *Nakhla hamra* and cv. Tijib) under salinity stresses. Adv Biosci Biotechnol 5: 64-72.

Dinnbier U, Limpinsel E, Schmid R, Bakker EP, 1988. Transient accumulation of potassium glutamate and its replacement by trehalose during adaptation of growing cells of *Escherichia coli* K-12 to elevated sodium chloride concentrations. Arch Microbiol 150: 348-357.

Diouf D, Duponnois R, Ba AT, Neyra M, Lesueur D, 2005. Symbiosis of *Acacia auriculiformis* and *Acacia mangium* with mycorrhizal fungi and *Bradyrhizobium* spp. improves salt tolerance in greenhouse conditions. Funct Plant Biol 32: 1143-1152.

Elhindi K, Sharaf El Din A, Abdel-Salam E, Elgorban A, 2016. Amelioration of salinity stress in different basil (*Ocimum basilicum* L.) varieties by vesicular-*Arbuscular mycorrhizal* fungi. Acta Agriculturae Scandinavica. Sect B-Soil Plant Sci 66: 583-592.

Elhindi KM, El-Din AS, Elgorban AM, 2017. The impact of *Arbuscular mycorrhizal* fungi in mitigating salt-induced adverse effects in sweet basil (*Ocimum basilicum* L.). Saudi J Biol Sci (1): 170-179.

Estrada B, Aroca R, Maathuis FJ, Barea JM, Ruiz-Lozano JM, 2013. *Arbuscular mycorrhizal* fungi native from a Mediterranean saline area enhance maize tolerance to salinity through improved ion homeostasis. Plant Cell Environ 36: 1771-1782.

Estrada-Luna AA, Davies FT, 2003. *Arbuscular mycorrhizal* fungi influence water relations, gas exchange, abscisic acid and growth of micropropagated chile ancho pepper (*Capsicum annuum*) plantlets during acclimatization and postacclimatization. J Plant Physiol 160: 1073-1083.

Evelin H, Giri B, Kapoor R, 2012. Contribution of *Glomus intraradices* inoculation to nutrient acquisition and mitigation of ionic imbalance in NaCl-stressed *Trigonella foenum-graecum*. Mycorrhiza 22: 203-217.

Evelin H, Giri B, Kapoor R, 2013. Ultrastructural evidence for AMF mediated salt stress mitigation in *Trigonella foenum-graecum*. Mycorrhiza 23: 71-86.

Evelin H, Kapoor R, Giri B, 2009. *Arbuscular mycorrhizal* fungi in alleviation of salt stress: a review. Ann Bot 104 (7): 1263-1280.

Felici C, Vettori L, Giraldi E, Forino LMC, Toffanin A, Tagliasacchi AM, Nuti M, 2008. Single and co-inoculation of *Bacillus subtilis* and *Azospirillum brasilense* on *Lycopersicum esculentum*: effects on plant growth and rhizosphere microbial community. Appl Soil Ecol 40: 260-270.

Feng G, Zhang F, Li X, Tian C, Tang C, Rengel Z, 2002. Improved tolerance of maize plants to salt stress by *Arbuscular mycorrhiza* is related to higher accumulation of soluble sugars in roots. Mycorrhiza 12 (4): 185-190.

Finlay RD, 2007. Ecological aspects of mycorrhizal symbiosis with special emphasis on the functional diversity of interactions involving the extraradical mycelium. J Exp Bot 59: 1115-1126.

Firth E, Carpenter SD, Sørensen HL, Collins RE, Deming JW, 2016. Bacterial use of choline to tolerate salinity shifts in sea-ice brines. Elementa Sci Anthropocene 4: 1-15.

Franche C, Lindstrom K, Elmerich C, 2009. Nitrogen - fixing bacteria associated with leguminous and non-leguminous plants. Plant Soil 321: 35-59.

Fukami J, de la Osa C, Ollero FJ, Megías M, Hungria M, 2018. Co-inoculation of maize with *Azospirillum brasilense* and *Rhizobium tropici* as a strategy to mitigate salinity stress. Funct Plant Biol 45: 328-339.

Galinski EA, 1995. Osmoadaptation in bacteria. Adv Microb Physiol 37: 273-328.

Galinski EA, Pfeiffer HP, Truper HG, 1985. 1, 4, 5, 6-Tetrahydro- 2-methyl-4- pyrimidinecarboxylic acid. A novel cyclic amino acid from halophilic phototrophic bacteria of the genus *Ectothiorhodospira*. Eur J Biochem 149 (135): 139.

Gamalero E, Berta G, Massa N, Glick BR, Lingua G, 2010. Interactions between *Pseudomonas putida* UW4 and *Gigaspora rosea* BEG9 and their consequences for the growth of cucumber under salt-stress conditions. J Appl Microbiol 108: 236-245.

García-Garrido JM, Ocampo JA, 2002. Regulation of the plant defence response in *Arbuscular mycorrhizal* symbiosis. J Exp Bot 53 (373): 1377-1386.

Garg N, Bhandari P, 2012. Influence of cadmium stress and *Arbuscular mycorrhizal* fungi on nodule senescence in *Cajanus cajan* (L.) Millsp. Int J Phytoremediation 14: 62-74.

Garg N, Bhandari P, 2016. Silicon nutrition and mycorrhizal inoculations improve growth, nutrient status, K^+/Na^+ ratio and yield of *Cicer arietinum* L. genotypes under salinity stress. Plant Growth Regul 78: 371-387.

Garg N, Chandel S, 2011. Effect of mycorrhizal inoculation on growth, nitrogenfixation, and nutrient uptake in *Cicer arietinum* (L.) under salt stress. Turk J Agric For 35: 205-214.

Garg N, Chandel S, 2015. Role of arbuscular mycorrhiza in arresting reactive oxygen species (ROS) and strengthening antioxidant defense in *Cajanus cajan* (L.) Millsp. nodules under salinity (NaCl) and cadmium (Cd) stress. Plant Growth Regul 75: 521-534.

Garg N, Manchanda G, 2008. Effect of *Arbuscular mycorrhizal* inoculation on salt - induced nodule senescence in *Cajanus cajan* (Pigeonpea). J Plant Growth Regul 27: 115-124.

Garg N, Pandey R, 2015. Effectiveness of native and exotic *Arbuscular mycorrhizal* fungi on nutrient uptake and ion homeostasis in saltstressed *Cajanus cajan* L. (Millsp.) genotypes. Mycorrhiza 25: 165-180.

Garg N, Pandey R, 2016. High effectiveness of exotic *Arbuscular mycorrhizal* fungi is reflected in improved rhizobial symbiosis and trehalose turnover in *Cajanus cajan* genotypes grown under salinity stress. Fungal Ecol 21: 57-67.

Ghoulam CA, Foursy A, Fares K, 2002. Effect of salt stress on growth, inorganic ions and proline accumulation in relation to osmotic adjustment infive sugar beet cultivars. Environ Exp Bot 47: 39-50.

Giri B, Kapoor R, Mukerji KG, 2002. VA mycorrhizal techniques/VAM technology in establishment of plants under salinity stress conditions. In: Mukerji KG, Manoharachary C, Chamola BP (eds) Techniques in mycorrhizal studies. Kluwer, Dordrecht, pp. 313-327.

Giri B, Kapoor R, Mukerji KG, 2007. Improved tolerance of *Acacia nilotica* to salt stress by *Arbuscular mycorrhizal*, *Glomus fasciculatum* may be partly related to elevated K/Na ratios in root and shoot tissues. Microb Ecol 54 (4): 753-760.

Giri B, Mukerji KG, 2004. Mycorrhizal inoculant alleviates salt stress in *Sesbania aegyptiaca* and *Sesbania grandiflora* under field conditions: evidence for reduced sodium and improved magnesium uptake. Mycorrhiza 14: 307-312.

Glick BR, 2010. Using soil bacteria to facilitate phytoremediation. Biotechnol Adv 28: 367-374.

Glick BR, Karaturovic DM, Newell PC, 1995. A novel procedure for rapid isolation of plant growth promoting *Pseudomonas*. Can J Microbiol 41: 533-536.

Glick BR, Patten CL, Holgin G, Penrose DM, 1999. Biochemical and genetic mechanisms used by plant growth promoting bacteria. Imperial College Press, London, p. 267.

Glick BR, Todorovic B, Czarny J, Cheng Z, Duan J, McConkey B, 2007. Promotion of plant growth by bacterial ACC deaminase. Crit Rev Plant Sci 26: 227-242.

Gouzou L, Burtin G, Philippy R, Bartoli F, Heulin T, 1993. Effect of inoculation with *Bacillus polymyxa* on soil aggregation in the wheat rhizosphere: preliminary examination. Geoderma 56: 479-491.

Govindarajan M, Balandreau J, Muthukumarasamy R, Revathi G, Lakshminarasimhan C, 2006. Improved yield of micropropagated sugarcane following inoculation by endophytic *Burkholderia vietnamiensis*. Plant Soil 280: 239-252.

Govindarajan M, Kwon SW, Weon HY, 2007. Isolation, molecular characterization and growth promoting activities of endophytic sugarcane diazotroph *Klebsiella* sp. GR9. World J Microbiol Biotechnol 23: 997-1006.

Govindarajulu M, Pfeffer PE, Jin HR et al., 2005. Nitrogen transfer in the *Arbuscular mycorrhizal* symbiosis. Nature 435: 819-823.

Gunde-Cimerman N, Zalar P, 2014. Extremely halotolerant and halophilic fungi inhabit brine in solar salterns around the globe. Food Technol Biotechnol 52: 170-179.

Gupta B, Huang B, 2014. Mechanism of salinity tolerance in plants: physiological, biochemical, and molecular characterization. Int J Genomics 2014: 701596.

Gupta V, Satyanarayana T, Garg S, 2000. General aspects of mycorrhiza. In: Mukerji KG, Chamola BP, Singh JE (eds) Mycorrhizal biology. Kluwer Academic/Plenum, New York, pp. 27-44.

Habib SH, Kausar H, Saud HM, 2016. Plant growth-promoting rhizobacteria enhance salinity stress tolerance in okra through ROS-scavenging enzymes. Biomed Res Int 1-10.

Hajbagheri S, Enteshari S, 2011. Effects of mycorrhizal fungi on photosynthetic pigments, root mycorrhizal colonization and morphological characteristics of salt stressed *Ocimum basilicum* L.

Iran J Plant Physiol 1: 215-222.

Hajiboland R, Aliasgharzadeh N, Laiegh SF, Poschenrieder C, 2010. Colonization with *Arbuscular mycorrhizal* fungi improves salinity tolerance of tomato (*Solanum lycopersicum* L.) plants. Plant Soil 331: 313-327.

Han HS, Lee KD, 2005. Plant growth promoting rhizobacteria effect on antioxidant status, photosynthesis, mineral uptake and growth of lettuce under soil salinity. Res J Agric Biol Sci 1: 210-215.

Hasegawa PM, Bressan RA, Zhu JK, Bohnert HJ, 2000. Plant cellular and molecular responses to high salinity. Plant Mol Biol 51: 463-499.

Hashem A, Abd Allah EF, Alqarawi AA, Al-Huqail AA, Shah MA, 2016b. Induction of osmoregulation and modulation of salt stress in *Acacia gerrardii* benth. by *Arbuscular mycorrhizal* fungi and *Bacillus subtilis* (BERA 71). Biomed Res Int 2016: 1-11.

Hashem A, Abd Allah EF, Alqarawi AA, Al-Huqail AA, Wirth S, Egamberdieva D, 2016c. The interaction between *Arbuscular mycorrhizal* fungi and endophytic bacteria enhances plant growth of *Acacia gerrardii* under salt stress. Front Microbiol 7: 1-15.

Hashem A, Abd-Allah EF, Alqarawi AA, Al Huqail AA, Egamberdieva D, Wirth S, 2016a. Alleviation of cadmium stress in *Solanum lycopersicum* L. by *Arbuscular mycorrhizal* fungi via induction of acquired systemic tolerance. Saudi J Biol Sci 23: 272-281.

Hashem A, Alqarawi AA, Radhakrishnan R, Al-Arjani A-BF, Aldehaish HA, Egamberdieva D, Abd_ Allah EF, 2018. *Arbuscular mycorrhizal* fungi regulate the oxidative system, hormones and ionic equilibrium to trigger salt stress tolerance in *Cucumis sativus* L. Saudi J Biol Sci 25 (6): 1102-1114.

He XH, Nara K, 2007. Element biofortification: can mycorrhizas potentially offer a more effective and sustainable pathway to curb human malnutrition? Trends Plant Sci 12: 331-333.

He Z, He C, Zhang Z, Zou Z, Wang H, 2007. Changes of antioxidative enzymes and cell membrane osmosis in tomato colonized by *Arbuscular mycorrhizae* under NaCl stress. Colloids Surf B: Biointerfaces 59: 128-133.

Hildebrandt U, Janetta K, Bothe H, 2002. Towards growth of *Arbuscular mycorrhizal* fungi independent of a plant host. Appl Environ Microbiol 68: 1919-1924.

Hildebrandt U, Janetta K, Ouziad F, Renne B, Nawrath K, Bothe H, 2001. *Arbuscular mycorrhizal* colonization of halophytes in Central European salt marshes. Mycorrhiza 10: 175-183.

Ibrahim AH, Abdel-Fattah GM, Eman FM, Abdel-Aziz MH, Shokr AE, 2011. *Arbuscular mycorrhizal* fungi and spermine alleviate the adverse effects of salinity stress on electrolyte leakage and productivity of wheat plants. Phyton 51: 261-276.

Jaderlund L, Arthurson V, Granhall U, Jansson JK, 2008. Specific interactions between *Arbuscular mycorrhizal* fungi and plant growth-promoting bacteria: as revealed by different combinations. FEMS Microbiol Lett 287: 174-180.

Jahromi F, Aroca R, Porcel R, Ruiz-Lozano JM, 2008. Influence of salinity on the in vitro development of *Glomus intraradices* and on the in vivo physiological and molecular responses of mycorrhizal lettuce plants. Microbial Ecol 55: 45-53.

Jeffries P, Gianinazzi S, Perotto S, Turnau K, Barea J, 2003. The contribution of *Arbuscular mycorrhizal* fungi in sustainable maintenance of plant health and soil fertility. Biol Fertil Soils

37: 1-16.

Jenab K, Moghimi H, Hamedi J, 2016. Bioremediation of crude oil and pyrene by halotolerant Fungus *Embellisia phragmospora* isolated from saline soil. International Conference and Iranian Congress of Microbiology, August 2016, University of Tehran, Tehran, Iran.

Jha Y, Subramanian RB, 2013. Paddy plants inoculated with PGPR show better growth physiology and nutrientcontent under saline conditions. Chil J Agric Res 73: 213-219.

Jindal V, Atwal A, Sekhon BS, Singh R, 1993. Effect of vesicular-*Arbuscular mycorrhizae* on metabolism of moong plants under NaCl salinity. Plant Physiol Biochem 31: 475-481.

Joe MM, Saravanan VS, Islam MR, Sa T, 2014. Development of alginate-based aggregate inoculants of *Methylobacterium* sp. and *Azospirillum brasilense* tested under in vitro conditions to promote plant growth. J Appl Microbiol 116 (2): 408-423.

Johannes AVV, Overbeek LSV, Elsas JDV, 1997. Fate and activity of microorganisms introduced into soil. Microbiol Mol Biol Rev 61: 121-135.

Johansson JF, Paul LR, Finlay RD, 2004. Microbial interactions in the mycorrhizosphere and their significance for sustainable agriculture. FEMS Microbiol Ecol 48: 1-13.

Joo HS, Chang CS, 2005. Oxidant and SDS-stable alkaline protease from a halo-tolerant *Bacillus clausii* I-52. 68: 772-783.

Juniper S, Abbott LK, 2006. Soil salinity delays germination and limits growth of hyphae from propagules of *Arbuscular mycorrhizal* fungi. Mycorrhiza 16: 371-379.

Kang SM, Khan AL, Waqas M, You YH, Kim JH, Kim JG, Hamayun M, Lee IJ (2014a) Plant growth-promoting rhizobacteria reduce adverse effects of salinity and osmotic stress by regulating phytohormones and antioxidants in *Cucumis sativus*. J Plant Interact 9: 673-682.

Kang SM, Radhakrishnan R, Khan AL, Kim MJ, Park JM, Kim BR, Shin DH, Lee IJ (2014b) Gibberellin secreting rhizobacterium, *Pseudomonas putida* H-2-3 modulates the hormonal and stress physiology of soybean to improve the plant growth under saline and drought conditions. Plant Physiol Biochem 84: 115-124.

Kapoor R, Sharma D, Bhatnagar AK, 2008. *Arbuscular mycorrhizae* in micropropagation systems and their potential applications. Sci Horticult 116: 227-239.

Kaya C, Ashraf M, Sonmez O, Aydemir S, Tuna LA, Cullu AM, 2009. The influence of *Arbuscular mycorrhizal* colonization on key growth parameters and fruit yield of pepper plants grown at high salinity. Sci Horticult 121: 1-6.

Kempf B, Bremer E, 1998. Uptake and synthesis of compatible solutes as microbial stress responses to highosmolality environments. Arch Microbiol 170 (319): 330.

Koc A, Balci G, Erturk Y, Keles H, Bakoglu N, Ercisli S, 2016. Influence of *Arbuscular mycorrhizae* and plant growth promoting rhizobacteria on proline, membrane permeability and growth of strawberry (*Fragaria* × *ananassa*) under salt stress. J Appl Bot Food Qual 89: 89-97.

Kohler J, Caravaca F, Carrasco L, Roldan A, 2006. Contribution of *Pseudomonas mendocina* and *Glomus intraradices* to aggregate stabilization and promotion of biological fertility in rhizosphere soil of lettuce plants under field conditions. Soil Use Manag 22: 298-304.

Kondepudi KK, Chandra TS, 2011. Identification of osmolytes from a moderately halophilic amylolytic *Bacillus* sp. strain TSCVKK. Eur J Exp Biol 1: 113-121.

Kong CC, Ren CG, Li RZ, Xie ZH, Wang JP, 2017. Hydrogen peroxide and strigolactones signaling are involved in alleviation of salt stress induced by *Arbuscular mycorrhizal* fungus in *Sesbania cannabina* seedlings. J Plant Growth Regul 36: 734-742.

Kumar A, Sharma S, Mishra S, Dames JF, 2015. *Arbuscular mycorrhizal* inoculation improves growth and antioxidative response of *Jatropha curcas* (L.) under Na_2SO_4 salt stress. Plant Biosyst 149 (2): 260-269.

Kumar K, Amaresan N, Madhuri K, 2017. Alleviation of the adverse effect of salinity stress by inoculation of plant growth promoting rhizobacteria isolated from hot humid tropical climate. Ecol Eng 102: 361-366.

Kunte HJ, 2012. Osmoregulation in halophilic bacteria. Extremophiles II: 263-277.

Lalnunhlimi S, Krishnaswamy V, 2016. Decolorization of azo dyes (direct blue 151 and direct red 31) by moderately alkaliphilic bacterial consortium. Braz J Microbiol 47: 39-46.

Landwehr M, Wilde P, Tóth T, Biró B, Hildebrandt U, Nawrath K, Bothe H, 2002. The *Arbuscular mycorrhizal* fungus *Glomus geosporum* in European saline, sodic and gypsum soils. Mycorrhiza 12 (4): 199-211.

Larkin RP, Fravel DR, 1998. Efficacy of various fungal and bacterial biocontrol organisms for control of fusarium wilt of tomato. Plant Dis 82: 1022-1028.

Lee Y, Krishnamoorthy R, Selvakumar G, Kim K, Sa T, 2015. Alleviation of salt stress in maize plant by co-inoculation of *Arbuscular mycorrhizal* fungi and *Methylobacterium oryzae* CB-MB20. J Korean Soc Appl Biol Chem 58: 533-540.

Li H, Lei P, Pang X, Li S, Xu H, Xu Z, Feng X, 2017. Enhanced tolerance to salt stress in canola (*Brassica napus* L.) seedlings inoculated with the halotolerant *Enterobacter cloacae* HSNJ4. Appl Soil Ecol 119: 26-34.

Lin J, Wang Y, Sun S, Mu C, Yan X, 2017. Effects of *Arbuscular mycorrhizal* fungi on the growth, photosynthesis and photosynthetic pigments of *Leymus chinensis* seedlings under salt-alkali stress and nitrogen deposition. Sci Total Environ 576: 234-241.

Linderman RG, 1992. Vesicular-*Arbuscular mycorrhizae* and soil microbial interactions. In: Bethlenfalvay GJ, Linderman RG (eds) Mycorrhizae in sustainable agriculture. ASA, Madison, pp. 1-26.

Liu C, Dai Z, Cui M, Lu W, Sun H, 2018. *Arbuscular mycorrhizal* fungi alleviate boron toxicity in *Puccinellia tenuiflora* under the combined stresses of salt and drought. Environ Pollut 240: 557-565.

Liu J, He H, Vitali M, Visentin I, Charnikhova T, Haider I, Schubert A, Ruyter-Spira C, Bouwmeester HJ, Lovisolo C, Cardinale F, 2015a. Osmotic stress represses strigolactone biosynthesis in *Lotus japonicus* roots: exploring the interaction between strigolactones and ABA under abiotic stress. Planta 241: 1435-1451.

Liu QL, Tang JC, Bai ZH, Hecker M, Giesy JP, 2015b. Distribution of petroleum degrading genes and factor analysis of petroleum contaminated soil from the Dagang Oilfield China. Sci Rep 5: 11068. https://doi.org/10.1038/srep11068.

Lowe SE, Jain MK, Zeikus JG, 1993. Biology, ecology, and biotechnological applications of anaerobic bacteria adapted to environmental stresses in temperature, pH, salinity, or substrates. Microbiol Biotechnol Rev 57: 451-509.

Lucht JM, Bremer E, 1994. Adaptation of Escherichia coli to high osmolarity environments: osmoregulation of the high-affinity glycine betaine transport system ProU. FEMS Microbiol Rev 14: 3-20.

Lugtenberg B, Kamilova F, 2009. Plant growth promoting rhizobacteria. Annu Rev Microbiol 63: 541-556.

Manchanda G, Garg N, 2011. Alleviation of salt-induced ionic, osmotic and oxidative stresses in *Cajanus Cajan* nodules by AM inoculation. Plant Biosyst 145: 88-97.

Mandal HK, 2014. Isolation of salt tolerant strains of *Rhizobium Trifolii*. Int J Agric Food Sci Technol 5: 325-332.

Manga A, Diop A, Diop TA, 2017. Functional diversity of mycorrhizal fungi has differential effects on salinity tolerance of *Acacia seyal* (Del.) seedlings. Open J Soil Sci 7: 315-332.

Mardukhi B, Rejali F, Daei G, Ardakani MR, Malakouti MJ, Miransari M, 2011. *Arbuscular mycorrhizas* enhance nutrient uptake in different wheat genotypes at high salinity levels under field and greenhouse conditions. C R Biol 334: 564-571.

Mardukhi B, Rejali F, Daei G, Ardakani MR, Malakouti MJ, Miransari M, 2015. Mineral uptake of mycorrhizal wheat (*Triticum aestivum* L.) under salinity stress. Commun Sci Plant Anal 46: 343-357.

Matamoros MA, Baird LM, Escuredo PR, Dalton DA, Minchin FR, Iturbe-Ormaetxe I, Rubio MC, Moran JF, Gordon AJ, Becana M, 1999. Stress-induced legume root nodule senescence: physiological, biochemical and structural alterations. Plant Physiol 121: 97-111.

Mayak S, Tirosh T, Glick BR, 2004. Plant growth-promoting bacteria confer resistance in tomato plants to salt stress. Plant Physiol Biochem 42: 565-572.

Mazhar S, Cohen JD, Hasnain S, 2013. Auxin producing non-heterocystous cyanobacteria and their impact on the growth and endogenous auxin homeostasis of wheat. J Basic Microbiol 53: 996-1003.

Mendpara J, Parekh V, Vaghela S, Makasana A, Kunjadia PD, Sanghvi G, Vaishnav D, Dave GS, 2013. Isolation and characterization of high salt tolerant bacteria from agricultural soil. Eur J Exp Biol 3: 351-358.

Meng L, Zhang A, Wang F, Han X, Wang D, Li S, 2015. *Arbuscular mycorrhizal* fungi and rhizobium facilitate nitrogen uptake and transfer in soybean/maize intercropping system. Front Plant Sci 6.

Meyer JR, Linderman RG, 1986. Response of subterranean clover to dual inoculationwithvesicular-*Arbuscular mycorrhizal* fungi and a plant growth-promoting bacterium, *Pseudomonas putida*. Soil Biol Biochem 18: 185-190.

Mitter B, Brader G, Afzal M, Compant S, Naveed M, Trognitz F, Sessitsch A, 2013. Advances in elucidating beneficial interactions between plants, soil and bacteria. Adv Agron 121: 381-445.

Mohamed HI, Gomaa EZ, 2012. Effect of plant growth promoting *Bacillus subtilis* and *Pseudomonas fluorescens* on growth and pigment composition of radish plants (*Raphanus sativus*) under NaCl stress. Photosynthetica 50: 263-272.

Munoz N, Soria-Diaz ME, Manyani M, Sanchez-Matamoros RC, Serrano AG, Megias M, Lascano R, 2014. Structure and biological activities of lipochitooligosaccharide nodulation signals produced by *Bradyrhizobium japonicum* USDA 138 under saline and osmotic stress. Biol Fertil Soils 50: 207-215.

Muthukumarasamy R, Kang UG, Park KD, Jeon WT, Park CY, Cho YS, Kwon SW, Song J, Roh DH, Revathi G, 2007. Enumeration, isolation and identification of diazotrophs from Korean wetland rice varieties grown with long-term application of N and compost and their short-term inoculation effect on rice plants. J Appl Microbiol 102: 981-991.

Nadeem SM, Ahmad M, Zahir ZA, Javaid A, Ashraf M, 2014. The role of mycorrhizae and plant growth promoting rhizobacteria (PGPR) in improving crop productivity under stressful environments. Biotechnol Adv 32: 429-448.

Nadeem SM, Zahair ZA, Naveed M, Asghar HN, Asghar M, 2010a. Rhizobacteria capable of producing ACC-deaminase may mitigate salt stress in wheat. J Soil Sci Soc Am 74: 533-542.

Nadeem SM, Zahir ZA, Naveed M, Ashraf M, 2010b. Microbial ACC-deaminase: prospects and applications for inducing salt tolerance in plants. Crit Rev Plant Sci 29: 360-393.

Najafi A, Ardakani MR, Rejali F, Sajedi N, 2012. Response of winter barley to co-inoculation with *Azotobacter* and Mycorrhiza fungi influenced by plant growth promoting rhizobacteria. Ann Biol Res 3: 4002-4006.

Navarro JM, Morte A, Rodríguez-Morán M, Pérez-Tornero O, 2015. Physiological response of Citrus macrophylla inoculated with *Arbuscular mycorrhizal* fungi under salt stress. Acta Hortic 1065: 1351-1358.

Nazareth SW, Gaitonde S, Marbaniang T, 2012. Metal resistance of halotolerant fungi from mangroves and salterns of Goa, India.

Nicholson CA, Fathepure BZ, 2004. Biodegradation of benzene by halophilic and halotolerant bacteria under aerobic conditions. Appl Environ Microbiol 70: 1222-1225.

Nishiyama R, Watanabe Y, Fujita Y, Le DT, Kojima M, Werner T, Vankova R, YamaguchiShinozaki K, Shinozaki K, Kakimoto T, Sakakibara H, Schmülling T, Tran LP, 2011. Analysis of cytokinin mutants and regulation of cytokinin metabolic genes reveals important regulatory roles of cytokinins in drought, salt and abscisic acid responses, and abscisic acid biosynthesis. Plant Cell 23: 2169-2183.

Nunkaew T, Kantachote D, Kanzaki H, Nitoda T, Ritchie RJ, 2014. Effects of 5-aminolevulinic acid (ALA) -containing supernatants from selected *Rhodopseudomonas palustris* strains on rice growth under NaCl stress, with mediating effects on chlorophyll, photosynthetic electron transport and antioxidative enzymes. Electron J Biotechnol 17: 4.

Okoro CC, 2010. Enhanced bioremediation of hydrocarbon contaminated mangrove swamp in the Nigerian oil-rich Niger-Delta using sea water microbial inocula amended with crude biosurfactants and micronutrients. Nature Sci 8: 195-206.

Oren A, 1999. Bioenergetic aspects of halophilism. Microbiol Mole Biol Rev 63: 334-348 Oren A, 2008. Microbial life at high salt concentrations: phylogenetic and metabolic diversity. Saline Syst 15: 4.

Oren A, Gurevich P, Azachi M, Henis Y, 1992. Microbial degradation of pollutants at high salt concentrations. Biodegradation 3: 387-398.

Osuji LC, Ezebuiro PE, 2006. Hydrocarbon contamination of a typical mangrovefloor in Niger Delta, Nigeria. Int J Environ Sci Tech 3: 313-332.

Palacio-Rodríguez R, Coria-Arellano JL, López-Bucio J, Sánchez-Salas J, Muro-Pérez G, Castañeda-Gaytán G, Sáenz-Mata J, 2017. Halophilic rhizobacteria from *Distichlis spicata* promote growth and improve salt tolerance in heterologous plant hosts. Symbiosis 73: 179-189.

Palaniyandi SA, Damodharan K, Yang SH, Suh JW, 2014. *Streptomyces* sp. strain PGPA39 alleviates salt stress and promotes growth of 'Micro-Tom' tomato plants. J Appl Microbiol 117: 766e773.

Pandey R, Garg N, 2017. High effectiveness of *Rhizophagus irregularis* is linked to superior modulation of antioxidant defence mechanisms in *Cajanus cajan* (L.) Millsp. genotypes grown under salinity stress. Mycorrhiza 27: 669-682.

Parshetti G, Kalme S, Saratale G, Govindwar S, 2006. Biodegradation of malachite green by *Kocuria rosea* MTCC 1532. Acta Chim Slov 53 (4): 492-498.

Patel VK, Srivastava R, Sharma A, Srivastava AK, Singh S, Srivastava AK, Kashyap PL, Chakdar H, Pandiyan K, Kalra A, Saxena AK, 2018. Halotolerant *Exiguobacterium profundum* PHM11 tolerate salinity by accumulating L-proline and fine-tuning gene expression profiles of related metabolic pathways. Front Microbiol 9: 423. https://doi.org/10.3389/fmicb.2018.00423.

Pedranzani H, Rodríguez-Rivera M, Gutiérrez M, Porcel R, Hause B, Ruiz-Lozano JM, 2016. *Arbuscular mycorrhizal* symbiosis regulates physiology and performance of *Digitaria eriantha* plants subjected to abiotic stresses by modulating antioxidant and jasmonate levels. Mycorrhiza 26: 141-152.

Perugini M, Visciano P, Giammarino A, Manera M, Di Nardo W, Amorena M, 2007. Polycyclic aromatic hydrocarbons in marine organisms from the Adriatic Sea, Italy. Chemosphere 66: 1904-1910.

Petrovic U, Gunde-Cimerman N, Plemenitaš A, 2002. Cellular responses to environmental salinity in the halophilic black yeast *Hortaea werneckii*. Mol Microbiol 45: 665-672.

Pinedo I, Ledger T, Greve M, Poupin MJ, 2015. *Burkholderia phytofirmans* PsJN induces longterm metabolic and transcriptional changes involved in *Arabidopsis thaliana* salt tolerance. Front Plant Sci 6 (466): 466.

Plemenitas A, Lenassi M, Konte T, Kejzar A, Zajc J, Gostincar C, Gunde-Cimerman N, 2014. Adaptation to high salt concentrations in halotolerant/halophilic fungi: a molecular perspective. Front Microbiol 5: 199.

Plenchette C, Duponnois R, 2005. Growth response of the saltbush *Atriplex nummularia* L. to inoculation with the *Arbuscular mycorrhizal* fungus *Glomus intraradices*. J Arid Environ 61: 535-540.

Pollastri S, Savvides A, Pesando M, Lumini E, Volpe MG, Ozudogru EA, Faccio A, De Cunzo F, Michelozzi M, Lambardi M, Fotopoulos V, 2017. Impact of two *Arbuscular mycorrhizal* fungi on *Arundo donax* L. response to salt stress. Planta 9: 1-3.

Porcel R, Aroca R, Azcon R, Ruiz-Lozano JM, 2016. Regulation of cation transporter genes by the *Arbuscular mycorrhizal* symbiosis in rice plants subjected to salinity suggests improved salt tolerance due to reduced Na^+ root-to-shoot distribution. Mycorrhiza 26: 673-684.

Porcel R, Aroca R, Azcón R, Ruiz-Lozano JM, 2006. *PIP* aquaporin gene expression in *Arbuscular mycorrhizal Glycine max* and *Lactuca sativa* plants in relation to drought stress tolerance. Plant Mol Biol 60: 389-404.

Porcel R, Aroca R, Ruíz-Lozano JM, 2012. Salinity stress alleviation using *Arbuscular mycorrhizal* fungi. A review. Agron Sustain Dev 32: 181-200.

Porcel R, Redondo-Gomez S, Mateos-Naranjo E, Aroca R, Garcia R, Ruiz-Lozano JM,

2015. Arbuscular mycorrhizal symbiosis ameliorates the optimum quantum yield of photosystem II and reduces non-photochemical quenching in rice plants subjected to salt stress. J Plant Physiol 185: 75-83.

Porcel R, Ruiz-Lozano JM, 2004. Arbuscular mycorrhizal influence on leaf water potential, solute accumulation, and oxidative stress in soybean plants subjected to drought stress. J Exp Bot 55: 1743-1750.

Porras-Soriano A, Soriano-Martín ML, Porras-Piedra A, Azcón R, 2009. Arbuscular mycorrhizal fungi increased growth, nutrient uptake and tolerance to salinity in olive trees under nursery conditions. J Plant Physiol 166 (13): 1350-1359.

Qin X, Tang JC, Li DS, Zhang QM, 2012. Effect of salinity on the bioremediation of petroleum hydrocarbons in a saline-alkaline soil. Lett Appl Microbiol 55: 210-217.

Querejeta JI, Egerton-Warburton LM, Allen MF, 2007. Hydraulic lift may buffer rhizosphere hyphae against the negative effects of severe soil drying in a California Oak savanna. Soil Biol Biochem 39: 409-417.

Qurashi AW, Sabri AN, 2012a. Bacterial exopolysaccharide and biofilm formation stimulate chickpea growth and soil aggregation under salt stress. Braz J Microbiol 11: 83-91.

Qurashi AW, Sabri AN, 2012b. Bacterial exopolysaccharide and biofilm formation stimulate chickpea growth and soil aggregation under salt stress. Braz J Microbiol 43: 1183-1191.

Rabie GH, Aboul-Nasr MB, Al-Humiany A, 2005. Increased salinity tolerance of cowpea plants by dual inoculation of an Arbuscular mycorrhizal fungus Glomus clarum and a nitrogen-fixer Azospirillum brasilense. Microbiology 33: 51-60.

Rabie GH, Almadini AM, 2005. Role of bioinoculants in development of salt-tolerance of Vicia faba plants under salinity stress. Afr J Biotechnol 4: 210-222.

Raja P, Uma S, Gopal H, Govindarajan K, 2006. Impact of bio inoculants consortium on rice root exudates, biological nitrogenfixation and plant growth. J Biol Sci 6: 815-823.

Rajput L, Imran A, Mubeen F, Hafeez FY, 2013. Salt-tolerant PGPR strain Planococcus rifietoensis promotes the growth and yield of wheat (Triticum aestivum L.) cultivated in saline soil. Pak J Bot 45: 1955-1962.

Rasooli I, SDA A, Borna H, Barchini KA, 2008. A thermostable Î-amylase producing natural variant of Bacillus spp. isolated from soil in Iran. Am J Agric Biol Sci 3: 591-596.

Reina-Bueno M, Argandoña M, Salvador M, Rodriguez-Moya J, Iglesias-Guerra F, Csonka LN, Nieto JJ, Vargas C, 2012. Role of trehalose in salinity and temperature tolerance in the model halophilic bacterium Chromohalobacter salexigens. PLoS One 7: 33587.

Ren CG, Bai YJ, Kong CC, Bian B, Xie ZH, 2016. Synergistic interactions between salt-tolerant rhizobia and Arbuscular mycorrhizal fungi on salinity tolerance of Sesbania cannabina plant. J Plant Growth Regul 35: 1098-1107.

Ruiz-Lozano JM, 2003. Arbuscular mycorrhizal symbiosis and alleviation of osmotic stress: new perspectives for molecular studies. Mycorrhiza 13: 309-317.

Ruiz-Lozano JM, Azcon R, 2000. Symbiotic efficiency and infectivity of an autochthonous Arbuscular mycorrhizal Glomus sp. from saline soils and Glomus deserticola under salinity. Mycorrhiza 10: 137-143.

Ruiz-Sanchez A, Armada E, Munoz Y, Garcia IE, Aroca R, Ruiz-Lozano JM, Azcon R, 2011. Azospirillum and Arbuscular mycorrhizal colonization enhance rice growth and physiological traits under well-watered and drought conditions. J Plant Physiol 168: 1031-1037.

Ruyter-Spira C, Al-Babili S, van der Krol S, Bouwmeester H, 2013. The biology of strigolactones. Trends Plant Sci 18: 72-83.

Sahoo RK, Gill SS, Tuteja N, 2012. Pea DNA helicase 45 promotes salinity stress tolerance in IR64 rice with improved yield. Plant Signal Behav 7: 1042-1046.

Salah UM, Zhou JT, Qu YY, Guo JB, Wang P, Zhao L, 2007. Biodecolorization of azo dye acid red B under high salinity condition. Bull Environ Contam Toxicol 79: 440-444.

Sanan-Mishra N, Pham XH, Sopory SK, Tuteja N, 2005. Pea DNA helicase 45 overexpression in tobacco confers high salinity tolerance without affecting yield. Proc Natl Acad Sci U S A 102: 509-514.

Sandhya V, Ali SZ, 2015. The production of exopolysaccharide by *Pseudomonas putida* GAP-P45 under various abiotic stress conditions and its role in soil aggregation. Microbiology 84: 512-519.

Sannazzaro AI, Echeverria M, Alberto EO, Ruiz OA, Menendez AB, 2007. Modulation of polyamine balance in *Lotus glaber* by salinity and arbuscular mycorrhiza. Plant Physiol Biochem 45: 39-46.

Sannazzaro AI, Ruiz OA, Alberto EO, Menendez AB, 2006. Alleviation of salt stress in *Lotus glaber* by *Glomus intraradices*. Plant Soil 285: 279-287.

Sarkar A, Ghosh PK, Pramanik K, Mitra S, Soren T, Pandey S, Mondal MH, Maiti TK, 2018. A halotolerant *Enterobacter* sp. displaying ACC deaminase activity promotes rice seedling growth under salt stress. Res Microbiol 169: 20-32.

Sarwat M, Hashem A, Ahanger MA, Abd Allah EF, Alqarawi AA, Alyemeni MN, Ahmad P, Gucel S, 2016. Mitigation of NaCl stress by *Arbuscular mycorrhizal* fungi through the modulation of osmolytes, antioxidants and secondary metabolites in mustard (*Brassica juncea* L.) plants. Front Plant Sci 7: 869. https://doi.org/10.3389/fpls.2016.00869.

Saum SH, Müller V, 2007. Salinity-dependent switching of osmolyte strategies in a moderately halophilic bacterium: glutamate induces proline biosynthesis in *Halobacillus halophilus*. J Bacteriol 189: 6968-6975.

Selvakumar G, Shagol CC, Kim K, Han S, Sa T, 2018. Spore associated bacteria regulates maize root K^+/Na^+ ion homeostasis to promote salinity tolerance during *Arbuscular mycorrhizal* symbiosis. BMC Plant Biol 18: 109. https://doi.org/10.1186/s12870-018-1317-2.

Sen S, Chandrasekhar CN, 2015. Effect of PGPR on enzymatic activities of rice (*Oryza sativa* L.) under salt stress. Asian J Plant Sci Rese 5: 44-48.

Sengupta A, Chaudhuri S, 1990. Vesicular *Arbuscular mycorrhizal* (VAM) in pioneer salt marsh plants of the Ganges river delta in West Bengal (India). Plant Soil 122: 111-113.

Shahid M, Akram MS, Khan MA, Zubair M, Shah SM, Ismail M, Shabir G, Basheer S, Aslam K, Tariq M, 2018. A phytobeneficial strain *Planomicrobium* sp. MSSA-10 triggered oxidative stress responsive mechanisms and regulated the growth of pea plants under induced saline environment. J Appl Microbiol 124: 1566-1579.

Shakirova FM, Sakhabutdinova AR, Bezrukova MV, Fatkhutdinova RA, Fatkhutdinova DR, 2003. Changes in the hormonal status of wheat seedlings induced by salicylic acid and salinity. Plant Sci 164: 317-322.

Sharifi M, Ghorbanli M, Ebrahimzadeh H, 2007. Improved growth of salinity-stressed soybean after inoculation with salt pre-treated mycorrhizal fungi. J Plant Physiol 164: 1144-1151.

Sharma N, Aggarwal A, Yadav K, 2017. Arbuscular *mycorrhizal* fungi enhance growth, physiological parameters and yield of salt stressed *Phaseolus mungo* (L.) Hepper. Eur J Environ Sci 7. doi: https://doi.org/10.14712/23361964.2017.1.

Sharma S, Kulkarni J, Jha B, 2016. Halotolerant rhizobacteria promote growth and enhance salinity tolerance in peanut. Front Microbiol. 13 7: 1600.

Sheng M, Tang M, Chen H, Yang BW, Zhang FF, Huang YH, 2008. Influence of *Arbuscular mycorrhizae* on photosynthesis and water status of maize plants under salt stress. Mycorrhiza 18: 287-296.

Shin W, Siddikee MA, Joe MM, Benson A, Kim K, Selvakumar G, Kang Y, Jeon S, Samaddar S, Chatterjee P, Walitang D, 2016. Halotolerant plant growth promoting bacteria mediated salinity stress amelioration in plants. Korean J Soil Sci Fertil 49: 355-367.

Shivan P, Jayaraman G, 2009. Production of extracellular protease from halotolerant bacterium, *Bacillus aquimaris* strain VITP4 isolated from Kumta coast. Process Biochem 44: 1088-1094.

Singh N, Samajpati N, Paul AK, 2011. Dual inoculation of salt tolerant *Bradyrhizobium* and *Glomus mosseae* for improvement of *Vigna radiata* L. cultivation in saline areas of West Bengal, India. Agric Sci 2: 413-423.

Singh RP, Jha P, Jha PN, 2015. The plant-growth-promoting bacterium *Klebsiella* sp. SBP-8 confers induced systemic tolerance in wheat (*Triticum aestivum*) under salt stress. J Plant Physiol 184: 57-67.

Singh RP, Jha PN, 2017a. Analysis of fatty acid composition of PGPR *Klebsiella* sp. SBP-8 and its role in ameliorating salt stress in wheat. Symbiosis 73: 213-222.

Singh RP, Jha PN, 2017b. The PGPR *Stenotrophomonas maltophilia* SBP-9 augments resistance against biotic and abiotic stress in wheat plants. Front Microbiol 8: 1945. https://doi.org/10.3389/fmicb.2017.01945.

Soliman AS, Shanan NT, Massoud ON, Swelim DM, 2012. Improving salinity tolerance of *Acacia saligna* (Labill.) plant by *Arbuscular mycorrhizal* fungi and Rhizobium inoculation. Afr J Biotechnol 11: 1259-1266.

Sowmya M, Rejula MP, Rejith PG, Mohan M, Karuppiah M, Hatha AM, 2014. Heavy metal tolerant halophilic bacteria from Vembanad Lake as possible source for bioremediation of lead and cadmium. J Environ Biol 35: 655.

Szabados L, Savouré A, 2010. Proline: a multifunctional amino acid. Trends Plant Sci 15: 89-97.

Tabak HH, Lens P, van Hullebusch ED, Dejonghe W, 2005. Developments in bioremediation of soils and sediments polluted with metals and radionuclides. Microbial processes and mechanisms affecting bioremediation of metal contamination and influencing metal toxicity and transport. Rev Environ Sci Biotechnol 4: 115-156.

Tajrishi MM, Vaid N, Tuteja R, Tuteja N, 2011. Overexpression of a pea DNA helicase 45 in bacteria confers salinity stress tolerance. Plant Signal Behav 6: 1271-1275.

Tan L, Qu YY, Zhou JT, Li A, Gou M, 2009. Identification and characteristics of a novel salttolerant *Exiguobacterium* sp. for azo dyes decolorization. Appl Biochem Biotechnol 159: 728-738.

Tester M, Davenport R, 2003. Na⁺ tolerant and Na⁺ transport in higher plants. Ann Bot 91: 503-527.

Tippannavar CM, Venkataramana M, Reddy V, Rajashekara E, 1989. Tolerance of *Azotobacter chroococcum* strains to different salt concentrations and pH levels. Farm Sys 5: 24-28.

Tiquia SM, Davis D, Hadid H, Kasparian S, Ismail M, Ahly S, Shim J, Singh S, Murray KS, 2007. Halophilic and halotolerant bacteria from river waters and shallow groundwater along the Rouge River of southeastern Michigan. Environ Technol 28: 297-307.

Tiwari S, Prasad V, Chauhan PS, Lata C, 2017. *Bacillus amyloliquefaciens* confers tolerance to various abiotic stresses and modulates plant response to phytohormones through osmoprotection and gene expression regulation in rice. Front Plant Sci. 29 8: 1510. https://doi.org/10.3389/fpls.2017.01510.

Torres MA, Dangl JL, 2005. Functions of the respiratory burst oxidase in biotic interactions, abiotic stress and development. Curr Opin Plant Biol 8: 397-403.

Tripathi AK, Nagarajan T, Verma SC, Le Rudulier D, 2002. Inhibition of biosynthesis and activity of nitrogenase in *Azospirillum brasilense* Sp7 under salinity stress. Curr Microbiol 44: 363-367.

Turan S, Cornish K, Kumar S, 2012. Salinity tolerance in plants: breeding and genetic engineering. AJCS 6: 1337-1348.

Turk M, Abramovic Z, Plemenitas A, Gunde-Cimerman N, 2007. Salt stress and plasma-membranefluidity in selected extremophilic yeasts and yeast-like fungi. FEMS Yeast Res 7: 550-557.

Turk M, Mejanelle L, Sentjurc M, Grimalt JO, Gunde-Cimerman N, Plemenitas A, 2004. Saltinduced changes in lipid composition and membranefluidity of halophilic yeast-like melanized fungi. Extremophiles 8: 53-61.

Tuteja N, 2007. Mechanisms of high salinity tolerance in plants. Methods Enzymol 428: 419-438.

Upadhyay SK, Maurya SK, Singh DP, 2012. Salinity tolerance in free living plant growth promoting rhizobacteria. Indian J Sci Res 3: 73-78.

Ventosa A, Nieto JJ, Oren A, 1998. Biology of moderately halophilic aerobic bacteria. Microbiol Mol Biol Rev 62: 504-544.

Vimal SR, Patel VK, Singh JS, 2018. Plant growth promoting *Curtobacterium albidum* strain SRV4: an agriculturally important microbe to alleviate salinity stress in paddy plants. Ecol Indic. https://doi.org/10.1016/j.ecolind.2018.05.014.

Vishnivetskaya TA, Sophia K, Tiedje JM, 2009. The Exiguobacterium genus: biodiversity and biogeography. Extremophiles 13: 541-555.

Viterbo A, Landau U, Kim S, Chernin L, Chet I, 2010. Characterization of ACC deaminase from the biocontrol and plant growth-promoting agent *Trichoderma asperellum* T203. FEMS Microbiol Lett 305: 42-48.

Wang FY, Liu RJ, Lin XG, Zhou JM, 2004. *Arbuscular mycorrhizal* status of wild plants in salinealkaline soils of the Yellow River Delta. Mycorrhiza 14: 133-137.

Welsh DT, 2000a. Ecological significance of compatible solute accumulation by micro-organisms: from single cells to global climate. FEMS Microbiol Rev 3 (263): 290.

Welsh DT, 2000b. Ecological significance of compatible solute accumulation by micro organisms:

from single cells to global climate. FEMS Microbiol Rev 24: 263-290.

Wu QS, Zou YN, He XH, 2010. Contributions of *Arbuscular mycorrhizal* fungi to growth, photosynthesis, root morphology and ionic balance of citrus seedlings under salt stress. Acta Physiol Planta 32: 297-304.

Wu ZS, Peng Y, Guo L, Li C, 2014. Root colonization of encapsulated Klebsiella oxytoca Rs-5 on cotton plants and its promoting growth performance under salinity stress. Eur J Soil Biol 60: 81-87.

Xia XJ, Wang YJ, Zhou YH, Tao Y, Mao WH, Shi K, Asami T, Chen ZX, Yu JQ, 2009. Reactive oxygen species are involved in brassinosteroid-induced stress tolerance in cucumber. Plant Physiol 150: 801-814.

Xie XN, Yoneyama K, Yoneyama K, 2010. The strigolactone story. Annu Rev Phytopathol 48: 93-117.

Xueming Z, Zhenping H, Yu Z, Huanshi Z, Pei Q, 2014. *Arbuscular mycorrhizal* fungi (AMF) and phosphate-solubilizing fungus (PSF) on tolerance of beach plum (*Prunus maritima*) under salt stress. Aust J Crop Sci 8: 945-950.

Xun F, Xie B, Liu S, Guo C, 2015. Effect of plant growth-promoting bacteria (PGPR) and *Arbuscular mycorrhizal* fungi (AMF) inoculation on oats in saline-alkali soil contaminated by petroleum to enhance phytoremediation. Environ Sci Pollut Res 22: 598-608.

Yang SJ, Zhang ZL, Xue YX, Zhang ZF, Shi SY, 2014. *Arbuscular mycorrhizal* fungi increase salt tolerance of apple seedlings. Bot Stud 55: 70.

Younesi O, Moradi A, 2014. Effects of plant growth-promoting rhizobacterium (PGPR) and *Arbuscular mycorrhizal* fungus (AMF) on antioxidant enzyme activities in salt-stressed bean (*Phaseolus vulgaris* L.). Agriculture. 2014 Mar 1 60 (1): 10-21.

Yu SH, Ke L, Wong YS, Tam NFY, 2005. Degradation of polycyclic aromatic hydrocarbons by a bacterial consortium enriched from mangrove sediments. Environ Int 31: 149-154.

Yusran Y, Volker R, Torsten M, 2009. Effects of plant prowth-promoting rhizobacteria andRhizobium on mycorrhizal development and drowth of *Paraserianthes falcataria* (L.) nielsen seedlings in two types of soils with contrasting levels of pH. In: Proc Int Plant Nutr Colloquium XVI. UC Davis: Dept Plant Sci.

Zhang H, Tang J, Wang L, Liu J, Gurav RG, Sun K, 2016. A novel bioremediation strategy for petroleum hydrocarbon pollutants using salt tolerant *Corynebacterium variabile* HRJ4 and biochar. J Environ Sci 47: 7-13.

Zhang HS, Qin CQ, Qin P, 2013. Effects of inoculation of *Arbuscular mycorrhizal* fungi and phosphate-solubilizing fungus with different proportion on P-uptake of *Castor Bean* (*Ricinus communis* L.) and rhizosphere soil enzyme activities in coastal saline soil. Chin Agric Sci Bull 29: 101-108.

Zhang YF, Wang P, Yang YF, Bi Q, Tian SY, Shi XW, 2011. *Arbuscular mycorrhizal* fungi improve reestablishment of *Leymus chinensis* in bare saline-alkaline soil: implication on vegetation restoration of extremely degraded land. J Arid Environ 75: 773-778.

Zhou N, Zhao S, Tian CY, 2017. Effect of halotolerant rhizobacteria isolated from halophytes on the growth of sugar beet (*Beta vulgaris* L.) under salt stress. FEMS Microbiol Lett 364. https://doi.org/https://doi.org/10.1093/femsle/fnx091.

Zhu J, FuX KYD, Zhu JK, Jenney JFE, Adams MW, Zhu Y, Shi H, Yun DJ, Hasegawa

PM, Bressan RA, 2007. An enhancer mutant of *Arabidopsis* salt overly sensitive 3 mediates both ion homeostasis and the oxidative stress response. Mol Cell Biol 27: 5214-5224.

Zhu RF, Tang F, Liu J, Liu FQ, Deng XY, Chen JS, 2016. Co-inoculation of *Arbuscular mycorrhizae* and nitrogen fixing bacteria enhance alfalfa yield under saline conditions. Pak J Bot 48: 763-769.

附表　主要名词中英文对照

英文名称	中文名称	简称
Food and Agriculture Organization	联合国粮农组织	FAO
Plant Growth-Promoting Rhizobacteria	植物生长促生菌	PGPR
nitrate reductase	硝酸还原酶	NR
reactive oxygen species	活性氧	ROS
nitric oxide	一氧化氮	NO
Induced Systemic Tolerance	系统性诱导抗性	IST
Indole-3-acetic acid	吲哚-3-乙酸	IAA
1-aminocyclopropane-1-carboxylic acid	1-氨基环丙烷-1-羧酸	ACC
catalase	过氧化氢酶	CAT
glutathione peroxidase	谷胱甘肽过氧化物酶	GPX
superoxide dismutase	超氧化物歧化酶	SOD
ascorbate peroxidase	抗坏血酸过氧化物酶	APX
monodehydroascorbate reductase	单脱氢抗坏血酸还原酶	MHDAR
glutathione reductase	谷胱甘肽还原酶	GR
groundwater-associated salinity	地下水伴生盐渍化	GAS
non-groundwater-associated salinity	非地下水伴生盐渍化	NAS
irrigation-associated salinity	灌溉水伴生盐渍化	IAS
exopolysaccharide	胞外多糖	EPS
lipopolysaccharide	脂多糖	LPS
S-adenosylmethionine	S-腺苷蛋氨酸	SAM
systemic acquired resistance	获得性系统性抗性	SAR
induced systemic resistance	诱导性系统性抗性	ISR
trimethylglycine	三甲基甘氨酸	GB
3-dimethylsulfonic acid	3-二甲基磺酸	DMSP
plant-probiotic microorganisms	植物益生菌	PPM

(续表)

英文名称	中文名称	简称
electroconductibility	电导率	EC
nitrate reductase	硝酸还原酶	NR
extracellular PGPR	胞外植物生长促生菌	ePGPR
intracellular PGPR	胞内植物生长促生菌	iPGPR
Aminocyclopropane-1-carboxylic acid deaminase	氨基环丙烷-1-羧酸脱氨酶	ACCD
abscisic acid	脱落酸	ABA
gibberellin	赤霉素	GA
halophyte rhizosphere-associated microbes	盐生植物根际相关微生物	HRAM
arbuscular mycorrhizal	丛枝菌根真菌	AM
plant growthpromoting microorganisms	植物生长促进微生物	PGPMs
plant growth-promoting	促进植物生长	PGP
water-use efficiency	水分利用效率	WUE
biological nitrogen fixation	生物固氮作用	BNF
phosphate-solubilizing microorganisms	溶磷菌	PSM
phosphate-solubilizing bacteria	溶磷细菌	PSB
phosphate-solubilizing fungi	溶磷真菌	PSF
potassium-solubilizing bacteria	溶钾细菌	KSB
high osmolarity glycerol	高渗甘油	HOG
strigolactone	粗内酯	SL
extracellular polymeric substances	胞外聚合物质	EPS
adenosine triphosphate	腺苷三磷酸	ATP